INCINERATION SYSTEMS

SYSTEMS

SELECTION AND DESIGN

INCINERATION SYSTEMS
SELECTION AND DESIGN

Calvin R. Brunner, P.E.

VNR VAN NOSTRAND REINHOLD COMPANY

Copyright © 1984 by Calvin R. Brunner

Library of Congress Catalog Card Number: 83-26124
ISBN: 0-442-21192-9

Manufactured in the United States of America

Published by Van Nostrand Reinhold Company Inc.
135 West 50th Street
New York, New York 10020

Van Nostrand Reinhold Company Limited
Molly Millars Lane
Wokingham, Berkshire RG11 2PY, England

Van Nostrand Reinhold
480 Latrobe Street
Melbourne, Victoria 3000, Australia

Macmillan of Canada
Division of Gage Publishing Limited
164 Commander Boulevard
Agincourt, Ontario MIS 3C7, Canada

15 14 13 12 11 10 9 8 7 6 5 4 3 2 1

Library of Congress Cataloging in Publication Data

Brunner, Calvin R.
    Incineration systems.

    Includes bibliographical references and index.
    1. Incineration. 2. Incinerators—Design and con-
struction. I. Title.
TD796.B78 1984   628.4′457   83-26124
ISBN 0-442-21192-9

In Celebration of Claire Helen Brunner

Preface

Incineration is increasingly looked to as a favorable means of waste disposal, especially when compared to alternative methods. As presented in this book, the field of incineration encompasses the destruction or processing of solid, sludge, liquid, gaseous, and radioactive wastes.

The text has been written to accommodate technical and nontechnical persons alike. It is meant to provide both a broad view of the subject as well as detailed system design techniques of primary interest to the specialist. References appear periodically to direct the reader to relevant publications and other sources of technical information.

The emphasis throughout is on detailed systems design. Before design can begin, however, it is obvious that the applicable satutory requirements must be understood. Two chapters are therefore alloted to the regulations that govern the incineration of hazardous and nonhazardous wastes. Six chapters then address analytical methods for systems design, from waste characterization to the prediction of air emissions. The various types of incinerator systems currently in use are discussed in the following eight chapters, which include design calculations, dimensional data, and other incinerator parameters.

Other chapters include one on energy recovery that presents a method for determining the heat recovery potential of an incinerator system, along with relevant design examples, and another on air pollution control equipment, which includes descriptions of the large variety of control devices available and their capacities, dimensions, and design parameters.

The reader has sufficient information in this single text to determine equipment selection, sizing, and parameters of operation for the incinerator equipment that burns the vast variety of wastes generated by municipal, commercial, industrial, and institutional sources.

CALVIN R. BRUNNER

Contents

INCINERATION SYSTEMS

SYSTEMS

SELECTION AND DESIGN

Chapter 1
Introduction

The term "waste" is a generalization for myriads of types, sizes, and configurations of materials which have no apparent utility. Household refuse, spent broth from the manufacture of penicillin, concrete chunks from a demolished structure, laboratory animal remains, rejects from a chemical product stream—these are all termed wastes. At one time off-gas generated in the drilling of oil wells in the United States was considered waste and was disposed of by burning on site (flaring). The price and availability of energy has made the collection and distribution of natural gas economical and today, instead of classification as a waste, it is considered a fuel source itself. In time, as natural resources are increasingly depleted, more and more means of reclamation and recycling will be developed removing materials from the category "waste." Industrial reclamation of combustible materials from waste streams, use of refuse and sewage sludge to generate gaseous fuel, hot water, steam and/or electric power, and recycling of waste paper, used cans and glass are all steps in the direction of reducing the damaging effect of an industrial society on the earth, the environment, and the people the society serves.

One of the most effective means of dealing with many wastes, to reduce their harmful potential and often to convert them to an energy form, is incineration. Comparing incineration (the destruction of a waste material by the application of heat) to other disposal options such as land burial, disposal at sea or in lagoons, the advantages of incineration are obvious:

- The volume and weight of the waste is reduced to a fraction of its original size.
- Waste reduction is immediate; it does not require long-term residence in a landfill or holding pond.
- Waste can be incinerated on-site, without having to be carted to a distant area.
- Air discharges can be effectively controlled for minimal impact on the atmospheric environment.
- The ash residue is usually non-putrescible, or sterile.

- Technology exists to completely destroy even the most hazardous of materials in a complete and effective manner.
- Incineration requires a relatively small disposal area, not the acres and acres required for lagoons or land burial.
- Using heat recovery techniques the cost of operation can often be reduced or offset by the use of or sale of energy.

Incineration will not solve all waste problems. Some disadvantages include:

- High capital cost.
- Skilled operators are required.
- Not all materials are incinerable, for example, high-aqueous wastes or noncombustible solids.
- Some materials require supplemental fuel to attain mandated efficiencies of destruction.

MATERIAL CLASSIFICATION

A substance will fall into one of three categories:

1. Gas: A state of matter where the substance completely fills a container in volume and in shape.
2. Liquid: The substance will fill a container in shape but not in volume.
3. Solid: The substance will neither fill a container in shape nor in volume.

LIQUID OR SOLID

The difference between a solid and a liquid is not always clear. Sludge, slurries, tars, and skimmings are terms describing states of matter lying somewhere between a true solid and true liquid. One quantitative measure of this difference is viscosity, which is a measure of the resistance of a fluid to shear. The resistance to shear in a solid is extremely high. In a "pure" liquid such as water it is almost zero. Flow and burning characteristics of a liquid are a function of the nature of and value of its viscosity. Types of viscosity and resultant fluid properties are defined as follows:

Newtonian fluid. A "pure" liquid such as water or light petroleum derivatives where the resistance to shear is initially zero and increases linearly with increase in shear, or velocity of flow.

Bingham plastic. A fluid which has an initial resistance to flow until a threshold is reached, at which point the fluid assumes Newtonian flow. Examples of such liquids are sewage sludge and aqueous mixtures of grain.

Pseudoplastic fluids and dilatant flow. These materials have a viscosity that increases rapidly, exponentially, with increasing velocity. These fluids include paper pulp, quicksand, and beach sand.

Thixotropic fluids. Initially these materials have a high viscosity. Under uniform shear, the viscosity decreases with time. Typical fluids are catsup, paints, inks, mayonnaise, and drilling muds.

Rheopectic fluids. These fluids display a rapid increase in viscosity with time. They will "set up" upon shaking or tapping. Examples are aqueous suspensions of vanadium pentoxide or gypsum.

Viscoelastic fluids. Fluids which exhibit elastic recovery from deformation which occurs during flow. This property causes the Weissenberg Effect, which is the tendency for the fluid to climb up a shaft rotating within it.

In general, a fluid can be pumped in a conventional manner if its viscosity is under 10,000 SSU. It can be atomized, i.e., burned in suspension, if its viscosity is below 750 SSU.

VISCOSITY

Figure 1-1 presents viscosities of a number of common fluids. Note the relative orders of magnitude, one substance to another, and the change in viscosity with respect to temperature.

INCINERABLE WASTE

The Incinerator Institute of America was a national organization attempting to quantify and standardize incinerator design parameters. It went out of business over ten years ago; however, a number of their standards are still in use. One such standard is given in Table 1-1, which is used by manufacturers of small and packaged incinerators in rating their equipment. The classifications in the table represent incinerable wastes, wastes which are combustible and are viable candidates for incineration.

Incinerability can be defined more specifically by consideration of the following factors:

Waste moisture content. The greater the moisture content the more fuel is required to destroy the waste. An aqueous waste with a moisture content greater than 95% or a sludge waste with less than 85% solids content would be considered poor candidates for incineration.

Heating value. Incineration is a thermal destruction process where the waste is degraded to non-putrescible form by the application and maintenance of a source of heat. With no significant heating value, incineration would not be a

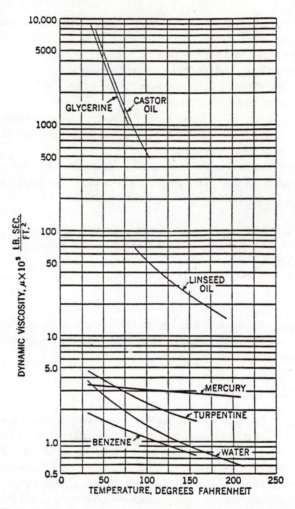

Fig. 1-1. Viscosity values for typical liquids. *Source:* Ref. 1-10.

practical disposal method. Generally, a waste with a heating value less than 1000 BTU/lb as received, such as concrete blocks or stone, is not applicable for incineration. There are instances, however, where an essentially inert material has a relatively small content (or coating) of combustibles and incineration would be a viable option even with a small heating value. Two such cases are incineration of empty drums with a residual coating of organic material on their inner surfaces, and incineration of grit from wastewater treatment plants. The grit adsorbs grease from within the wastewater flow which results in a slight heating value to the grit material, normally less than 500 BTU/lb.

Inorganic salts. Wastes rich in inorganic, alkaline salts are troublesome to dispose of in a conventional incineration system. A significant fraction of the salt will become airborne. It will collect on furnace surfaces creating a slag, or cake, which severely reduces the ability of an incinerator to function properly.

High sulfur or halogen content. The presence of chlorides or sulfides in a waste will normally result in the generation of acid forming compounds in the off-gas. The cost of protecting equipment from acid attack must be balanced against the cost of alternative disposal methods for the waste in question.

Radioactive waste. Incinerators have been developed specifically for the destruction of radioactive waste materials, as described in a later chapter. Unless designed specifically for radioactive waste disposal, however, an incinerator should not be used for the destruction of a radioactive waste.

STANDARD INDUSTRIAL CLASSIFICATION

The Department of Commerce has established a classification of industries known as the Standard Industrial Classification Code, SIC. A listing of the SIC's of manufacturing industries which produce substantive quantities of incinerable wastes is presented in Table 1-2. Table 1-3 lists the number of such manufacturing establishments in the United States (as of 1978) related to SIC codes 20 through 39.

DISPOSAL OPTIONS

There is no simple answer to the disposal problem. Proper waste management involves the examination of the entire breadth of available options for each type of material. Disposal options for waste materials include:

1. Physical treatment processes:
   - Gas cleaning.
   - Liquid/solid separation.
   - Removal of specific components.
   - Blending of wastes.
2. Chemical treatment processes:
   - Absorption.
   - Chemical oxidation.
   - Chemical reduction.
3. Biological treatment processes.
4. Ultimate disposal processes:
   - Deep well disposal.
   - Sanitary landfill.
   - Composting.

Table 1-1.   Classification of Wastes to be Incinerated.

Classification of Wastes Type Description	Principal Components	Approximate Composition % by Weight
*0 Trash	Highly combustible waste, paper, wood, cardboard cartons, including up to 10% treated papers, plastic or rubber scraps; commercial and industrial sources	Trash 100%
*1 Rubbish	Combustible waste, paper, cartons, rags, wood scraps, combustible floor sweepings; domestic, commercial, and industrial sources	Rubbish 80% Garbage 20%
*2 Refuse	Rubbish and garbage; residential sources	Rubbish 50% Garbage 50%
*3 Garbage	Animal and vegetable wastes, restaurants, hotels, markets; institutional, commercial, and club sources	Garbage 65% Rubbish 35%
4 Animal solids and organic wastes	Carcasses, organs, solid organic wastes; hospital, laboratory, abattoirs, animal pounds, and similar sources	100% Animal and Human Tissue
5 Gaseous, liquid or semi-liquid wastes	Industrial process wastes	Variable
6 Semi-solid and solid wastes	Combustibles requiring hearth, retort, or grate burning equipment	Variable

*The above figures on moisture content, ash, and B.T.U. as fired have been determined by analysis of many samples. They are recommended for use in computing

Moisture Content %	Incombustible Solids %	B.T.U. Value/lb. of Refuse as Fired	B.T.U. of Aux. Fuel Per Lb. of Waste to be included in Combustion Calculations	Recommended Min. B.T.U./hr. Burner Input per lb. Waste
10%	5%	8500	0	0
25%	10%	6500	0	0
50%	7%	4300	0	1500
70%	5%	2500	1500	3000
85%	5%	1000	3000	8000 (5000 Primary) (3000 Secondary)
Dependent on predominant components	Variable according to wastes survey	Variable according to wastes survey	Variable according to wastes survey	Variable according to wastes survey
Dependent on predominant components	Variable according to wastes survey	Variable according to wastes survey	Variable according to wastes survey	Variable according to wastes survey

heat release, burning rate, velocity, and other details of incinerator designs. Any design based on these calculations can accommodate minor variations.

Source: Ref. 1-2.

Table 1-2. Sources and Types of Industrial Wastes.

Code	SIC Group Classification	Waste Generating Processes	Expected Specific Wastes
17	Plumbing, heating, air conditioning Special trade contractors	Manufacturing and installation in homes, buildings, factories	Scrap metal from piping and duct work; rubber, paper, insulating materials, misc. construction, demolition debris
19	Ordnance and accessories	Manufacturing and assembling	Metals, plastic, rubber, paper, wood, cloth, chemical residues
20	Food and kindred products	Processing, packaging, shipping	Meats, fats, oils, bones, offal vegetables, fruits, nuts and shells, cereals
22	Textile mill products	Weaving, processing, dyeing and shipping	Cloth and fiber residues
23	Apparel and other finished products	Cutting, sewing, sizing, pressing	Cloth, fibers, metals, plastics, rubber
24	Lumber and wood products	Sawmills, mill work plants, wooden container, misc. wood products, manufacturing	Scrap wood, shavings, sawdust; in some instances metals, plastics, fibers, glues, sealers, paints solvents
25	Furniture, wood	Manufacture of household and office furniture, partitions, office and store fixtures, mattresses	Those listed under Code 24; in addition, cloth and padding residues
25	Furniture, metal	Manufacture of household and office furniture, lockers, bedsprings, frames	Metals, plastics, resins, glass, wood, rubber, adhesives, cloth, paper
26	Paper and allied products	Paper manufacture, conversion of paper and paperboard, manufacture of paperboard boxes and containers	Paper and fiber residues, chemicals, paper coatings and fillers, inks, glues, fasteners
27	Printing and publishing	Newspaper publishing, printing lithography, engraving, printing, and bookbinding	Paper, newsprint, cardboard, metals, chemicals, cloth, inks, glues
28	Chemicals and related products	Manufacture and preparation of inorganic chemicals (ranges from drugs and soups to paints and varnishes, and explosives)	Organic and inorganic chemicals, metals, plastics, rubber, glass, oils, pigments
29	Petroleum refining and related industries	Manufacture of paving and roofing materials	Asphalt and tars, felts, asbestos, paper, cloth, fiber
30	Rubber and miscellaneous plastic products	Manufacture of fabricated rubber and plastic products	Scrap rubber and plastics, lampblack, curing compounds, dyes

8

Table 1-2. (Continued)

Code	SIC Group Classification	Waste Generating Processes	Expected Specific Wastes
31	Leather and leather products	Leather tanning and finishing; manufacture of leather belting and packing	Scrap leather, thread, dyes, oils, processing and curing compounds
32	Stone, clay, and glass products	Manufacture of flat glass, fabrication or forming of glass; manufacture of concrete, gypsum, and plaster products; forming and processing of stone and stone products, abrasives, asbestos, and misc. nonmineral products	Glass, cement, clay, ceramics, gypsum, asbestos, stome, paper, abrasives
33	Primary metal industries	Melting, casting, forging, drawing, rolling, forming, extruding operations	Ferrous and nonferrous metals scrap, slag, sand, cores, patterns, bonding agents
34	Fabricated metal products	Manufacture of metal cans, hand tools, general hardware, nonelectric heating apparatus, plumbing fixtures, fabricated structural products, wire, farm machinery and equipment, coating and engraving of metal	Metals, ceramics, sand, slag, scale, coatings, solvents, lubricants, pickling liquors
35	Machinery (except electrical)	Manufacture of equipment for construction, mining, elevators, moving stairways, conveyors, industrial trucks, trailers, stackers, machine tools, etc.	Slag, sand, cores, metal scrap, wood, plastics, resins, rubber, cloth, paints, solvents, petroleum products
36	Electrical	Manufacture of electric equipment, appliances, and communication apparatus, machining, drawing, forming, welding, stamping, winding, painting, plating, baking, firing, operations	Metal scrap, carbon, glass, exotic metals, rubber, plastics, resins, fibers, cloth residues
37	Transportation equipment	Manufacture of motor vehicles, truck and bus bodies, motor vehicle parts and accessories, aircraft and parts, ship and boat building and	Metal scrap, glass, fiber, wood, rubber, plastics, cloth, paints, solvents, petroleum products

Table 1-2. (Continued)

Code	SIC Group Classification	Waste Generating Processes	Expected Specific Wastes
		repairing motorcycles and bicycles and parts, etc.	
38	Professional, scientific controlling instruments	Manufacture of engineering, laboratory, and research instruments and associated equipment	Metals, plastics, resins, glass, wood, rubber, fibers, abrasives
39	Miscellaneous manufacturing	Manufacture of jewelry, silverware, plated ware, toys, amusement, sport- and athletic goods, costume novelties, buttons, brooms, brushes, signs, advertising displays	Metals, glass, plastics, resins, leather, rubber, composition, bone, cloth, straw, adhesives, paints, solvents

Source: Ref. 1-1.

Table 1-3. Manufacturing Establishments in the United States by SIC Number.

SIC Code	Number of Establishments in Nation (1978)
20 Food	18,195
21 Tobacco	167
22 Textile	5,990
23 Apparel	20,016
24 Lumber	28,881
25 Furniture	7,639
26 Paper	4,259
27 Printing	45,528
28 Chemical	8,407
29 Petroleum	1,494
30 Rubber, plastic	12,450
31 Leather	2,428
32 Stone, glass, and clay	13,385
33 Primary metal	5,477
34 Fabricated metal	27,453
35 Machinery	42,139
36 Electric machinery	13,144
37 Transportation equipment	9,078
38 Instruments	6,308
39 Miscellaneous manufacturing	16,052
Total	288,490

- Land burial.
- Encapsulation and solidification.
- Secure landfill.
- Ocean dumping.
- Dilute and disperse.
- Recycle/reuse.
- Incineration.

Degradation options, as opposed to disposal, include the following:

- *Destruction* is the conversion of the waste materials to innocuous by-products (i.e., conversion of organic phosphate to $CO_2 + H_2O$ + phosphoric acid).
- *Detoxification* is removal or destruction of the toxic component; this process renders the material non-hazardous (e.g., dechlorination of PCBs).

Degradation options include non-oxidative and oxidative techniques, as follows:

1. Non-oxidative techniques include:
   - Microbiological.
   - Reductive dechlorination.
   - High energy radiation; microwaves.
2. Oxidative techniques include:
   - Wet catalytic oxidation; aqueous phase reaction.
   - Ozonation.
   - Photochemical decomposition.
   - Chemical oxidation (i.e., hydrogen peroxide, potassium permanganate and chlorine dioxide oxidants).
   - Electrochemical oxidation.
   - Flameless oxidation; low temperature vapor combustion.
   - Incineration.
3. *Refractory* is the inverse of destructible, i.e., it refers to the degree to which a material resists destruction. For instance, PCBs are highly refractory.

LOAD ESTIMATING

The quantity of solid waste generated in the United States, industrial and municipal, is approximately 300 million tons per year. Of this figure approximately 2000 pounds of household refuse is produced per year per capita.

The estimation of incinerator loading, where the waste quantity is not known, usually requires a survey of the area in question including a study of past records, demographic trends, etc. Table 1-4 can be used as a guide in determining the solid waste produced from various sources.

Table 1-4. Incinerator Capacity Chart.

CLASSIFICATION	BUILDING TYPES	QUANTITIES OF WASTE PRODUCED
INDUSTRIAL BUILDINGS	Factories Warehouses	Survey must be made 2 lbs. per 100 sq. ft. per day
COMMERCIAL BUILDINGS	Office Buildings Department Stores Shopping Centers Supermarkets Restaurants Drug Stores Banks	1 lb. per 100 sq. ft. per day 4 lbs. per 100 sq. ft. per day Study of plans or survey required 9 lbs. per 100 sq. ft. per day 2 lbs. per meal per day 5 lbs. per 100 sq. ft. per day Study of plans or survey required
RESIDENTIAL	Private Homes Apartment Buildings	5 lbs. basic & 1 lb. per bedroom 4 lbs. per sleeping room per day
SCHOOLS	Grade Schools High Schools Universities	10 lbs. per room & ½ lb. per pupil per day 8 lbs. per room & ½ lb. per pupil per day Survey required
INSTITUTIONS	Hospitals Nurses or Interns Homes Homes for Aged Rest Homes	15 lbs. per bed per day 3 lbs. per person per day 3 lbs. per person per day 3 lbs. per person per day
HOTELS, ETC.	Hotels—1st Class Hotels—Medium Class Motels Trailer Camps	3 lbs. per room and 2 lbs. per meal per day 1½ lbs. per room & 1 lb. per meal per day 2 lbs. per room per day 6 to 10 lbs. per trailer per day
MISCELLANEOUS	Veterinary Hospitals Industrial Plants Municipalities	Study of plans or survey required

Do not estimate more than 7 hours operation per shift of industrial installations.

Do not estimate more than 6 hours operation per day for commercial buildings, institutions, and hotels.

Do not estimate more than 4 hours operation per day for schools.

Do not estimate more than 3 hours operation per day for apartment buildings.

Whenever possible an actual survey of the amount and nature of refuse to be burned should be carefully taken. The data herein is of value in estimating capacity of the incinerator where no survey is possible and also to double check against an actual survey.

Source: Ref. 1-2.

Table 1-5. Average Weight of Solid Waste.

Type	Lbs. per cu. ft.
Type 0 Waste	8 to 10
Type 1 Waste	8 to 10
Type 2 Waste	15 to 20
Type 3 Waste	30 to 35
Type 4 Waste	45 to 55
Garbage (70% H_2O)	40 to 45
Magazines and packaged paper	35 to 50
Loose paper	5 to 7
Scrap wood and sawdust	12 to 15
Wood shavings	6 to 8
Wood sawdust	10 to 12

Source: Ref. 1-2.

Table 1-6. Average Solid Waste Collected.
[Pounds per person per day]

Solid Wastes	Urban	Rural	National
Household	1.26	0.72	1.14
Commercial	0.46	0.11	0.38
Combined	2.63	2.60	2.63
Industrial	0.65	0.37	0.59
Demolition, construction	0.23	0.02	0.18
Street and alley	0.11	0.03	0.09
Miscellaneous	0.38	0.08	0.31
Totals	5.72	3.93	5.32

Source: Ref. 1-3.

Table 1-7. Industrial Solid Waste Density Data.

Waste	Lb/Cubic Yard as Discarded
Department store waste	80
Hospital waste (not research)	100
School waste w/lunch program	110
Supermarket waste	100
Bakalite	600
Bitumen waste	1500
Brown paper	135
Cardboard	180
Cork	320
Corn cobs	300
Corrugated paper (loose)	100
Disposable hospital plastics	120
Grass, green	120
Hardboard	900
Latex	1200
Magazines	945
Meat scraps	400
Milk cartons, coated	80
Nylon	200
Paraffin—wax	1400
Plastic coated paper	135
Polyethylene film	20
Polystyrene	175
Polyurethane (foamed)	55
Resin bonded fiberglass	990
Rubber—synthetics	1200
Shoe leather	540
Tar paper	450
Textile waste (non-synthetic)	280
Textile waste (synthetic)	240
Vegetable food waste	375
Wax paper	150
Wood	300

Source: Ref. 1-12.

Table 1-8.   Typical Moisture Content of
Municipal Solid Waste Components.

Component	Moisture, percent	
	Range	Typical
Food wastes	50–80	70
Paper	4–10	6
Cardboard	4–8	5
Plastics	1–4	2
Textiles	6–15	10
Rubber	1–4	2
Leather	8–12	10
Garden trimmings	30–80	60
Wood	15–40	20
Glass	1–4	2
Tin cans	2–4	3
Nonferrous metals	2–4	2
Ferrous metals	2–6	3
Dirt, ashes, brick, etc.	6–12	8
Municipal solid waste	15–40	20

Source: Ref. 1-11.

Table 1-9.   Typical Heating Value of MSW
Components.

Component	Energy, BTU/lb	
	Range	Typical
Food wastes	1500–3000	2000
Paper	5000–8000	7200
Cardboard	6000–7500	7000
Plastics	12000–16000	14000
Textiles	6500–8000	7500
Rubber	9000–12000	10000
Leather	6500–8500	7500
Garden trimmings	1000–8000	2800
Wood	7500–8500	8000
Glass	50–100	60
Tin cans	100–500	300
Nonferrous metals	–	–
Ferrous metals	100–500	300
Dirt, ashes, brick, etc.	1000–5000	3000
Municipal solid wastes	4000–6500	4500

Source: Ref. 1-12.

Table 1-5 lists the average weight of various solid wastes and Table 1-6 lists per capita waste generation in the United States.

Another major waste, sewage sludge, can be estimated to be generated at the rate of 0.2 pounds of sludge solids per day per capita.

ESTIMATING SOLID WASTE QUALITY

While a general figure for waste generation can be obtained as noted in the previous sections, a more accurate means of determining the quality of a solid waste stream is by use of Table 1-7, Table 1-8 and/or Table 1-9. By a visual inspection of the waste, a percentage of each waste component as listed in these tables can be established. Multiplying the moisture percent or heating value or density of each of these components by the indicated moisture, heating value or density, a more accurate figure for the total waste quality can be estimated. (A more detailed analysis of heating value of wastes is included in a later chapter.)

As an example, to estimate the heating value of a particular municipal solid waste, with the waste components as listed below, using the heating value listed in Table 1-9, the total waste heating value is calculated as follows:

Component	Solid Wastes, %	Inherent Energy BTU/lb	Total Energy Contribution, BTU/lb
Food wastes	15	2,000	300
Paper	40	7,200	2,880
Cardboard	5	7,000	350
Plastics	5	14,000	700
Wood	15	8,000	1,200
Glass	10	60	6
Tin cans	10	300	30
Total	100		5,466

Total energy content is therefore 5,466 BTU/lb.

Chapter 2
Regulatory Requirements

Each political entity in the United States shows concern for discharges from incineration by deferring to a higher authority or by establishing their own set of standards. Local zoning boards are often used to regulate the external effects of an incinerator facility. In this chapter the regulations of federal and state governments will be discussed.

FEDERAL REGULATIONS

The Federal Government has established an intensive program for regulating the incineration of hazardous wastes. Chapter 3 describes the Federal hazardous waste incineration regulations. Non-hazardous waste incineration is regulated only if the waste is generated from municipal sources. Non-hazardous waste generated and incinerated by industry is not presently regulated by the federal government except for the provisions of the National Ambient Air Quality Standards, discussed later in this chapter.

The second Federal regulation established for control of emissions from municipal incinerators is subpart O of the above standard. This standard establishes the emissions from incinerators burning sewage sludge generated from municipal sewage treatment facilities. It limits particulate emissions from the facility to 1.3 pounds per dry ton of sludge charged. It also limits visible emissions to 20 percent opacity.

Solid waste incinerator standards apply to incinerators constructed after August 17, 1971. The sewage sludge incinerator standards apply to incinerators constructed after June 11, 1973.

STATE PARTICULATE EMISSION STANDARDS

State emission standards are presented as grains per unit volume emitted or pounds per unit weight charged. Table 2-1 is a list of factors which allow con-

Table 2-1.   Emissions Standards Conversion Factors.

	lb/ton Refuse (As Received)	lb/1000 lb Flue Gas at 50% Excess Air	lb/1000 lb Flue Gas at 12% CO_2	gr/st ft^3 at 50% Excess Air	gr/st ft^3 at 12% CO_2	G/Nm^3 at ntp, 7% CO_2
lb/ton refuse (as received)	1	0.089	0.10	0.047	0.053	0.067
lb/1000 lb flue gas at 50% excess air	11.27	1	1.12	0.52	0.585	0.74
lb/1000 lb flue gas at 12% CO_2	10.0	0.89	1	0.46	0.52	0.66
gr/st ft^3 at 50% excess air	21.31	1.93	2.16	1	1.12	1.42
gr/st ft^3 SCF at 12% CO_2	18.85	1.71	1.92	0.89	1	1.26
g/Nm^3 at ntp, 7% CO_2	15.0	1.36	1.53	0.704	0.79	1

Source: Ref. 2-13.

version from one form of a standard to another. These conversions are a function of the constituents of the feed and for this particular chart the waste feed chosen was 50 percent cellulose ($C_6H_{10}O_5$, the prime constituent of paper), 50 percent water, with a heating value of 4462 BTU/lb as charged.

Table 2-2 is a listing of the particulate emission regulations for each of the fifty states. The last column of this chart is an equivalent discharge in grains per cubic foot, in accordance with the conversion factors listed in Table 2-1 (as applicable) which provide a common basis of comparison for one state with another. These standards apply to all incineration facilities, e.g., municipal refuse, municipal sludge, industrial, and agricultural wastes.

STATE OPACITY STANDARDS

Opacity is a visual measure of particulate emission. The greater the particulate loading, the darker (or more opaque) the exhaust. In general, the standards governing particulate emissions have an effect on stack opacity. With opacity a visible quality of a plume, however, it was incumbent upon public officials to establish a standard that can be easily identified and verified by the layman. Fig. 2-1 is a typical opacity chart. The chart is normally printed on a transparent surface

Table 2-2. Particulate Emission Limitations for New and Existing Incinerators.

State	Value	Units	Regulation Corrected to	Process Conditions	Validity	Equivalent Common Regulation (gr/dscf @ 12% CO_2)
1 Alabama	0.1	lbs/100 lbs charged		>50 TPD		0.12
	0.2	lbs/100 lbs charged		≤50 TPD		0.24
2 Alaska	0.3	gr/dscf	12% CO_2	<200 lbs/hr		0.3
	0.2	gr/dscf	12% CO_2	200–1000 lbs/hr		0.2
	0.1	gr/dscf	12% CO_2	>1000 lbs/hr		0.1
3 Arizona	0.1	gr/dscf	12% CO_2			0.1
4 Arkansas	0.2	gr/dscf	12% CO_2	>200 lbs/hr		0.2
	0.3	gr/dscf	12% CO_2	<200 lbs/hr		0.3
5 California	0.3	gr/dscf	12% CO_2	typical of the 43 APCD's		0.3
6 Colorado	0.1	gr/dscf	12% CO_2		designated control areas	0.1
	0.15	gr/dscf	12% CO_2		other areas	0.15
7 Connecticut	0.08	gr/dscf	12% CO_2		built after 6/1/72	0.08
	0.4	lbs/1000 lbs	50% excess air		built before 6/1/72	0.26
8 Delaware	0.2	lbs/hr		100 lbs/hr		0.24
	1.0	lbs/hr		500 lbs/hr		0.24
	2.0	lbs/hr		1000 lbs/hr		0.24
	5.0	lbs/hr		3000 lbs/hr		0.2
9 Florida	0.08	gr/dscf	50% excess air	>50 TPD	built after 2/11/72	0.1
	0.1	gr/dscf	50% excess air	>50 TPD	built before 2/11/72	0.12
10 Georgia	0.1	gr/dscf	12% CO_2	≤50 TPD—type 0, 1, 2 waste	new (built after 1/1/72)	0.1
	0.2	gr/dscf	12% CO_2	≤50 TPD—type 3, 4, 5, 6 waste	new (built after 1/1/72)	0.2
	0.2	gr/dscf	12% CO_2	type 0, 1, 2 waste	existing before 1/1/72	0.2
	0.3	gr/dscf	12% CO_2	type 3, 4, 5, 6 waste	existing before 1/1/72	0.3

No.	State	Value	Units	Basis	Capacity	Applicability	Value
11	Hawaii	0.08	gr/dscf	12% CO_2	≥50 TPD	new (built after 1/1/72)	0.08
12	Idaho	0.2	lbs/100 lbs charged				0.24
		0.2	lbs/100 lbs charged				0.24
13	Illinois	0.08	gr/dscf	12% CO_2	2000–60,000 lbs/hr		0.08
		0.2	gr/dscf	12% CO_2	<2000 lbs/hr	built before 4/15/72	0.2
		0.1	gr/dscf	12% CO_2	<2000 lbs/hr	built after 4/15/72	0.1
14	Indiana	0.3	lbs/1000 lbs gas	50% excess air	≥200 lbs/hr		0.19
		0.5	lbs/1000 lbs gas	50% excess air	<200 lbs/hr		0.32
15	Iowa	0.2	gr/dscf	12% CO_2	≥1000 lbs/hr		0.2
		0.35	gr/dscf	12% CO_2	<1000 lbs/hr		0.35
16	Kansas	0.3	gr/dscf	12% CO_2	<200 lbs/hr		0.3
		0.2	gr/dscf	12% CO_2	200–20,000 lbs/hr		0.2
		0.1	gr/dscf	12% CO_2	>20,000 lbs/hr		0.1
17	Kentucky	0.2	gr/dscf	12% CO_2	≤50 TPD		0.2
		0.08	gr/dscf	12% CO_2	>50 TPD		0.08
18	Louisiana	0.2	gr/dscf	12% CO_2			0.2
19	Maine	0.2	gr/dscf	12% CO_2			0.2
20	Maryland	0.1	gr/dscf	12% CO_2	<2000 lbs/hr	built after 1/17/72	0.1
		0.03	gr/dscf	12% CO_2	>2000 lbs/hr	built after 1/17/72	0.03
		0.3	gr/dscf	12% CO_2	<200 lbs/hr	built before 1/17/72	0.3
		0.2	gr/dscf	12% CO_2	>200 lbs/hr	built before 1/17/72	0.2
21	Massachusetts	0.1	gr/dscf	12% CO_2		existing	0.1
		0.05	gr/dscf	12% CO_2		new	0.05
22	Michigan	0.65	lbs/1000 lbs gas	50% excess air	0–100 lbs/hr		0.42
		0.3	lbs/100 lbs gas	50% excess air	>100 lbs/hr		0.19
23	Minnesota	0.3	gr/dscf	12% CO_2	<200 lbs/hr	existing before 8/17/71	0.3
		0.2	gr/dscf	12% CO_2	200–2000 lbs/hr	existing before 8/17/71	0.2
		0.1	gr/dscf	12% CO_2	<200 lbs/hr	existing before 8/17/71	0.1
		0.2	gr/dscf	12% CO_2	200–2000 lbs/hr	new (built after 8/17/71)	0.2
		0.15	gr/dscf	12% CO_2	>2000 lbs/hr	new (built after 8/17/71)	0.15
		0.1	gr/dscf	12% CO_2	>2000 lbs/hr	new (built after 8/17/71)	0.1
24	Mississippi	0.2	gr/dscf	12% CO_2	Design capacity		0.2

Table 2-2. (*Continued*)

State	Regulation					Equivalent
	Value	Units	Corrected to	Process Conditions	Validity	Common Regulation (gr/dscf @ 12% CO_2)
	0.1	gr/dscf	12% CO_2	New sources near residential areas		0.1
25 Missouri	0.2	gr/dscf	12% CO_2	≥200 lbs/hr		0.2
	0.3	gr/dscf	12% CO_2	<200 lbs/hr		0.3
26 Montana	0.2	gr/dscf	12% CO_2	>200 lbs/hr	existing before 9/5/75	0.2
	0.3	gr/dscf	12% CO_2	≤200 lbs/hr	existing before 9/5/75	0.3
	0.1	gr/dscf	12% CO_2		all others	0.1
27 Nebraska	0.2	gr/dscf	12% CO_2	<2000 lbs/hr		0.2
	0.1	gr/dscf	12% CO_2	≥2000 lbs/hr		0.1
28 Nevada	3.0	lbs/ton charged	12% CO_2	<2000 lbs/hr		0.18
	variable	$E = 40.7 \times 10^{-5}\ C$	$C, E = $ lbs/hr	>2000 lbs/hr		0.05
29 New Hampshire	0.3	gr/dscf	12% CO_2	≤200 lbs/hr		0.3
	0.2	gr/dscf	12% CO_2	>200 lbs/hr		0.2
	0.08	gr/dscf	12% CO_2	>50 TPD	built after 4/20/74	0.08
30 New Jersey	0.2	gr/dscf	12% CO_2	<2000 lbs/hr	type 0, 1, 2, 3 waste only	0.2
	0.2	gr/dscf	12% CO_2	all others		0.1
31 New Mexico	only opacity	regulations		≤50 TPD		–
	0.08	gr/dscf	12% CO_2	>50 TPD	new (built after 8/17/71)	0.08
32 New York	0.5	lbs/100 lbs charged		>2000 lbs/hr	built between 4/1/62 and 1/1/70	0.6
	0.5	lbs/100 lbs charged		≤2000 lbs/hr	built between 4/1/62 and 1/1/68	0.6
	variable (e.g., 0.3)	lbs/hr		<100 lbs/hr	built after 1/1/68	0.36
	variable (e.g., 3.0)	lbs/hr		@1000 lbs/hr	built after 1/1/68	0.36

State						
	variable (e.g., 7.5)	lbs/hr		@3000 lbs/hr	built after 1/1/70	0.3
33 North Carolina	0.2	lbs/hr		0–100 lbs/hr		0.24
	0.4	lbs/hr		@200 lbs/hr		0.24
	1.0	lbs/hr		@500 lbs/hr		0.24
	2.0	lbs/hr		@1000 lbs/hr		0.24
	4.0	lbs/hr		>2000 lbs/hr		0.24
34 North Dakota	variable	lbs/hr		@100 lbs/hr		0.4
				@1000 lbs/hr		0.31
				@3000 lbs/hr		0.24
35 Ohio	0.1	lbs/100 lbs charged		>100 lbs/hr		0.12
	0.2	lbs/100 lbs charged		<100 lbs/hr		0.24
36 Oklahoma	variable	lbs/hr		@100 lbs/hr		0.48
				@1000 lbs/hr		0.31
				@3000 lbs/hr		0.21
37 Oregon	0.3	gr/dscf		>100 lbs/hr		0.3
	0.2	gr/dscf		>200 lbs/hr	built before 6/1/70	0.2
	0.1	gr/dscf		>200 lbs/hr	built after 6/1/70	0.1
38 Pennsylvania	0.1	gr/dscf	12% CO_2			0.16
39 Rhode Island	0.16	gr/dscf	12% CO_2	<2000 lbs/hr		0.08
	0.08	gr/dscf	12% CO_3	>2000 lbs/hr		0.27
40 South Carolina	0.5	lbs/10^6 Btu		@10 mm Btu/hr		0.24
41 South Dakota	0.2	lbs/100 lbs charged				0.12
42 Tennessee	0.2	% of charge		<2000 lbs/hr		0.41
	0.1	% of charge		>2000 lbs/hr		0.27
43 Texas	variable	lbs/hr		@1000 lbs/hr		
				@3000 lbs/hr		
44 Utah	0.08	gr/dscf	12% CO_2	>50 TPD		0.08
45 Vermont	0.1	lbs/100 lbs charged				0.12
46 Virginia	0.14	gr/dscf	12% CO_2			0.14
47 Washington	0.1	gr/dscf	7% O_2			0.11
48 West Virginia	8.25	lbs/ton		<200 lbs/hr		0.5
	5.43	lbs/ton		>200 lbs/hr		0.33

Table 2-2. (*Continued*)

| State | Value | Units | Regulation | | | Equivalent |
			Corrected to	Process Conditions	Validity	Common Regulation (gr/dscf @ 12% CO_2)
49 Wisconsin	0.2	lbs/1000 lbs exhaust gas	12% CO_2	500–4000 lbs/hr	built after 4/1/72	0.11
	0.3	lbs/1000 lbs exhaust gas	12% CO_2	≤500 lbs/hr	built after 4/1/72	0.17
	0.5	lbs/1000 lbs exhaust gas	12% CO_2	>500 lbs/hr	built before 4/1/72	0.28
	0.6	lbs/1000 lbs exhaust gas	12% CO_2	≤500 lbs/hr	built before 4/1/72	0.34
	0.15	lbs/1000 lbs exhaust gas	12% CO_2	≥4000 lbs/hr	built after 4/1/72	0.08
50 Wyoming	0.2	lbs/100 lbs charged				0.24

Source: Ref. 2-11.

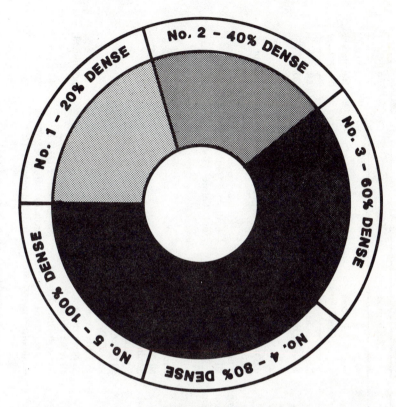

INSTRUCTIONS

1. Hold chart at arms length and view smoke through circle provided.
2. Observer should not be less than 100 ft, nor more than 1/4 mile from the stack.
3. Line of observation should be at right angles to the direction the smoke travels and viewed in the same light.
4. Do not try to observe smoke into direct sun or setting sun. Sun should always be overhead.
5. Match smoke with corresponding color on chart—note density number and time.

Fig. 2-1. Opacity measurement. Alken smoke chart (Ringelmann type).

and the degree of opacity of the exhaust is matched to the chart reading to obtain a qualitative identification of degree of opacity.

Table 2-3 lists opacity requirements for each state. The percent opacity is that opacity noted in Fig. 2-1.

Table 2-3. Opacity Regulations for New and Existing Commercial and Industrial Incinerators.

States	Value	Units	Regulation — Process Conditions	Regulation — Validity	Equivalent Common Regulation (% opacity)
1 Alabama	60	% opacity	3 min discharge/60 min		60
	20	% opacity	all other times		20
2 Alaska	40	% opacity		installed before 7/1/72	40
	20	% opacity		installed after 7/1/72	20
3 Arizona	exempt				exempt
	20	% opacity	0.5 min discharge/60 min		20
			all other times		
4 Arkansas	No. 3	Ringelmann	5 min discharge/60 min		60
	No. 1	Ringelmann	all other times	built after 7/30/73	20
	No. 2	Ringelmann		built before 7/30/73	40
5 California					
6 Colorado	20	% opacity			20
7 Connecticut	40	% opacity	5 min discharge/60 min		40
	20	% opacity	all other times		20
8 Delaware	20	% opacity	3 min discharge/60 min		20
9 District of Columbia	20	% opacity	2 min discharge/60 min	existing	20
			all other times	existing	
10 Florida	prohibited				prohibited
	20	% opacity	≤50 TPD, 3 min discharge/60 min		20
			all other times		
11 Georgia	prohibited				prohibited
	20	% opacity	6 min discharge/60 min	installed after 1/1/72	20
	40	% opacity	all other times	installed after 1/1/72	40
	40	% opacity	6 min discharge/60 min	installed before 1/1/72	40
	60	% opacity	all other times	installed before 1/1/72	60
12 Hawaii	40	% opacity			40
13 Idaho	No. 2	Ringelmann	3 min discharge/60 min	built before 4/1/72	40

No.	State	Level	Method	Condition	Note	Value
14	Illinois	No. 1	Ringelmann	3 min discharge/60 min	built after 4/1/72	20
		30	% opacity	all other times		30
15	Indiana	30–40	% opacity	3 min discharge/60 min		30–60
16	Iowa	40	% opacity	15 min discharge/60 min		40
		40	% opacity			40
		60	% opacity	3 min discharge/60 min during breakdowns, etc.		60
17	Kansas	20	% opacity			20
18	Kentucky	20	% opacity	all other times		20
19	Louisiana	No. 1	Ringelmann			20
20	Maine	>No. 1	Ringelmann	4 min discharge/60 min		>20
21	Maryland	No. 1	Ringelmann			20
22	Massachusetts	No. 1	Ringelmann	3 min discharge/60 min		20
23	Michigan	40	% opacity	all other times		40
		20	% opacity			20
24	Minnesota	20	% opacity			20
25	Mississippi	40	% opacity			40
26	Missouri	No. 1	Ringelmann		built after 2/10/72	20
		No. 2	Ringelmann		built before 2/10/72	40
27	Montana	10	% opacity			10
28	Nebraska	20	% opacity			20
29	Neveda	20	% opacity			20
30	New Hampshire	No. 1	Ringelmann	1 min discharge/60 min		20
31	New Jersey	No. 2	Ringelmann	3 min discharge/60 min		40
		No. 1	Ringelmann	3 consecutive minutes		20
32	New Mexico	No. 1	Ringelmann	2 min discharge/60 min		20
33	New York (state)	40	% opacity	all other times	built before 1/26/67	40
	(state)	20	% opacity		built after 1/26/67	20
	(city)	No. 1	Ringelmann	3 min discharge/60 min		20
34	North Carolina	No. 3	Ringelmann	4 min discharge/60 min		60
35	North Dakota	No. 1	Ringelmann	all other times		20

Table 2-3. (*Continued*)

States	Value	Units	Regulation Process Conditions	Validity	Equivalent Common Regulation (% opacity)
36 Ohio	60	% opacity	3 min discharge/60 min		60
	20	% opacity	all other times		20
37 Oklahoma	No. 1	Ringelmann	all other times		20
	No. 3	Ringelmann			60
38 Oregon	40	% opacity	5 min discharge/60 min	built before 5/1/70	40
	20	% opacity	3 min discharge/60 min	built after 6/1/70	20
39 Pennsylvania	20	% opacity	3 min discharge/60 min		20
40 Puerta Rica	20	% opacity	all other times		20
	60	% opacity	6 min discharge/60 min		60
41 Rhode Island	20	% opacity	3 min discharge/60 min		20
42 South Carolina	No. 1	Ringelmann	3 min discharge/60 min		20
43 South Dakota	20	% opacity	all other times		20
	60	% opacity	3 min discharge/60 min		60
44 Tennessee	20	% opacity	5 min discharge/60 min		20
45 Texas	30	% opacity	5 min average	built before 1/31/72	30
	20	% opacity	5 min average	built after 1/31/72	20
46 Utah	No. 1	Ringelmann			20
47 Vermont	40	% opacity	6 min discharge/60 min	built before 4/30/70	40
	20	% opacity	6 min discharge/60 min	built after 4/30/70	20
48 Virginia	20	% opacity			20
49 Washington	20	% opacity	3 min discharge/60 min		20
	>20	% opacity	15 min/8 hr		>20
50 West Virginia	No. 1	Ringelmann		built after 4/1/72	20
51 Wisconsin	20	% opacity			20
52 Wyoming	20	% opacity			20

Source: Ref. 2-11.

THE NATIONAL AMBIENT AIR QUALITY STANDARDS

The Federal Government has established National Ambient Air Quality Standards (NAAQS) which attempt to define the desired, or permissible, maximum levels of pollutants in the air throughout the country. The NAAQS establishes clean air standards for minimal (Class I), moderate (Class II) and extensive (Class III) growth areas. In conjunction with the NAAQS geographical locations are designated as either attainment or non-attainment areas:

> *Attainment area.* Air in this geographical location is presently considered clean, i.e., within the definition of clean air in the NAAQS for the designated growth area classification.
>
> *Non-attainment area.* Air quality in a non-attainment area is below the quality established in the NAAQS for the designated growth area classification.

PREVENTION OF SIGNIFICANT DETERIORATION

A set of regulations have been established by the federal government to help achieve the air quality standards designated in the NAAQS.

Implementation of NSR and BACT procedures is required, as noted above, for significant sources of air pollution. The PSD defines a significant source as either of the following:

- Any source which has the potential to emit over 250 tons per year of pollutants into the atmosphere. Pollutants are any of those pollutants listed in Table 2-4.
- Any one of the 28 source categories listed in Table 2-5 with the potential to discharge more than 100 tons per year of pollutants into the atmosphere.

The pollutants are measured at the stack. The term "potential" refers to the total amount of pollutants which may be discharged by the process in question. For instance, although a unit may be in operation only 30 weeks a year, it can be operated for 52 weeks a year. The potential emissions therefore, refer to, in this illustration, 52 weeks per year operation, although the process may only operate a fraction of the year.

Table 2-4.   De Minimis Emission Values and
Monitoring Exemption.

Pollutant	Emission Rate (tons/year)	Air Quality Impact ($\mu g/m^3$) and Averaging Time
Carbon monoxide	100.	575, 8-hour
Nitrogen oxides	40.	14, 24-hour
Sulfur dioxide	40.	13, 24-hour
Total suspended particulates	25.	10, 24-hour
Ozone (volatile organic compounds)	40.	—[1]
Lead	0.6	0.1, 24-hour
Asbestos	0.007	—[2]
Beryllium	0.0004	0.0005, 24-hour
Mercury	0.1	0.25, 24-hour
Vinyl chloride	1.0	15, 24-hour
Fluorides	3.	0.25, 24-hour
Sulfuric acid mist	7.	—[2]
Total reduced sulfur (including H_2S)	10.	10, 1-hour
Reduced sulfur (including H_2S)	10.	10, 1-hour
Hydrogen sulfide	10.	0.023, 1-hour

[1] All cases where VOC emissions are less than 100 tons per year.
[2] No satisfactory monitoring technique available at this time.
Source: Ref. 2-12.

THE NEW SOURCE REVIEW PROCESS

The chart designated Fig. 2-2 illustrates the New Source Review (NSR) process. Following this chart:

If an installation does not constitute a major new or modified source it is not subject to federal PSD review. It must, however, comply with applicable emissions regulations of the state in which it is located. Some states have adopted modified provisions of the National Ambient Air Quality Standards (NAAQS) and may require simplified source modeling. A public hearing for a non-major source may or may not be necessary and a decision to hold a hearing is governed by state statute.

If a source is a major new or modified source, as defined previously, a determination must first be made as to its location. If it is not located in a nonattainment area the PSD requires that an assessment of the Best Available Con-

Table 2-5. Major Stationary Sources of Air Pollution.

Coal cleaning plants (with thermal dryers)
Kraft pulp mills
Portland cement plants
Primary zinc smelters
Iron and steel mills
Primary aluminum ore reduction plants
Primary copper smelters
Municipal incinerators capable of charging more than 250 tons of refuse per day
Hydrofluoric acid plants
Sulfuric acid plants
Nitric acid plants
Petroleum refineries
Lime plants
Phosphate rock processing plants
Coke oven batteries
Sulfur recovery plants
Carbon black plant (furnace process)
Primary lead smelters
Fuel conversion plants
Sintering plants
Secondary metal production facilities
Chemical process plants
Fossil-fuel boilers of more than 250,000,000 BTU/hr heat input
Petroleum storage and transfer facilities with a capacity exceeding 300,000 bbl
Taconite ore processing facilities
Glass fiber processing plants
Charcoal production facilities

Source: Ref. 2-12.

trol Technology (BACT) be applied to each pollutant emission in excess of the de minimis value. The BACT analysis evaluates the state-of-the art in air pollution control equipment for the application in question and considers the economic and energy implications of that equipment.

A screening model is applied. This is a relatively simple mathematical modeling technique to determine the order of magnitude of the resultant emissions on the NAAQS. If the result of the screening model is a discharge less than 50% of what is allowable under the NAAQS, further source study is not necessary and the permitting question can go to a public hearing. If the screening model indicates that a discharge equivalent to greater than 50% of the NAAQS allowable amount will occur, then further analysis must be performed. Ambient air monitoring is necessary and a complete computer modeling procedure must be performed. This procedure utilizes inputs from the anticipated source (pollutants discharged, stack temperature, stack gas velocity, etc.), meteorological data

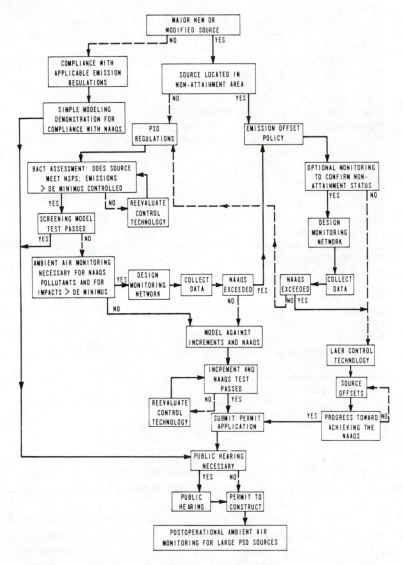

Fig. 2-2. Principal steps in the new source review process. *Source:* Ref. 2-12.

(prevailing winds, latitude, storm frequency, mean temperature, precipitation, etc.), and receptor data (topography, i.e., presence of hills or mountains and their height, water surfaces, location of other structures, etc.). If this modeling indicates that the NAAQS will not be met the emissions offset policy (EOP) pro-

Table 2-6. Air Emissions Glossary.

AQCR	Air Quality Control Region
BACT	Best Available Control Technology
CAA	Clean Air Act
CAAA	Clean Air Act Amendments
CTG	Control Technology Guideline
DEC	(New York State) Department of Environmental Conservation
EOP	Emission Offset Policy
EPA	Environmental Protection Agency
GEP	Good Engineering Policy
LAER	Lowest Achievable Emission Rate
NA	Non-Attainment
NAAQS	National Ambient Air Quality Standards
NESHAPs	National Emission Standards for Hazardous Air Pollutants
NSPS	New Source Performance Standards
NSR	New Source Review
ORD	Office of Research and Development
PSD	Prevention of Significant Deterioration
RACT	Reasonable Available Control Technology
RCRA	Resource Conservation and Recovery Act
RDF	Resource-Derived Fuel
SIP	State Implementation Plan
TSCA	Toxic Substances Control Act
VOC	Volatile Organic Compound

cedures must be instituted. If the NAAQS will be satisfied the way is open for a public hearing.

If the new (or modified) source is located in a non-attainment area, procedures of the emissions offset policy (EOP) must be followed. The EOP requires the following:

- Source emissions must be controlled to the greatest degree possible.
- More than equivalent offsetting emission reductions must be secured for existing sources on the same site.
- Progress must be made toward achieving the NAAQS.

After the EOP is satisfied the adequacy of the new source with respect to emissions can be demonstrated by a modeling procedure or by implementation of the lowest achievable emissions rate (LAER) review. This review is similar to the BACT review except that economic or energy factors are not taken into account.

If the modeling procedure is utilized and it is found that the NAAQS is exceeded, the LAER must be instituted. If the NAAQS is not exceeded, the source can be considered as if it were located in an "attainment" area.

Application of the LAER procedure must demonstrate that there is progress towards meeting the NAAQS before a public hearing can be advertised.

An agency review of the above procedure must prove successful before a public hearing is called. A permit is granted after a public hearing and a designated comment period.

Table 2-6 is a glossary of the acronyms used in this section.

Chapter 3
Hazardous Waste

As a result of an increased public awareness of the effect of industrialization on the environment, the quality of our natural resources and our national health, the Federal Government has established statutory control over discharges into the environment. In this chapter the nature of these requirements will be discussed with regard to incineration.

BACKGROUND

As population increases the quantity of wastes generated will increase. With technological advances the quality of wastes will change. As energy costs increase, many so-called wastes are no longer discarded but are used as fuels. Not all wastes are hazardous to the environment, or to the population. Most wastes generated, such as agricultural waste, domestic sewage and household waste, are not hazardous, although they are objectionable and must be effectively disposed of. Of the thousands of millions of tons of wastes generated in this country annually, approximately 60 million tons are classified as hazardous. The Federal Government, as a first step toward control of hazardous waste disposal, has established definitions and classifications of hazardous waste.

HAZARDOUS WASTE CLASSIFICATION

The Environmental Protection Agency through regulation 40 CFR Part 261, dated May 19, 1980, and subsequent revisions, has established a set of definitions of hazardous wastes:

1. Ignitable Waste (ignitability), Hazard Code "I," will have at least one of the following properties:

   a. A liquid having a flash point less than 140°F. An aqueous solution containing less than 20% alcohol by volume is excluded from this definition.
   b. A substance, other than a liquid, which can cause fire through friction, ab-

sorption of moisture or spontaneous chemical changes, under standard temperature and pressure. When this substance burns it does so vigorously and persistently.

c. An ignitable compressed gas (see 49 CFR 173.300 for further definition).

d. An oxidizer (see 40 CFR 173.151 for further definition).

An ignitable waste which is not listed elsewhere as a hazardous waste is given an EPA Hazardous Waste Number of D001.

2. Corrosive Waste (Corrosivity), Hazard Code "C," will have either or both of the following properties:

a. An aqueous waste with pH equal to or less than 2.0 or a pH equal to or greater than 12.5.

b. A liquid that corrodes carbon steel (grade SAE 1030) at a rate greater than 0.250 inch per year.

A corrosive waste which is not listed elsewhere as a hazardous waste is given an EPA Hazardous Waste Number of D002.

3. Reactive Waste (Reactivity), Hazard Code "R," will have at least one of the following properties:

a. A substance which is normally unstable and undergoes violent physical and/or chemical change without detonating.

b. A substance that reacts violently with water.

c. A waste which forms a potentially explosive mixture when wetted with water.

d. A substance which can generate harmful gases, vapors or fumes when mixed with water.

e. A cyanide or sulfide bearing waste which can generate harmful gases, vapors, or fumes when exposed to pH conditions between 2 and 12.5.

f. A waste which, when subjected to a strong initiating source or when heated in confinement, will detonate and/or generate an explosive reaction.

g. A substance which is readily capable of detonation at standard temperature and pressure.

h. An explosive listed as Class A, Class B, or "forbidden" in accordance with 49 CFR 173.

A reactive waste which is not listed elsewhere as a hazardous waste is given an EPA Hazardous Waste Number of D003.

4. EP Toxic Waste (EP Toxicity), Hazard Code "E." If the extract from a representative sample of this waste (EP, extract procedure) contains contamination in excess of that allowed in Table 3-1 it is classified as a hazardous waste.

Table 3-1.   Allowable Contaminants.

EPA hazardous waste number	Contaminant	Maximum concentration (milligrams per liter)
D004	Arsenic	5.0
D005	Barium	100.0
D006	Cadmium	1.0
D007	Chromium	5.0
D008	Lead	5.0
D009	Mercury	0.2
D010	Selenium	1.0
D011	Silver	5.0
D012	Endrin (1,2,3,4,10,10-hexachloro-1,7-epoxy-1,4,4a,5,6,7,8,8a-octahydro-1,4-endo, endo-5,8-dimethano naphthalene.	0.02
D013	Lindane (1,2,3,4,5,6-hexachlorocyclohexane, gamma isomer.	0.4
D014	Methoxychlor (1,1,1-Trichloro-2,2-bis [p-methoxyphenyl]ethane).	10.0
D015	Toxaphene ($C_{10}H_{10}Cl_6$, Technical chlorinated camphene, 67–69 percent chlorine).	0.5
D016	2,4-D, (2,4-Dichlorophenoxyacetic acid).	10.0
D017	2,4,5-TP Silvex (2,4,5-Trichlorophenoxypropionic acid).	1.0

Source: Ref. 3-9.

An EP Toxic Waste which is not listed elsewhere as a hazardous waste is given an EPA Hazardous Waste Number corresponding to that contaminant listed in Table 3-1 causing it to be hazardous.

5. Acute Hazardous Waste, Hazard Code "H." A substance which has been found to be fatal to humans in low doses, or in the absence of data on human toxicity, has been found to be fatal in corresponding human concentrations in laboratory animals.

6. Toxic Waste, Hazard Code "T." Wastes that have been found, through laboratory studies, to have a carcinogenic, mutagenic, or teratogenic effect on human or other life forms. Definitions of these terms are as follows:

- *Carcinogenic:* producing or tending to produce cancer.
- *Mutagenic:* capable of inducing mutations in future offspring.
- *Teratogenic:* producing abnormal growth in fetuses.

HAZARDOUS WASTE LISTINGS

Based on the above definitions a set of lists have been included in the RCRA regulation identifying hazardous wastes.

1. Nonspecific Sources. The listing in Table 3-2 identifies nonspecific sources the wastes of which are hazardous. The hazard code at the right of the listing refers to the quality of the waste creating its hazardous nature.

2. Specific Sources. Table 3-3 lists specific processes where the wastes generated are classified as hazardous. The waste hazard code identifies the reason(s) for its hazard classification.

3. Acute Hazardous Wastes. The substances identified in Table 3-4 when discarded are classified as acute hazardous wastes.

4. Toxic Wastes. The substances listed in Table 3-5 when discarded are classified as toxic hazardous wastes.

5. Hazardous Constitutents. A waste containing any of the substances listed in Table 3-6 is considered a hazardous waste.

ANCILLARY MATERIALS

The container or container liner in contact with a hazardous waste is itself hazardous. In addition, clothing, debris, soil, etc., which has become contaminated with a hazardous waste is considered hazardous. These hazardous waste definitions are subject to the small quantity exclusion noted later in this chapter.

NON-HAZARDOUS WASTES

A number of wastes are specifically excluded from classification as hazardous wastes under RCRA, including the following:

1. Domestic sewage.
2. Irrigation return flows.
3. Nuclear waste.
4. Household waste.
5. Wastes generated from the growing of crops and the raising of animals (manure) and which are returned to the soil as fertilizers.
6. Mining overburden returned to the mine site.
7. Fly ash, bottom ash, slag waste and waste from flue gas emissions control systems when generated from the burning of coal or other fossil fuels.
8. Wastes associated with the exploration, development, or production of crude oil, natural gas or geothermal energy such as drilling fluids and oil laden waters.

Table 3-2. Hazardous Waste from Nonspecific Sources.

Industry and EPA hazardous waste No.	Hazardous waste	Hazard code
Generic:		
F001	The spent halogenated solvents used in degreasing, tetrachloroethylene, trichloroethylene, methylene chloride, 1,1,1-trichloroethane, carbon tetrachloride, and the chlorinated fluorocarbons; and sludges from the recovery of these solvents in degreasing operations.	(T)
F002	The spent halogenated solvents, tetrachloroethylene, methylene chloride, trichloroethylene, 1,1,1-trichloroethane, chlorobenzene, 1,1,2-trichloro-1,2,2-trifluoroethane, o-dichlorobenzene, trichlorofluoromethane and the still bottoms from the recovery of these solvents.	(T)
F003	The spent non-halogenated solvents, xylene, acetone, ethyl acetate, ethyl benzene, ethyl ether, n-butyl alcohol, cyclohexanone, and the still bottoms from the recovery of these solvents.	(I)
F004	The spent non-halogenated solvents, cresols and cresylic acid, nitrobenzene, and the still bottoms from the recovery of these solvents.	(T)
F005	The spent non-halogenated solvents, methanol, toluene, methyl ethyl ketone, methyl isobutyl ketone, carbon disulfide, isobutanol, pyridine and the still bottoms from the recovery of these solvents.	(I, T)
F006	Wastewater treatment sludges from electroplating operations.	(T)
F007	Spent plating bath solutions from electroplating operations.	(R, T)
F008	Plating bath sludges from the bottom of plating baths from electroplating operations.	(R, T)
F009	Spent stripping and cleaning bath solutions from electroplating operations.	(R, T)
F010	Quenching bath sludge from oil baths from metal heat treating operations.	(R, T)
F011	Spent solutions from salt bath pot cleaning from metal heat treating operations.	(R, T)
F012	Quenching wastewater treatment sludges from metal heat treating operations.	(T)
F013	Flotation tailings from selective flotation from mineral metals recovery operations.	(T)
F014	Cyanidation wastewater treatment tailing pond sediment from mineral metals recovery operations.	(T)
F015	Spent cyanide bath solutions from mineral metals recovery operations.	(R, T)
F016	Dewatered air pollution control scrubber sludges from coke ovens and blast furnaces.	(T)

Source: Ref. 3-9.

Table 3-3.   Hazardous Waste from Specific Sources.

Industry and EPA hazardous waste No.	Hazardous waste	Hazard code
Wood Preservation: K001	Bottom sediment sludge from the treatment of wastewaters from wood preserving processes that use creosote and/or pentachlorophenol	(T)
Inorganic Pigments:		
K002	Wastewater treatment sludge from the production of chrome yellow and orange pigments	(T)
K003	Wastewater treatment sludge from the production of molybdate orange pigments	(T)
K004	Wastewater treatment sludge from the production of zinc yellow pigments	(T)
K005	Wastewater treatment sludge from the production of chrome green pigments	(T)
K006	Wastewater treatment sludge from the production of chrome oxide green pigments (anhydrous and hydrated)	(T)
K007	Wastewater treatment sludge from the production of iron blue pigments	(T)
K008	Oven residue from the production of chrome oxide green pigments	(T)
Organic Chemicals:		
K009	Distillation bottoms from the production of acetaldehyde from ethylene	(T)
K010	Distillation side cuts from the production of acetaldehyde from ethylene	(T)
K011	Bottom stream from the wastewater stripper in the production of acrylonitrile	(R,T)
K012	Still bottoms from the final purification of acrylonitrile in the production of acrylonitrile	(T)
K013	Bottom stream from the acetonitrile column in the production of acrylonitrile	(R,T)
K014	Bottoms from the acetonitrile purification column in the production of acrylonitrile	(T)
K015	Still bottoms from the distillation of benzyl chloride	(T)
K016	Heavy ends or distillation residues from the production of carbon tetrachloride	(T)
K017	Heavy ends (still bottoms) from the purification column in the production of epichlorohydrin	(T)
K018	Heavy ends from fractionation in ethyl chloride production	(T)
K019	Heavy ends from the distillation of ethylene dichloride in ethylene dichloride production	(T)
K020	Heavy ends from the distillation of vinyl chloride in vinyl chloride monomer production	(T)
K021	Aqueous spent antimony catalyst waste from fluoromethanes production	(T)
K022	Distillation bottom tars from the production of phenol/acetone from cumene	(T)
K023	Distillation light ends from the production of phthalic anhydride from naphthalene	(T)
K024	Distillation bottoms from the production of phthalic anhydride from naphthalene	(T)
K025	Distillation bottoms from the production of nitrobenzene by the nitration of benzene	(T)
K026	Stripping still tails from the production of methyl ethyl pyridines	(T)
K027	Centrifuge residue from toluene diisocyanate production	(R,T)
K028	Spent catalyst from the hydrochlorinator reactor in the production of 1,1,1-trichloroethane	(T)
K029	Waste from the product stream stripper in the production of 1,1,1-trichloroethane	(T)
K030	Column bottoms or heavy ends from the combined production of trichloroethylene and perchloroethylene	(T)
Pesticides:		
K031	By-products salts generated in the production of MSMA and cacodylic acid	(T)
K032	Wastewater treatment sludge from the production of chlordane	(T)
K033	Wastewater and scrub water from the chlorination of cyclopentadiene in the production of chlordane	(T)
K034	Filter solids from the filtration of hexachlorocyclopentadiene in the production of chlordane	(T)
K035	Wastewater treatment sludges generated in the production of creosote	(T)
K036	Still bottoms from toluene reclamation distillation in the production of disulfoton	(T)

Code	Description	
K037	Wastewater treatment sludges from the production of disulfoton	(E)
K038	Wastewater from the washing and stripping of phorate production	(E)
K039	Filter cake from the filtration of diethylphosphorodithoic acid in the production of phorate	(E)
K040	Wastewater treatment sludge from the production of phorate	(E)
K041	Wastewater treatment sludge from the production of toxaphene	(E)
K042	Heavy ends or distillation residues from the distillation of tetrachlorobenzene in the production of 2,4,5-T	(E)
K043	2,6-Dichlorophenol waste from the production of 2,4-D	(E)
Explosives:		
K044	Wastewater treatment sludges from the manufacturing and processing of explosives	(R)
K045	Spent carbon from the treatment of wastewater containing explosives	(R)
K046	Wastewater treatment sludges from the manufacturing, formulation and loading of lead-based initiating compounds	
K047	Pink/red water from TNT operations	(R)
Petroleum Refining:		
K048	Dissolved air flotation (DAF) float from the petroleum refining industry	(E)
K049	Slop oil emulsion solids from the petroleum refining industry	(E)
K050	Heat exchanger bundle cleaning sludge from the petroleum refining industry	(E)
K051	API separator sludge from the petroleum refining industry	(E)
K052	Tank bottoms (leaded) from the petroleum refining industry	(E)
Leather Tanning Finishing:		
K053	Chrome (blue) trimmings generated by the following subcategories of the leather tanning and finishing industry: hair pulp/chrome tan/retan/wet finish; hair save/chrome tan/retan/wet finish; retan/wet finish; no beamhouse; through-the-blue; and shearling.	(T)
K054	Chrome (blue) shavings generated by the following subcategories of the leather tanning and finishing industry: hair pulp/chrome tan/retan/wet finish; hair save/chrome tan/retan/wet finish; retan/wet finish; no beamhouse; through-the-blue; and shearling.	(T)
K055	Buffing dust generated by the following subcategories of the leather tanning and finishing industry: hair pulp/chrome tan/retan/wet finish; hair save/chrome tan/retan/wet finish; retan/wet finish; no beamhouse; and through-the-blue.	(T)
K056	Sewer screenings generated by the following subcategories of the leather tanning and finishing industry: hair pulp/chrome tan/retan/wet finish; hair save/chrome tan/retan/wet finish; retan/wet finish; no beamhouse; through-the-blue; and shearling.	(T)
K057	Wastewater treatment sludges generated by the following subcategories of the leather tanning and finishing industry: hair pulp/chrome tan/retan/wet finish; hair save/chrome tan/retan/wet finish; retan/wet finish; no beamhouse; through-the-blue and shearling.	(T)
K058	Wastewater treatment sludges generated by the following subcategories of the leather tanning and finishing industry: hair pulp/chrome tan/retan/wet finish; hair save/chrome tan/retan/wet finish; and through-the-blue.	(R, T)
K059	Wastewater treatment sludges generated by the following subcategory of the leather tanning and finishing industry: hair save/non-chrome tan/retan/wet finish.	(R)
Iron and Steel:		
K060	Ammonia still lime sludge from coking operations	(E)
K061	Emission control dust/sludge from the electric furnace production of steel	(E)
K062	Spent pickle liquor from steel finishing operations	(C, T)
K063	Sludge from lime treatment of spent pickle liquor from steel finishing operations	(E)
Primary Copper: K064	Acid plant blowdown slurry/sludge resulting from the thickening of blowdown slurry from primary copper production	(E)
Primary Lead: K065	Surface impoundment solids contained in and dredged from surface impoundments at primary lead smelting facilities	(E)
Primary Zinc:		
K066	Sludge from treatment of process wastewater and/or acid plant blowdown from primary zinc production	(E)
K067	Electrolytic anode slimes/sludges from primary zinc production	(E)
K068	Cadmium plant leach residue (iron oxide) from primary zinc production	(E)
Secondary Lead: K069	Emission control dust/sludge from secondary lead smelting	(E)

Source: Ref. 3-9.

Table 3-4.   Acute Hazardous Wastes.

Hazardous waste No.	Substance [1]	Hazardous waste No.	Substance [1]
	1080 see P058	P031.........	Cyanogen
	1081 see P057	P032.........	Cyanogen bromide
	(Acetato)phenylmercury see P092	P033.........	Cyanogen chloride
	Acetone cyanohydrin see P069		Cyclodan see P050
P001.........	3-(alpha-Acetonylbenzyl)-4-hydroxycoumarin and salts	P034.........	2-Cyclohexyl-4,6-dinitrophenol
			D-CON see P001
P002.........	1-Acetyl-2-thiourea		DETHMOR see P001
P003.........	Acrolein		DETHNEL see P001
	Agarin see P007		DFP see P043
	Agrosan GN 5 see P092	P035.........	2,4-Dichlorophenoxyacetic acid (2,4-D)
	Aldicarb see P069	P036.........	Dichlorophenylarsine
	Aldifen see P048		Dicyanogen see P031
P004.........	Aldrin	P037.........	Dieldrin
	Algimycin see P092		DIELDREX see P037
P005.........	Allyl alcohol	P038.........	Diethylarsine
P006.........	Aluminum phosphide (R)	P039.........	0,0-Diethyl-S-(2-(ethylthio)ethyl)ester of phosphorothioic acid
	ALVIT see P037		
	Aminoethylene see P054	P040.........	0,0-Diethyl-0-(2-pyrazinyl)phosphorothioate
P007.........	5-(Aminomethyl)-3-isoxazolol	P041.........	0,0-Diethyl phosphoric acid, 0-p-nitrophenyl ester
P008.........	4-Aminopyridine	P042.........	3,4-Dihydroxy-alpha-(methylamino)-methyl benzyl alcohol
	Ammonium metavanadate see P119		
P009.........	Ammonium picrate (R)	P043.........	Di-isopropylfluorophosphate
	ANTIMUCIN WDR see P092		DIMETATE see P044
	ANTURAT see P073		1,4:5,8-Dimethanonaphthalene, 1,2,3,4,10,10-hexachloro-1,4,4a,5,8,8a-hexahydro endo, endo see P060
	AQUATHOL see P088		
	ARETIT see P020		
P010.........	Arsenic acid	P044.........	Dimethoate
P011.........	Arsenic pentoxide	P045.........	3,3-Dimethyl-1-(methylthio)-2-butanone-O-[(methylamino)carbonyl] oxime
P012.........	Arsenic trioxide		
	Athrombin see P001	P046.........	alpha,alpha-Dimethylphenethylamine
	AVITROL see P008		Dinitrocyclohexylphenol see P034
	Aziridene see P054	P047.........	4,6-Dinitro-o-cresol and salts
	AZOFOS see P061	P048.........	2,4-Dinitrophenol
	Azophos see P061		DINOSEB see P020
	BANTU see P072		DINOSEBE see P020
P013.........	Barium cyanide		Disulfoton see P039
	BASENITE see P020	P049.........	2,4-Dithiobiuret
	BCME see P016		DNBP see P020
P014.........	Benzenethiol		DOLCO MOUSE CEREAL see P108
	Benzoepin see P050		DOW GENERAL see P020
P015.........	Beryllium dust		DOW GENERAL WEED KILLER see P020
P016.........	Bis(chloromethyl) ether		DOW SELECTIVE WEED KILLER see P020
	BLADAN-M see P071		DOWICIDE G see P090
P017.........	Bromoacetone		DYANACIDE see P092
P018.........	Brucine		EASTERN STATES DUOCIDE see P001
P019.........	2-Butanone peroxide		ELGETOL see P020
	BUFEN see P092	P050.........	Endosulfan
	Butaphene see P020	P051.........	Endrin
P020.........	2-sec-Butyl-4,6-dinitrophenol		Epinephrine see P042
P021.........	Calcium cyanide	P052.........	Ethylcyanide
	CALDON see P020	P053.........	Ethylenediamine
P022.........	Carbon disulfide	P054.........	Ethyleneimine
	CERESAN see P092		FASCO FASCRAT POWDER see P001
	CERESAN UNIVERSAL see P092		FEMMA see P091
	CHEMOX GENERAL see P020	P055.........	Ferric cyanide
	CHEMOX P.E. see P020	P056.........	Fluorine
	CHEM-TOL see P090	P057.........	2-Fluoroacetamide
P023..........	Chloroacetaldehyde	P058.........	Fluoroacetic acid, sodium salt
P024.........	p-Chloroaniline		FOLODOL-80 see P071
P025.........	1-(p-Chlorobenzoyl)-5-methoxy-2-methylindole-3-acetic acid		FOLODOL M see P071
			FOSFERNO M 50 see P071
P026.........	1-(o-Chlorophenyl)thiourea		FRATOL see P058
P027.........	3-Chloropropionitrile		Fulminate of mercury see P065
P028.........	alpha-Chlorotoluene		FUNGITOX OR see P092
P029.........	Copper cyanide		FUSSOF see P057
	CRETOX see P108		GALLOTOX see P092
	Coumadin see P001		GEARPHOS see P071
	Coumafen see P001		GERUTOX see P020
P030.........	Cyanides	P059.........	Heptachlor

Table 3-4. (Continued)

Hazardous waste No.	Substance [1]	Hazardous waste No.	Substance [1]
P060	1,2,3,4,10,10-Hexachloro-1,4,4a,5,8,8a-hexahydro-1,4:5,8-endo, endo-dimethanonaphthalene	P085	Octamethylpyrophosphoramide
			OCTAN see P092
	1,4,5,6,7,7-Hexachloro-cyclic-5-norbornene-2,3-dimethanol sulfite see P050	P086	Oleyl alcohol condensed with 2 moles ethylene oxide
P061	Hexachloropropene		OMPA see P085
P062	Hexaethyl tetraphosphate		OMPACIDE see P085
	HOSTAQUICK see P092		OMPAX see P085
	HOSTAQUIK see P092	P087	Osmium tetroxide
	Hydrazomethane see P068	P088	7-Oxabicyclo[2.2.1]heptane-2,3-dicarboxylic acid
P063	Hydrocyanic acid		PANIVARFIN see P001
	ILLOXOL see P037		PANORAM D-31 see P037
	INDOCI see P025		PANTHERINE see P007
	Indomethacin see P025		PANWARFIN see P001
	INSECTOPHENE see P050	P089	Parathion
	Isodrin see P060		PCP see P090
P064	Isocyanic acid, methyl ester		PENNCAP-M see P071
	KILOSEB see P020		PENOXYL CARBON N see P048
	KOP-THIODAN see P050	P090	Pentachlorophenol
	KWIK-KIL see P108		Pentachlorophenate see P090
	KWIKSAN see P092		PENTA-KILL see P090
	KUMADER see P001		PENTASOL see P090
	KYPFARIN see P001		PENWAR see P090
	LEYTOSAN see P092		PERMICIDE see P090
	LIQUIPHENE see P092		PERMAGUARD see P090
	MALIK see P050		PERMATOX see P090
	MAREVAN see P001		PERMITE see P090
	MAR-FRIN see P001		PERTOX see P090
	MARTIN'D MAR-FRIN see P001		PESTOX III see P085
	MAVERAN see P001		PHENMAD see P092
	MEGATOX see P005		PHENOTAN see P020
P065	Mercury fulminate	P091	Phenyl dichloroarsine
	MERSOLITE see P092		Phenyl mercaptan see P014
	METACID 50 see P071	P092	Phenylmercury acetate
	METAFOS see P071	P093	N-Phenylthiourea
	METAPHOR see P071		PHILIPS 1861 see P008
	METAPHOS see P071		PHIX see P092
	METASOL 30 see P092	P094	Phorate
P066	Methomyl	P095	Phosgene
P067	2-Methylaziridine	P096	Phosphine
	METHYL-E 605 see P071	P097	Phosphorothioic acid, 0,0-dimethyl ester, 0-ester with N,N-dimethyl benzene sulfonamide
P068	Methyl hydrazine		Phosphorothioic acid 0,0-dimethyl-0-(p-nitro-phenyl) ester see P071
	Methyl isocyanate see P064		
P069	2-Methyllactonitrile		PIED PIPER MOUSE SEED see P108
P070	2-Methyl-2-(methylthio)propionaldehyde-o-(methylcarbonyl) oxime	P098	Potassium cyanide
		P099	Potassium silver cyanide
	METHYL NIRON see P042		PREMERGE see P020
P071	Methyl parathion	P100	1,2-Propanediol
	METRON see P071		Propargyl alcohol see P102
	MOLE DEATH see P108	P101	Propionitrile
	MOUSE-NOTS see P108	P102	2-Propyn-1-o1
	MOUSE-RID see P108		PROTHROMADIN See P001
	MOUSE-TOX see P108		QUICKSAM see P092
	MUSCIMOL see P007		QUINTOX see P037
P072	1-Naphthyl-2-thiourea		RAT AND MICE BAIT see P001
P073	Nickel carbonyl		RAT-A-WAY see P001
P074	Nickel cyanide		RAT-B-GON see P001
P075	Nicotine and salts		RAT-O-CIDE #2 see P001
P076	Nitric oxide		RAT-GUARD see P001
P077	p-Nitroaniline		RAT-KILL see P001
P078	Nitrogen dioxide		RAT-MIX see P001
P079	Nitrogen peroxide		RATS-NO-MORE see P001
P080	Nitrogen tetroxide		RAT-OLA see P001
P081	Nitroglycerine (R)		RATOREX see P001
P082	N-Nitrosodimethylamine		RATTUNAL see P001
P083	N-Nitrosodiphenylamine		RAT-TROL see P001
P084	N-Nitrosomethylvinylamine		RO-DETH see P001
	NYLMERATE see P092		RO-DEX see P108
	OCTALOX see P037		ROSEX see P001

Table 3-4.   (Continued)

Hazardous waste No.	Substance [1]	Hazardous waste No.	Substance [1]
	ROUGH & READY MOUSE MIX see P001		Thallium peroxide see P113
	SANASEED see P108	P114	Thallium selenite
	SANTOBRITE see P090	P115	Thallium (I) sulfate
	SANTOPHEN see P090		THIFOR see P092
	SANTOPHEN 20 see P090		THIMUL see P092
	SCHRADAN see P085		THIODAN see P050
P103	Selenourea		THIOFOR see P050
P104	Silver Cyanide		THIOMUL see P050
	SMITE see P105		THIONEX see P050
	SPARIC see P020		THIOPHENIT see P071
	SPOR-KIL see P092	P116	Thiosemicarbazide
	SPRAY-TROL BRAND RODEN-TROL see P001		Thiosulfan tionel see P050
	SPURGE see P020	P117	Thiuram
P105	Sodium azide		THOMPSON'S WOOD FIX see P090
	Sodium coumadin see P001		TIOVEL see P050
P106	Sodium cyanide	P118	Trichloromethanethiol
	Sodium fluoroacetate see P056		TWIN LIGHT RAT AWAY see P001
	SODIUM WARFARIN see P001		USAF RH–8 see P069
	SOLFARIN see P001		USAF EK–4890 see P002
	SOLFOBLACK BB see P048	P119	Vanadic acid, ammonium salt
	SOLFOBLACK SB see P048	P120	Vanadium pentoxide
P107	Strontium sulfide		VOFATOX see P071
P108	Strychnine and salts		WANADU see P120
	SUBTEX see P020		WARCOUMIN see P001
	SYSTAM see P085		WARFARIN SODIUM see P001
	TAG FUNGICIDE see P092		WARFICIDE see P001
	TEKWAISA see P071		WOFOTOX see P072
	TEMIC see P070		YANOCK see P057
	TEMIK see P070		YASOKNOCK see P058
	TERM-I-TROL see P090		ZIARNIK see P092
P109	Tetraethyldithiopyrophosphate	P121	Zinc cyanide
P110	Tetraethyl lead	P122	Zinc phosphide (R,T)
P111	Tetraethylpyrophosphate		ZOOCOUMARIN see P001
P112	Tetranitromethane		
	Tetraphosphoric acid, hexaethyl ester see P062		[1] The Agency included those trade names of which it was
	TETROSULFUR BLACK PB see P048		aware; an omission of a trade name does not imply that the
	TETROSULPHUR PBR see P048		omitted material is not hazardous. The material is hazardous
P113	Thallic oxide		if it is listed under its generic name.

Source: Ref. 3-9.

USABLE HAZARDOUS WASTE

Regulations applicable to disposal of hazardous wastes are not currently applicable to a hazardous waste that meets any of the following criteria:

1. The waste is being recycled or reclaimed.
2. The waste is in storage or is being treated prior to its reclamation.

Reclamation includes the use of a waste for the generation of heat energy. Of note, however, is that the waste must have significant heating value. The regulations state specifically that a waste cannot be fired to avoid regulation under this provision unless it can legitimately be used for heat generation and recovery.

QUANTITY EXCLUSION

Provisions are included in the RCRA regulations to exempt small generators of hazardous waste from the rigorous procedures necessary for compliance with the disposal statutes. The small generator exclusion is as follows:

Table 3-5. Toxic Wastes.

Hazardous Waste No.	Substance[1]	Hazardous Waste No.	Substance[1]
	AAF see U005	U055.........	Cumene
U001.........	Acetaldehyde		Cyanomethane see U003
U002.........	Acetone (I)	U056.........	Cyclohexane (I)
U003.........	Acetonitrile (I,T)	U057.........	Cyclohexanone (I)
U004.........	Acetophenone	U058.........	Cyclophosphamide
U005.........	2-Acetylaminoflourene	U059.........	Daunomycin
U006.........	Acetyl chloride (C,T)	U060.........	DDD
U007.........	Acrylamide	U061.........	DDT
	Acetylene tetrachloride see U209	U062.........	Diallate
	Acetylene trichloride see U228	U063.........	Dibenz[a,h]anthracene
U008.........	Acrylic acid (I)		Dibenzo[a,h]anthracene see U063
U009.........	Acrylonitrile	U064.........	Dibenzo[a,i]pyrene
	AEROTHENE TT see U226	U065.........	Dibromochloromethane
	3-Amino-5-(p-acetamidophenyl)-1H-1,2,4-triazole, hydrate see U011	U066.........	1,2-Dibromo-3-chloropropane
		U067.........	1,2-Dibromoethane
U010.........	6-Amino-1,1a,2,8,8a,8b-hexahydro-8-(hydroxymethyl)8-methoxy-5-methylcarbamate azirino(2′,3′:3,4) pyrrolo(1,2-a) indole-4, 7-dione (ester)	U068.........	Dibromomethane
		U069.........	Di-n-butyl phthalate
		U070.........	1,2-Dichlorobenzene
		U071.........	1,3-Dichlorobenzene
		U072.........	1,4-Dichlorobenzene
U011.........	Amitrole	U073.........	3,3′-Dichlorobenzidine
U012.........	Aniline (I)	U074.........	1,4-Dichloro-2-butene
U013.........	Asbestos		3,3′-Dichloro-4,4′-diaminobiphenyl see U073
U014.........	Auramine	U075.........	Dichlorodifluoromethane
U015.........	Azaserine	U076.........	1,1-Dichloroethane
U016.........	Benz[c]acridine	U077.........	1,2-Dichloroethane
U017.........	Benzal chloride	U078.........	1,1-Dichloroethylene
U018.........	Benz[a]anthracene	U079.........	1,2-trans-dichloroethylene
U019.........	Benzene	U080.........	Dichloromethane
U020.........	Benzenesulfonyl chloride (C,R)		Dichloromethylbenzene see U017
U021.........	Benzidine	U081.........	2,4-Dichlorophenol
	1,2-Benzisothiazolin-3-one, 1,1-dioxide see U202	U082.........	2,6-Dichlorophenol
	Benzo[a]anthracene see U018	U083.........	1,2-Dichloropropane
U022.........	Benzo[a]pyrene	U084.........	1,3-Dichloropropene
U023.........	Benzotrichloride (C,R,T)	U085.........	Diepoxybutane (I,T)
U024.........	Bis(2-chloroethoxy)methane	U086.........	1,2-Diethylhydrazine
U025.........	Bis(2-chloroethyl) ether	U087.........	O,O-Diethyl-S-methyl ester of phosphorodithioic acid
U026.........	N,N-Bis(2-chloroethyl)-2-naphthylamine		
U027.........	Bis(2-chloroisopropyl) ether		
U028.........	Bis(2-ethylhexyl) phthalate	U088.........	Diethyl phthalate
U029.........	Bromomethane	U089.........	Diethylstilbestrol
U030.........	4-Bromophenyl phenyl ether	U090.........	Dihydrosafrole
U031.........	n-Butyl alcohol (I)	U091.........	3,3′-Dimethoxybenzidine
U032.........	Calcium chromate	U092.........	Dimethylamine (I)
	Carbolic acid see U188	U093.........	p-Dimethylaminoazobenzene
	Carbon tetrachloride see U211	U094.........	7,12-Dimethylbenz[a]anthracene
U033.........	Carbonyl fluoride	U095.........	3,3′-Dimethylbenzidine
U034.........	Chloral	U096.........	alpha,alpha-Dimethylbenzylhydroperoxide (R)
U035.........	Chlorambucil	U097.........	Dimethylcarbamoyl chloride
U036.........	Chlordane	U098.........	1,1-Dimethylhydrazine
U037.........	Chlorobenzene	U099.........	1,2-Dimethylhydrazine
U038.........	Chlorobenzilate	U100.........	Dimethylnitrosoamine
U039.........	p-Chloro-m-cresol	U101.........	2,4-Dimethylphenol
U040.........	Chlorodibromomethane	U102.........	Dimethyl phthalate
U041.........	1-Chloro-2,3-epoxypropane	U103.........	Dimethyl sulfate
	CHLOROETHENE NU see U226	U104.........	2,4-Dinitrophenol
U042.........	Chloroethyl vinyl ether	U105.........	2,4-Dinitrotoluene
U043.........	Chloroethene	U106.........	2,6-Dinitrotoluene
U044.........	Chloroform (I,T)	U107.........	Di-n-octyl phthalate
U045.........	Chloromethane (I,T)	U108.........	1,4-Dioxane
U046.........	Chloromethyl methyl ether	U109.........	1,2-Diphenylhydrazine
U047.........	2-Chloronaphthalene	U110.........	Dipropylamine (I)
U048.........	2-Chlorophenol	U111.........	Di-n-propylnitrosamine
U049.........	4-Chloro-o-toluidine hydrochloride		EBDC see U114
U050.........	Chrysene		1,4-Epoxybutane see U213
	C.I. 23060 see U073	U112.........	Ethyl acetate (I)
U051.........	Cresote	U113.........	Ethyl acrylate (I)
U052.........	Cresols	U114.........	Ethylenebisdithiocarbamate
U053.........	Crotonaldehyde	U115.........	Ethylene oxide (I,T)
U054.........	Cresylic acid		

Table 3-5.   (Continued)

Hazardous Waste No.	Substance[1]	Hazardous Waste No.	Substance[1]
U116.........	Ethylene thiourea		Nitrobenzol see U169
U117.........	Ethyl ether (I,T)	U170.........	4-Nitrophenol
U118.........	Ethylmethacrylate	U171.........	2-Nitropropane (I)
U119.........	Ethyl methanesulfonate	U172.........	N-Nitrosodi-n-butylamine
	Ethylnitrile see U003	U173.........	N-Nitrosodiethanolamine
	Firemaster T23P see U235	U174.........	N-Nitrosodiethylamine
U120.........	Fluoranthene	U175.........	N-Nitrosodi-n-propylamine
U121.........	Fluorotrichloromethane	U176.........	N-Nitroso-n-ethylurea
U122.........	Formaldehyde	U177.........	N-Nitroso-n-methylurea
U123.........	Formic acid (C,T)	U178.........	N-Nitroso-n-methylurethane
U124.........	Furan (I)	U179.........	N-Nitrosopiperidine
U125.........	Furfural (I)	U180.........	N-Nitrosopyrrolidine
U126.........	Glycidylaldehyde	U181.........	5-Nitro-o-toluidine
U127.........	Hexachlorobenzene	U182.........	Paraldehyde
U128.........	Hexachlorobutadiene		PCNB see U185
U129.........	Hexachlorocyclohexane	U183.........	Pentachlorobenzene
U130.........	Hexachlorocyclopentadiene	U184.........	Pentachloroethane
U131.........	Hexachloroethane	U185.........	Pentachloronitrobenzene
U132.........	Hexachlorophene	U186.........	1,3-Pentadiene (I)
U133.........	Hydrazine (R,T)		Perc see U210
U134.........	Hydrofluoric acid (C,T)		Perchlorethylene see U210
U135.........	Hydrogen sulfide	U187.........	Phenacetin
	Hydroxybenzene see U188	U188.........	Phenol
U136.........	Hydroxydimethyl arsine oxide	U189.........	Phosphorous sulfide (R)
	4,4'-(Imidocarbonyl)bis(N,N-dimethyl)aniline see U014	U190.........	Phthalic anhydride
U137.........	Indeno(1,2,3-cd)pyrene	U191.........	2-Picoline
U138.........	Iodomethane	U192.........	Pronamide
U139.........	Iron Dextran	U193.........	1,3-Propane sultone
U140.........	Isobutyl alcohol	U194.........	n-Propylamine (I)
U141.........	Isosafrole	U196.........	Pyridine
U142.........	Kepone	U197.........	Quinones
U143.........	Lasiocarpine	U200.........	Reserpine
U144.........	Lead acetate	U201.........	Resorcinol
U145.........	Lead phosphate	U202.........	Saccharin
U146.........	Lead subacetate	U203.........	Safrole
U147.........	Maleic anhydride	U204.........	Selenious acid
U148.........	Maleic hydrazide	U205.........	Selenium sulfide (R,T)
U149.........	Malononitrile		Silvex see U233
	MEK Peroxide see U160	U206.........	Streptozotocin
U150.........	Melphalan		2,4,5-T see U232
U151.........	Mercury	U207.........	1,2,4,5-Tetrachlorobenzene
U152.........	Methacrylonitrile	U208.........	1,1,1,2-Tetrachloroethane
U153.........	Methanethiol	U209.........	1,1,2,2-Tetrachloroethane
U154.........	Methanol	U210.........	Tetrachloroethene
U155.........	Methapyrilene		Tetrachloroethylene see U210
	Methyl alcohol see U154	U211.........	Tetrachloromethane
U156.........	Methyl chlorocarbonate	U212.........	2,3,4,6-Tetrachlorophenol
	Methyl chloroform see U226	U213.........	Tetrahydrofuran (I)
U157.........	3-Methylcholanthrene	U214.........	Thallium (I) acetate
	Methyl chloroformate see U156	U215.........	Thallium (I) carbonate
U158.........	4,4'-Methylene-bis-(2-chloroaniline)	U216.........	Thallium (I) chloride
U159.........	Methyl ethyl ketone (MEK) (I,T)	U217.........	Thallium (I) nitrate
U160.........	Methyl ethyl ketone peroxide (R)	U218.........	Thioacetamide
	Methyl iodide see U138	U219.........	Thiourea
U161.........	Methyl isobutyl ketone	U220.........	Toluene
U162.........	Methyl methacrylate (R,T)	U221.........	Toluenediamine
U163.........	N-Methyl-N'-nitro-N-nitrosoguanidine	U222.........	o-Toluidine hydrochloride
U164.........	Methylthiouracil	U223.........	Toluene diisocyanate
	Mitomycin C see U010	U224.........	Toxaphene
U165.........	Naphthalene		2,4,5-TP see U233
U166.........	1,4-Naphthoquinone	U225.........	Tribromomethane
U167.........	1-Naphthylamine	U226.........	1,1,1-Trichloroethane
U168.........	2-Naphthylamine	U227.........	1,1,2-Trichloroethane
U169.........	Nitrobenzene (I,T)	U228.........	Trichloroethene
			Trichloroethylene see U228

Table 3-5. (Continued)

Hazardous Waste No.	Substance[1]	Hazardous Waste No.	Substance[1]
U229.........	Trichlorofluoromethane	U237.........	Uracil mustard
U230..........	2,4,5-Trichlorophenol	U238.........	Urethane
U231..........	2,4,6-Trichlorophenol		Vinyl chloride see U043
U232.........	2,4,5-Trichlorophenoxyacetic acid		Vinylidene chloride see U078
U233.........	2,4,5-Trichlorophenoxypropionic acid alpha, alpha, alpha- Trichlorotoluene see U023 TRI-CLENE see U228	U239.........	Xylene
U234.........	Trinitrobenzene (R,T)		
U235.........	Tris(2,3-dibromopropyl) phosphate		
U236.........	Trypan blue		

[1] The Agency included those trade names of which it was aware; an omission of a trade name does not imply that it is not hazardous. The material is hazardous if it is listed under its generic name.

Source: Ref. 3-9.

Table 3-6. Hazardous Constituents.

Acetaldehyde
(Acetato)phenylmercury
Acetonitrile
3-(alpha-Acetonylbenzyl)-4-hydroxycoumarin
   and salts
2-Acetylaminofluorene
Acetyl chloride
1-Acetyl-2-thiourea
Acrolein
Acrylamide
Acrylonitrile
Aflatoxins
Aldrin
Allyl alcohol
Aluminum phosphide
4-Aminobiphenyl
6-Amino-1,1a,2,8,8a,8b-hexahydro-8-
   (hydroxymethyl)-8a-methoxy-5-
   methylcarbamate azirino(2′,3′:3,4)
   pyrrolo(1,2-a)indole-4,7-dione (ester)
   (Mitomycin C)
5-(Aminomethyl)-3-isoxazolol
4-Aminopyridine
Amitrole
Antimony and compounds, N.O.S.[1]
Aramite
Arsenic and compounds, N.O.S.
Arsenic acid
Arsenic pentoxide
Arsenic trioxide
Auramine

Azaserine
Barium and compounds, N.O.S.
Barium cyanide
Benz[c]acridine
Benz[a]anthracene
Benzene
Benzenearsonic acid
Benzenethiol
Benzidine
Benzo[a]anthracene
Benzo[b]fluoranthene
Benzo[j]fluoranthene
Benzo[a]pyrene
Benzotrichloride
Benzyl chloride
Beryllium and compounds, N.O.S.
Bis(2-chloroethoxy)methane
Bis(2-chloroethyl) ether
N,N-Bis(2-chloroethyl)-2-naphthylamine
Bis(2-chloroisopropyl) ether
Bis(chloromethyl) ether
Bis(2-ethylhexyl) phthalate
Bromoacetone
Bromomethane
4-Bromophenyl phenyl ether
Brucine
2-Butanone peroxide
Butyl benzyl phthalate
2-sec-Butyl-4,6-dinitrophenol (DNBP)
Cadmium and compounds, N.O.S.
Calcium chromate
Calcium cyanide
Carbon disulfide
Chlorambucil
Chlordane (alpha and gamma isomers)

[1] The abbreviation N.O.S. signifies those members of the general class "not otherwise specified" by name in this listing.

Table 3-6.   (Continued)

Chlorinated benzenes, N.O.S.
Chlorinated ethane, N.O.S.
Chlorinated naphthalene, N.O.S.
Chlorinated phenol, N.O.S.
Chloroacetaldehyde
Chloroalkyl ethers
p-Chloroaniline
Chlorobenzene
Chlorobenzilate
1-(p-Chlorobenzoyl)-5-methoxy-2-
    methylindole-3-acetic acid
p-Chloro-m-cresol
1-Chloro-2,3-epoxybutane
2-Chloroethyl vinyl ether
Chloroform
Chloromethane
Chloromethyl methyl ether
2-Chloronaphthalene
2-Chlorophenol
1-(o-Chlorophenyl)thiourea
3-Chloropropionitrile
alpha-Chlorotoluene
Chlorotoluene, N.O.S.
Chromium and compounds, N.O.S.
Chrysene
Citrus red No. 2
Copper cyanide
Creosote
Crotonaldehyde
Cyanides (soluble salts and complexes),
    N.O.S.
Cyanogen
Cyanogen bromide
Cyanogen chloride
Cycasin
2-Cyclohexyl-4,6-dinitrophenol
Cyclophosphamide
Daunomycin
DDD
DDE
DDT
Diallate
Dibenz[a,h]acridine
Dibenz[a,j]acridine
Dibenz[a,h]anthracene(Dibenzo[a,h]
    anthracene)
7H-Dibenzo[c,g]carbazole
Dibenzo[a,e]pyrene
Dibenzo[a,h]pyrene
Dibenzo[a,i]pyrene
1,2-Dibromo-3-chloropropane
1,2-Dibromoethane
Dibromomethane
Di-n-butyl phthalate
Dichlorobenzene, N.O.S.

3,3'-Dichlorobenzidine
1,1-Dichloroethane
1,2-Dichloroethane
trans-1,2-Dichloroethane
Dichloroethylene, N.O.S.
1,1-Dichloroethylene
Dichloromethane
2,4-Dichlorophenol
2,6-Dichlorophenol
2,4-Dichlorophenoxyacetic acid (2,4-D)
Dichloropropane
Dichlorophenylarsine
1,2-Dichloropropane
Dichloropropanol, N.O.S.
Dichloropropene, N.O.S.
1,3-Dichloropropene
Dieldrin
Diepoxybutane
Diethylarsine
0,0-Diethyl-S-(2-ethylthio)ethyl ester of
    phosphorothioic acid
1,2-Diethylhydrazine
0,0-Diethyl-S-methylester phosphorodithioic
    acid
0,0-Diethylphosphoric acid, 0-p-nitrophenyl
    ester
Diethyl phthalate
0,0-Diethyl-0-(2-pyrazinyl)phosphorothioate
Diethylstilbestrol
Dihydrosafrole
3,4-Dihydroxy-alpha-(methylamino)-methyl
    benzyl alcohol
Di-isopropylfluorophosphate (DFP)
Dimethoate
3,3'-Dimethoxybenzidine
p-Dimethylaminoazobenzene
7,12-Dimethylbenz[a]anthracene
3,3'-Dimethylbenzidine
Dimethylcarbamoyl chloride
1,1-Dimethylhydrazine
1,2-Dimethylhydrazine
3,3-Dimethyl-1-(methylthio)-2-butanone-0-
    ((methylamino) carbonyl)oxime
Dimethylnitrosoamine
alpha,alpha-Dimethylphenethylamine
2,4-Dimethylphenol
Dimethyl phthalate
Dimethyl sulfate
Dinitrobenzene, N.O.S.
4,6-Dinitro-o-cresol and salts
2,4-Dinitrophenol
2,4-Dinitrotoluene
2,6-Dinitrotoluene Di-n-octyl phthalate
1,4-Dioxane
1,2-Diphenylhydrazine

Table 3-6. (Continued)

Di-n-propylnitrosamine
Disulfoton
2,4-Dithiobiuret
Endosulfan
Endrin and metabolites
Epichlorohydrin
Ethyl cyanide
Ethylene diamine
Ethylenebisdithiocarbamate (EBDC)
Ethyleneimine
Ethylene oxide
Ethylenethiourea
Ethyl methanesulfonate
Fluoranthene
Fluorine
2-Fluoroacetamide
Fluoroacetic acid, sodium salt
Formaldehyde
Glycidylaldehyde
Halomethane, N.O.S.
Heptachlor
Heptachlor epoxide (alpha, beta, and gamma
   isomers)
Hexachlorobenzene
Hexachlorobutadiene
Hexachlorocyclohexane (all isomers)
Hexachlorocyclopentadiene
Hexachloroethane
1,2,3,4,10,10-Hexachloro-1,4,4a,5,8,8a-
   hexahydro-1,4:5,8-endo,endo-
   dimethanonaphthalene
Hexachlorophene
Hexachloropropene
Hexaethyl tetraphosphate
Hydrazine
Hydrocyanic acid
Hydrogen sulfide
Indeno(1,2,3-c,d)pyrene
Iodomethane
Isocyanic acid, methyl ester
Isosafrole
Kepone
Lasiocarpine
Lead and compounds, N.O.S.
Lead acetate
Lead phosphate
Lead subacetate
Maleic anhydride
Malononitrile
Melphalan
Mercury and compounds, N.O.S.
Methapyrilene
Methomyl
2-Methylaziridine
3-Methylcholanthrene

4,4'-Methylene-bis-(2-chloroaniline)
Methyl ethyl ketone (MEK)
Methyl hydrazine
2-Methyllactonitrile
Methyl methacrylate
Methyl methanesulfonate
2-Methyl-2-(methylthio)propionaldehyde-o-
   (methylcarbonyl) oxime
N-Methyl-N'-nitro-N-nitrosoguanidine
Methyl parathion
Methylthiouracil
Mustard gas
Naphthalene
1,4-Naphthoquinone
1-Naphthylamine
2-Naphthylamine
1-Naphthyl-2-thiourea
Nickel and compounds, N.O.S.
Nickel carbonyl
Nickel cyanide
Nicotine and salts
Nitric oxide
p-Nitroaniline
Nitrobenzene
Nitrogen dioxide
Nitrogen mustard and hydrochloride salt
Nitrogen mustard N-oxide and hydrochloride
   salt
Nitrogen peroxide
Nitrogen tetroxide
Nitroglycerine
4-Nitrophenol
4-Nitroquinoline-1-oxide
Nitrosamine, N.O.S.
N-Nitrosodi-N-butylamine
N-Nitrosodiethanolamine
N-Nitrosodiethylamine
N-Nitrosodimethylamine
N-Nitrosodiphenylamine
N-Nitrosodi-N-propylamine
N-Nitroso-N-ethylurea
N-Nitrosomethylethylamine
N-Nitroso-N-methylurea
N-Nitroso-N-methylurethane
N-Nitrosomethylvinylamine
N-Nitrosomorpholine
N-Nitrosonornicotine
N-Nitrosopiperidine
N-Nitrosopyrrolidine
N-Nitrososarcosine
5-Nitro-o-toluidine
Octamethylpyrophosphoramide
Oleyl alcohol condensed with 2 moles
   ethylene oxide
Osmium tetroxide

Table 3-6.   (Continued)

7-Oxabicyclo[2.2.1]heptane-2,3-dicarboxylic acid
Parathion
Pentachlorobenzene
Pentachloroethane
Pentachloronitrobenzene (PCNB)
Pentacholorophenol
Phenacetin
Phenol
Phenyl dichloroarsine
Phenylmercury acetate
N-Phenylthiourea
Phosgene
Phosphine
Phosphorothioic acid, O,O-dimethyl ester, O-ester with N,N-dimethyl benzene sulfonamide
Phthalic acid esters, N.O.S.
Phthalic anhydride
Polychlorinated biphenyl, N.O.S.
Potassium cyanide
Potassium silver cyanide
Pronamide
1,2-Propanediol
1,3-Propane sultone
Propionitrile
Propylthiouracil
2-Propyn-1-ol
Pryidine
Reserpine
Saccharin
Safrole
Selenious acid
Selenium and compounds, N.O.S.
Selenium sulfide
Selenourea
Silver and compounds, N.O.S.
Silver cyanide
Sodium cyanide
Streptozotocin
Strontium sulfide
Strychnine and salts
1,2,4,5-Tetrachlorobenzene
2,3,7,8-Tetrachlorodibenzo-p-dioxin (TCDD)
Tetrachloroethane, N.O.S.
1,1,1,2-Tetrachloroethane
1,1,2,2-Tetrachloroethane
Tetrachloroethene (Tetrachloroethylene)

Tetrachloromethane
2,3,4,6-Tetrachlorophenol
Tetraethyldithiopyrophosphate
Tetraethyl lead
Tetraethylpyrophosphate
Thallium and compounds, N.O.S.
Thallic oxide
Thallium (I) acetate
Thallium (I) carbonate
Thallium (I) chloride
Thallium (I) nitrate
Thallium selenite
Thallium (I) sulfate
Thioacetamide
Thiosemicarbazide
Thiourea
Thiuram
Toluene
Toluene diamine
o-Toluidine hydrochloride
Tolylene diisocyanate
Toxaphene
Tribromomethane
1,2,4-Trichlorobenzene
1,1,1-Trichloroethane
1,1,2-Trichloroethane
Trichloroethene (Trichloroethylene)
Trichloromethanethiol
2,4,5-Trichlorophenol
2,4,6-Trichlorophenol
2,4,5-Trichlorophenoxyacetic acid (2,4,5-T)
2,4,5-Trichlorophenoxypropionic acid (2,4,5-TP) (Silvex)
Trichloropropane, N.O.S.
1,2,3-Trichloropropane
0,0,0-Triethyl phosphorothioate
Trinitrobenzene
Tris(1-azridinyl)phosphine sulfide
Tris(2,3-dibromopropyl) phosphate
Trypan blue
Uracil mustard
Urethane
Vanadic acid, ammonium salt
Vanadium pentoxide (dust)
Vinyl chloride
Vinylidene chloride
Zinc cyanide
Zinc phosphide

Source: Ref. 3-9.

1. Generation of less than 1000 kilograms (2200 pounds) per month of hazardous wastes, unless otherwise specified.
2. Accumulation of less than 1000 kilograms (2200 pounds) of hazardous waste in any one calendar month, unless otherwise specified.
3. A waste containing no more than one kilogram (2.2 pounds) of any product or chemical listed in Table 3-4 (acute hazardous waste).
4. Containers not greater than 20 liters (5.3 gallons) or liners no greater than 10 kilograms (22 pounds) in weight, holding materials listed in Table 3-4 (acute hazardous waste).
5. No greater than 100 kilograms (220 pounds) of residue, soil contaminated from a spill, etc. from a material listed in Table 3-4 (acute hazardous waste).

The regulations allow that a hazardous waste subject to the small quantity exclusion may be mixed with non-hazardous waste and remain subject to this exclusion even though the resultant mixture exceeds the small quantity limitations.

GENERAL CONSIDERATIONS

It is the responsibility of the waste generator to determine if his waste is hazardous, and then to classify it with an appropriate hazardous waste code. He is then obligated to report the existence of this waste, its transport and method of disposal using a manifest which is described in 40 CFR Part 260.

These regulations will undoubtedly be subject to intense review as they are applied. Litigation and subsequent revisions are to be expected and these regulations should be monitored with regularity because of these expected changes.

HAZARDOUS WASTE INCINERATION REGULATIONS

If a hazardous waste is hazardous only because it has characteristics of ignitability or of corrosivity or if the hazardous waste will be reused, reclaimed or used for heat recovery its incineration does not have to comply with hazardous waste incineration criteria. In addition, certain wastes which are hazardous solely because of reactivity may be incinerated without compliance with these criteria. These criteria require that a test burn be run on the principal organic hazardous constitutent (POHC). The POHC is that compound listed in Table 3-6 that is present in the waste. If more than one compound from Table 3-6 can be identified, the selection of the POHC will be a matter of judgement based on the difficulty of incineration, quantity present and toxicity of that compound. More than one of the compounds present may be designated as the POHC.

The test burn is one phase of a four phase permit process. Permitting provides for a one-month shake-down phase prior to performance of a test burn. The test burn is the second phase of the permitting process. The third phase comprises limited operation while results of the test burn are evaluated. The fourth phase is the final or permanent operating phase. The permit requirements, promulgated in Federal regulations 40 CFR Parts 122, 264 and 265, include the following provisions:

1. Identification, by the generator, of the POHCs.
2. Operation of incineration equipment to achieve a Destruction and Removal Efficiency (DRE) of at least 99.99%. The DRE is defined, with W_{in} the POHC mass rate into the system and W_{out} the POHC mass rate leaving in the incinerator exhaust gas stream, as follows:

$$DRE = \frac{(W_{in} - W_{out}) \times 100\%}{W_{in}}$$

3. If the hydrogen chloride exiting the stack is less than 4 lbs/hr no HCl removal is necessary. If the stack emission contains in excess of 4 lbs/hr of hydrogen chloride, 99% of the hydrogen chloride produced must be removed from the exhaust gas stream.
4. Particulate emissions into the atmosphere must not exceed 0.08 grain per dry standard cubic foot when corrected to 50% excess air. A correction to 50% excess air is provided by the following formula:

$$Pc = Pm \times \frac{14}{21 - Y}$$

where Pm is the measured particulate, Y is the percent oxygen by volume and Pc is the corrected particulate concentration.
5. Continuous monitoring is required for combustion temperature, waste feed rate, combustion gas flow rate and carbon monoxide (CO in the exhaust gas stream).

In lieu of a test burn, which requires EPA approval prior to its implementation, operating criteria can be established on the basis of published data on disposal of the POHC's in question. The published data, however, must have been generated from incinerator equipment similar in type to the proposed incinerator.

WHEN CRITERIA ARE NOT AVAILABLE

For purposes of initial design, where test data are not immediately available for a particular waste, a reasonable estimate of combustion criteria can be assumed, as follows:

1. For a non-halogenated hazardous waste provide 1000°C (1832°F) in the combustion chamber, with a retention time of 2 seconds and 2% oxygen in the exhaust.
2. For halogenated hazardous waste (one containing at least 0.5% chlorine) provide a temperature of 1200°C (2192°F) in the combustion chamber, with a retention time of 2 seconds, and 3% oxygen in the exhaust.

Note that these criteria are not stated within RCRA or other federal statutory requirements. They are only suggested here as criteria to be used as a guide to equipment design where no other data are available and are not presented in lieu of a test burn.

PCB INCINERATION

Incineration of substances containing PCB's (Polychlorinated Biphenyls) is covered by the Toxic Substances Control Act (TSCA) 44FR106 paragraph 761.41. The incineration of PCB's is subject to the following criteria, exerpted from TSCA.

1. Combustion at 1200°C (2192°F), with a 2 second retention time at 3% excess oxygen in the exhaust gas or 1600°C (2912°F), with a $1\frac{1}{2}$ second retention time at 2% excess oxygen in the exhaust.

2. Combustion efficiency (CE of 99.9%. With C_{CO_2} and C_{CO} the concentration of CO_2 and CO in the exhaust gas, respectively, the CE is calculated as follows:

$$CE = \frac{C_{CO_2}}{C_{CO_2} + C_{CO}} \times 100\%$$

3. The PCB charging rate and total feed must be monitored at least every 15 minutes.

4. The combustion temperature must be monitored on a continuous basis.

5. Upon a drop in temperature below 1200°C (or 1600°C) the flow of PCBs shall automatically cease. PCB flow shall also cease if there is a failure in the monitoring operations or if excess oxygen falls below that required.

6. When an incinerator is initially used for disposal of PCBs the following stack emissions must be monitored:

a. Oxygen (O_2).
b. Carbon monoxide (CO).
c. Carbon dioxide (CO_2).
d. Oxides of nitrogen (NO_x).

    e. Hydrogen chloride (HCl).

    f. Total chlorinated organic component (RCl).

    g. PCBs.

    h. Total particulate matter.

7. During normal operation of the incinerator the CO_2 concentration in the exhaust gas shall be monitored on a periodic basis. On a continuous basis the O_2 and CO component of the exhaust shall be monitored.

8. Water scrubbers or equivalent gas cleaning equipment shall be used to control the HCl emission in the exhaust gas. Spent scrubber water effluent must be monitored and shall be in compliance with applicable effluent standards.

9. If the PCBs to be incinerated are non-liquid in addition to the above requirements the mass air emissions shall not be greater than 1 pound PCB per million pounds of PCBs charged into the furnace.

COMMENTARY ON PCB INCINERATOR REGULATIONS

The EPA intends to include PCB regulations in the RCRA statutes at a later date, removing these regulations from TSCA. In this transferral of authority changes may be made to these regulations.

Other considerations regarding the present PCB incineration regulations include the following:

1. Instrumentation for incinerator monitoring of low levels of CO (for combustion efficiency calculations) for PCB and HCl detection from incinerator systems is not readily available. They are being developed on an application by application basis, advancing the present state of the art.

2. A requirement of 15 minute monitoring is impractical for manual recordkeeping. It would require a full time operator who does nothing but record data. This requirement mandates use of continuous automatic monitoring equipment.

Chapter 4
Basic Thermodynamics

Thermodynamics is defined as the science concerned with the conversion of heat to energy. Many texts deal with heat conversion, thermodynamic properties of materials, and the production of energy. This chapter is not intended as a substitution for these texts but a concise explanation of basic thermodynamic concepts relative to the incineration process including enthalpy, energy levels and psychometric analysis.

ELEMENTAL LAWS OF THERMODYNAMICS

The two basic laws of thermodynamics can be simply stated as follows:

1. In any reaction the energy output must equal the energy into the system.
2. In any process the flow of heat will always be from a region of higher to a region of lower temperature.

UNITS OF ENERGY

The unit of energy in common use in the United States is the British Thermal Unit, BTU. One BTU is that amount of heat required to raise the temperature of one pound of water one degree Fahrenheit. The calorie is an equivalent unit, that amount of heat which will raise the temperature of one gram of water one degree centigrade.

In accordance with the first law of thermodynamics the energy input to a system must be accounted for within that system. A corollary of this law is the equivalence of heat and work. For instance, a fixed quantity of water stirred by a mixer (mechanical work) will experience an equivalent rise in temperature (heat energy). The constants of conversion of mechanical and heat energy, as well as other thermodynamic conversions, are presented as follows:

Heat Conversions

1 BTU	$\equiv$	778 Foot-Pounds
1 BTU	$\equiv$	1055 Joules
1 BTU	$\equiv$	252 calories
1 BTU	$\equiv$	0.252 Calories
1 BTU	$\equiv$	0.0002931 Kilowatt-Hours
1 horsepower-hour	$\equiv$	2544 BTU
1 kilowatt-hour	$\equiv$	3412 BTU
1 Calorie	$\equiv$	3.968 BTU
1 Calorie	$\equiv$	1 kilocalorie (1000 calories)

ENTHALPY

To facilitate quantification if these laws a number of material properties have been defined. Of interest to an understanding of combustion and related processes the total energy, or enthalpy, is of particular concern. Enthalpy h, BTU/lb, is the total energy of a substance at one temperature relative to the energy of the same substance at another temperature. It is a relative property and whenever it is used it must be related to a base or datum point.

Table 4-1 lists the enthalpy of water vapor. The enthalpy of air is listed relative to the enthalpy of air at 60°F and at 80°F. The enthalpy of water vapor (steam) is listed relative to liquid water at both 60°F and 80°F.

Table 4-2 lists the enthalpy of gases other than air and moisture (steam) normally found in flue gas, i.e., CO_2, N_2, and O_2.

REACTION TEMPERATURES

The temperature of a reaction can be calculated knowing the heat produced by the reaction and the quantities of products generated in the reaction. For example:

If 15 pounds of water vapor (steam) is produced in a reaction generating 20000 BTU and all of this heat is absorbed by the water vapor, and assuming that the water vapor entered the reaction as liquid water at 60°F, what is the temperature of the water vapor (steam) exiting the reaction?

In this case the steam carries 20000 BTU ÷ 15 pounds or 1333 BTU/lb, a definition of its enthalpy. Consulting Table 4-1, an enthalpy of 1333 BTU/lb corresponds to a temperature of approximately 650°F, based on a 60°F datum.

This simple example illustrates that the temperature of a substance is that

Table 4-1.   Enthalpy, Air and Moisture.

Relative to 60°F		Temp., °F	Relative to 80°F	
H_{Air}, BTU/lb	H_{H_2O} BTU/lb		H_{Air}, BTU/lb	H_{H_2O}, BTU/lb
21.61	1091.92	150	16.82	1071.91
33.65	1116.62	200	28.86	1096.61
45.71	1140.72	250	40.92	1120.71
57.81	1164.52	300	53.02	1144.51
69.98	1188.22	350	65.19	1168.21
82.19	1211.82	400	77.40	1191.81
94.45	1235.47	450	89.66	1215.46
106.79	1259.22	500	102.00	1239.21
119.21	1283.07	550	114.42	1263.06
131.69	1307.12	600	126.90	1287.11
144.25	1331.27	650	139.46	1311.26
156.87	1355.72	700	152.08	1335.71
169.59	1380.27	750	164.80	1360.26
187.38	1405.02	800	177.59	1385.01
195.26	1430.02	850	190.47	1410.01
208.21	1455.32	900	203.42	1435.31
221.25	1480.72	950	216.46	1460.71
234.36	1506.42	1000	229.57	1486.41
247.55	1532.40	1050	242.76	1512.40
260.81	1558.32	1100	256.02	1538.31
274.15	1584.80	1150	264.36	1564.80
287.55	1611.22	1200	282.76	1591.21
301.02	1638.26	1250	296.23	1618.20
314.56	1665.12	1300	309.77	1645.11
328.17	1692.15	1350	323.38	1672.15
341.85	1719.82	1400	337.06	1699.81
355.58	1747.70	1450	364.58	1727.70
369.37	1775.52	1500	364.58	1755.51
397.17	1832.12	1600	392.33	1812.11
425.08	1890.11	1700	420.29	1870.10
453.24	1948.02	1800	448.45	1928.01
481.57	2007.17	1900	476.78	1987.70
510.07	2067.42	2000	505.28	2047.41
538.72	2128.70	2100	533.93	2108.70
567.52	2189.92	2200	562.73	2169.91
596.45	2252.60	2300	591.66	2232.60
625.52	2315.32	2400	620.73	2295.31
654.70	2377.80	2500	649.91	2357.80
684.01	2443.30	2600	679.22	2423.30
713.42	2511.88	2700	708.63	2491.80

Source: Ref. 4-7 and 4-9.

Table 4-2. Enthalpies of Various
Gases Expressed in BTU/lb of Gas.

Temp, °F	CO_2	N_2	O_2
100	5.8	6.4	8.8
150	17.6	20.6	19.8
200	29.3	34.8	30.9
250	40.3	47.7	42.1
300	51.3	59.8	53.4
350	63.1	73.3	64.8
400	74.9	84.9	76.2
450	87.0	97.5	87.8
500	99.1	110.1	99.5
550	111.8	122.9	111.3
600	124.5	135.6	123.2
700	150.2	161.4	147.2
800	176.8	187.4	171.7
900	204.1	213.8	196.5
1000	231.9	240.5	221.6
1100	260.2	267.5	247.0
1200	289.0	294.9	272.7
1300	318.0	326.1	298.5
1400	347.6	350.5	324.6
1500	377.6	378.7	350.8
1600	407.8	407.3	377.3
1700	438.2	435.9	403.7
1800	469.1	464.8	430.4
1900	500.1	493.7	457.3
2000	531.4	523.0	484.5
2100	562.8	552.7	511.4
2200	594.3	582.0	538.6
2300	626.2	612.3	566.1
2400	658.2	642.3	593.5
2500	690.2	672.3	621.0
3000	852.3	823.8	760.1
3500	1017.4	978.0	901.7

Source: Ref. 4-14.

temperature at which its enthalpy is equal to the total heat contained within that substance.

For a more complex example:

A liquid at 60°F is combusted and releases 700,000 BTU. The products of combustion are 1000 pounds of dry gas, which has properties identical to that of air, and 250 pounds of moisture. Using Table 4-1, assuming a reaction temperature of 1200°F:

Air: 287.55 BTU/lb × 1000 lb = 287,550 BTU
H_2O: 1611.22 BTU/lb × 250 lb = 402,805 BTU
                    Total heat content = 690,355 BTU

The heat release, 700,000 BTU, is greater than the heat contained by the gas mixture at 1200°F. Therefore, the mixture temperature must be greater than 1200°F. Trying 1250°F:

Air: 301.02 BTU/lb × 1000 lb = 301,020 BTU
          1638.26 BTU/lb × 250 lb = 409,565 BTU
                    Total heat content = 710,585 BTU

The heat content at 1250°F is greater than that of the heat released, 700,000 BTU. The mixture temperature x must lie between 1200°F and 1250°F, as follows;

Temperature	Heat Content
1200	690,355
x	700,000
1250	710,585

By interpolation:

$$x = 1200 + (1250 - 1200)\, \frac{700,000 - 690,355}{710,585 - 690,355} = 1224°F$$

The mixture temperature is 1224°F.

The determination of temperature is a trial and error calculation, based on component enthalpies and heat release.

Note that the moisture enthalpy values in Table 4-2 are based on moisture in the liquid phase at 60°F or 80°F. If moisture enters a reaction in the vapor phase its heat of vaporization, that amount of heat required to change water from a liquid to a gas (steam) without a change in temperature, must be subtracted .from the moisture value in Table 4-1. At 60°F the heat of vaporization is 1059.6 BTU/lb and at 80°F the heat of vaporization is 1048.3 BTU/lb.

In the previous example, with moisture entering the reaction at 60°F, in the vapor phase, the temperature of the mixture, y, is calculated as follows:

At 1800°F:

Air: 435.24 BTU/lb × 1000 lb = 453,240 BTU
H_2O: (1948.02 - 1059.6) BTU/lb × 250 lb = 222,104 BTU
                            Total heat content = 675,344 BTU

At 1900°F:

Air:	481.57 BTU/lb × 1000 lb = 481,570 BTU
H_2O:	(2007.17 - 1059.6) BTU/lb × 250 lb = 236,893 BTU
	Total heat content = 718,463 BTU

To summarize:

Temperature	Heat Content
1800	675,344
y	700,000
1900	718,463

By interpolation:

$$y = 1800 + (1900 - 1800) \frac{700,000 - 675,345}{718,463 - 675,345} = 1857°F$$

The mixture temperature is 1857°F. Note that the heat of vaporization is a significant factor in the calculated heat level. When present as liquid water, the resultant temperature is 1224°F. Without inclusion of the heat of vaporization, i.e., if moisture enters the reaction as steam or vapor, the temperature is 1857°F.

THE PERFECT GAS LAW

In general, gases at low (atmospheric) pressures will behave in accordance with the perfect gas law, as follows:

$$144P = WRT$$

where

P = Pressure, psia
W = Specific weight, lb/ft^3
T = Absolute temperature, degrees Rankine
   (the absolute temperature is equal to $t + 460$ where t is degrees farenheit)
R = Universal gas constant divided by the molecular weight of the gas, ft/°F

The universal gas constant, 1545, is unvarying for any gas. The molecular weight and the R value of some common gases are listed in Table 4-3.

The gas law is convenient for calculating W, the specific weight of a gas at a particular temperature and pressure. For example:

Table 4-3.

	Gas Molecular Weight	R Value (ft/°R)
Air	28.9	53.3
Oxygen	32.00	48.3
Hydrogen	2.02	764.9
Nitrogen	28.02	55.1
Carbon monoxide	28.01	55.2
Carbon dioxide	44.01	35.1
Sulfur dioxide	64.06	24.01
Water vapor	18.02	85.83

What is the specific weight of nitrogen at 14.7 psia and 100°F? Using the perfect gas law, with $R = 55.1$ from Table 4-3, $P = 14.7$ and $T = 100F° + 460° = 560°R°$,

$$W = \frac{144P}{RT} = \frac{144 \times 14.7}{55.1 \times 560} = 0.0686 \text{ lb/ft}^3$$

Table 4-4 lists specific volumes, ft^3/lb, the reciprocal of specific weight, for air and water vapor at atmospheric pressure, 14.7 psia.

Table 4-4. Specific Volume.

T, °F	Air, ft^3/lb	H_2O, ft^3/lb	T, °F	Air, ft^3/lb	H_2O, ft^3/lb
70	13.3	21.5	2000	61.9	99.7
100	14.1	22.7	2100	64.5	103.8
200	16.6	26.8	2200	67.0	107.9
300	19.1	30.8	2300	69.5	111.9
400	21.7	34.9	2400	72.0	116.0
500	24.2	38.9	2500	74.5	120.0
600	26.7	43.0	2600	77.0	124.1
700	29.2	47.0	2700	79.6	128.1
800	31.7	51.1	2800	82.1	132.2
900	34.2	55.1	2900	84.6	136.2
1000	36.8	59.2	3000	87.1	140.3
1100	39.3	63.3	3100	89.6	144.3
1200	41.8	67.3	3200	92.2	148.4
1300	44.3	71.4	3300	94.7	152.5
1400	46.8	75.4	3400	97.2	156.5
1500	49.4	79.5	3500	99.7	160.6
1600	51.9	83.5	3600	102.2	164.6
1700	54.4	87.6	3700	104.7	168.7
1800	56.9	91.6	3800	107.3	172.7
1900	59.4	95.7	3900	109.8	176.8

A MIXTURE OF GASES

A mixture of two or more gases has the following properties:

- The temperature of each gas is the same as that of the mixture.
- The weight of the mixture is equal to the sum of the weights of each of the component gases.
- The volume of the mixture is equal to the sum of the volumes of each gas each calculated at the mixture pressure.
- The enthalpy of the mixture is equal to the sum of the total enthalpies of each of the component gases.

An illustration of calculations involving these mixture properties is as follows:

A mixture of 1000 pounds of air plus 200 pounds of steam is contained, at a pressure of 14.7 psia at 1200°F. Determine the total enthalpy of the mixture, the weight and volume percentage of each component.

The enthalpy of the mixture is equal to the sum of the component enthalpies (from Table 2, 60°F datum):

$$\begin{aligned}
\text{Air at } 1200°F: \quad & 287.55 \text{ BTU/lb} \times 1000 \text{ lb} = 287{,}550 \text{ BTU} \\
\text{H}_2\text{O at } 1200°F: \quad & 1611.22 \text{ BTU/lb} \times 200 \text{ lb} = \underline{322{,}244 \text{ BTU}} \\
& \text{Total enthalpy} = 609{,}794 \text{ BTU}
\end{aligned}$$

The total weight of the mixture is equal to the sum of the component weights:

$$\begin{aligned}
\text{Weight of Air} &= 1000 \text{ lb} \\
\text{Weight of H}_2\text{O} &= \underline{\ 200 \text{ lb}} \\
\text{Total weight} &= 1200 \text{ lb}
\end{aligned}$$

Therefore, the enthalpy is

$$609{,}794 \div 1200 \text{ lb} = 508 \text{ BTU/lb mixture}$$

The weight percentage:

$$\begin{aligned}
\text{Air:} \quad & 1000 \text{ lb}/1200 \text{ lb} = 83\% \\
\text{H}_2\text{O:} \quad & 200 \text{ lb}/1200 \text{ lb} = 17\%
\end{aligned}$$

From Table 4-4, the mixture volume is:

$$\begin{aligned}
\text{Air at } 1200°F: \quad & 41.8 \text{ ft}^3/\text{lb} \times 1000 \text{ lb} = 41{,}800 \text{ ft}^3 \\
\text{H}_2\text{O at } 1200°F: \quad & 67.3 \text{ ft}^3/\text{lb} \times 200 \text{ lb} = \underline{13{,}460 \text{ ft}^3} \\
& \text{Total} = 55{,}260 \text{ ft}^3
\end{aligned}$$

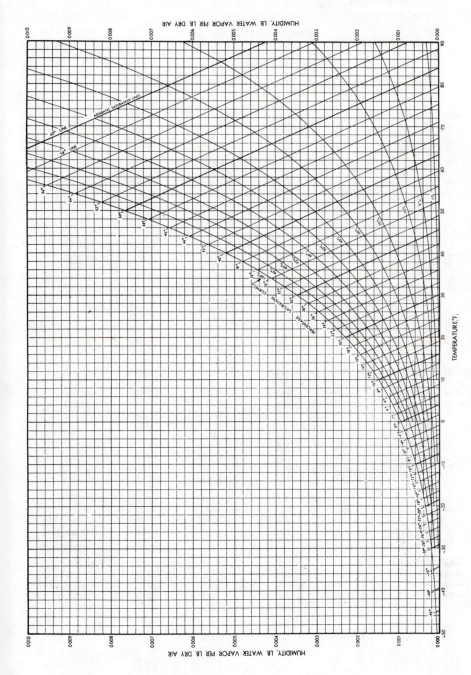

Fig. 4-1. Adiabatic saturation lines and percentage saturation curves. *Source:* Ref. 4-10.

Table 4-5. Properties of Dry Air, Water Vapor, and Saturated Air-Water Vapor Mixtures.
Temperature Range, –100 to +211°F; Pressure, 29.921″ Hg

Temp., °F.	Saturation pressure		Saturation humdity, Wt. water vapor/lb. dry air		Saturation moisture content, Wt. water vapor/cu. ft. sat. mixture	
	Lb./sq. in. $\times 10^4$	In. of Hg $\times 10^4$	Pounds $\times 10^5$	Grains	Pounds $\times 10^7$	Grains
−100	0.47315	0.9633	0.2002	0.01402	2.213	0.001549
−99	.51235	1.0431	.2168	.01518	2.390	.001673
−98	.55446	1.1289	.2347	.01643	2.579	.001806
−97	.59979	1.2212	.2538	.01777	2.783	.001948
−96	.64852	1.3204	.2745	.01922	3.000	.002100
−95	0.70090	1.4270	0.2966	0.02077	3.234	0.002264
−94	.75721	1.5417	.3205	.02243	3.484	.002439
−93	.81760	1.6646	.3460	.02422	3.751	.002626
−92	.88249	1.7968	.3735	.02614	4.038	.002827
−91	.95209	1.9385	.4029	.02821	4.345	.003041
−90	1.0267	2.0904	0.4345	0.03042	4.672	0.003271
−89	1.1068	2.2534	.4684	.03279	5.023	.003516
−88	1.1925	2.4279	.5047	.03533	5.397	.003778
−87	1.2844	2.6150	.5436	.03805	5.797	.004058
−86	1.3828	2.8153	.5852	.04095	6.225	.004357
−85	1.4880	3.0296	0.6298	0.04408	6.680	0.004676
−84	1.6007	3.2590	.6774	.04742	7.167	.005017
−83	1.7212	3.5043	.7284	.05099	7.686	.005380
−82	1.8500	3.7665	.7829	.05481	8.239	.005767
−81	1.9876	4.0468	.8412	.05888	8.828	.006180
−80	2.1346	4.3460	0.9034	0.06324	9.456	0.006619
−79	2.2914	4.6653	0.9698	.06788	10.124	.007087
−78	2.4592	5.0070	1.0408	.07286	10.837	.007586
−77	2.6379	5.3709	1.1164	.07815	11.593	.008115
−76	2.8266	5.7590	1.1971	.08380	12.399	.008679
−75	3.0320	6.1732	1.283	0.08983	13.26	0.009279
−74	3.2486	6.6142	1.375	.09624	14.17	.009910
−73	3.4793	7.0838	1.473	.10308	15.13	.010592
−72	3.7254	7.5850	1.577	.11037	16.16	.011312
−71	3.9869	8.1173	1.687	.11811	17.25	.012074
−70	4.2653	8.6842	1.805	0.1264	18.41	0.01289
−69	4.5619	9.2882	1.931	.1352	19.64	.01375
−68	4.8769	9.9295	2.064	.1445	20.94	.01466
−67	5.2118	10.6114	2.206	.1544	22.32	.01562
−66	5.5678	11.3362	2.357	.1650	23.78	.01595
−65	5.9460	12.106	2.517	0.1762	25.33	0.01773
−64	6.3477	12.924	2.687	.1881	26.98	.01888
−63	6.7741	13.792	2.867	.2007	28.71	.02010
−62	7.2262	14.713	3.058	.2141	30.55	.02139
−61	7.7061	15.690	3.262	.2283	32.50	.02275
−60	8.2152	16.726	3.477	0.2434	34.56	0.02419
−59	8.7545	17.824	3.705	.2594	36.74	.02572
−58	9.3265	18.989	3.947	.2763	39.04	.02733
−57	9.9325	20.223	4.204	.2943	41.47	.02903
−56	10.5745	21.530	4.476	.3133	44.04	.03083

Saturation density, Wt. air plus water vapor/cu. ft. sat. mixture		Volume			Enthalpy			Temp., °F
Pounds	Grains	Dry air, cu. ft./lb.	Water vapor, cu. ft./lb.*	Saturated mixture, cu. ft./lb. dry air	Dry air, B.t.u./lb.	Water vapor, B.t.u./lb.	Saturated mixture, B.t.u./lb. dry air	
0.1105	773.8	9.047	14.580	9.047	−31.691	1016.6	−31.689	−100
.1102	771.6	9.072	14.620	9.072	−31.451	1017.1	−31.449	−99
.1099	769.4	9.098	14.661	9.098	−31.211	1017.5	−31.209	−98
.1096	767.3	9.123	14.702	9.123	−30.971	1018.0	−30.968	−97
.1093	765.2	9.148	14.742	9.148	−30.731	1018.4	−30.728	−96
0.1090	763.1	9.173	14.783	9.173	−30.491	1018.8	−30.488	−95
.1087	761.0	9.199	14.823	9.199	−30.251	1019.3	−30.248	−94
.1084	758.9	9.224	14.864	9.224	−30.011	1019.7	−30.007	−93
.1081	756.8	9.249	14.904	9.249	−29.771	1020.2	−29.767	−92
.1078	754.7	9.275	14.944	9.275	−29.531	1020.6	−29.527	−91
0.1075	752.7	9.300	14.985	9.300	−29.291	1021.1	−29.287	−90
.1072	750.6	9.325	15.026	9.326	−29.051	1021.5	−29.046	−89
.1069	748.6	9.351	15.066	9.351	−28.811	1021.9	−28.806	−88
.1067	746.6	9.376	15.107	9.376	−28.571	1022.4	−28.565	−87
.1064	744.6	9.402	15.147	9.402	−28.331	1022.8	−28.325	−86
0.1061	742.6	9.427	15.188	9.427	−28.091	1023.3	−28.085	−85
.1058	740.6	9.452	15.229	9.452	−27.851	1023.7	−27.844	−84
.1055	738.6	9.478	15.269	9.478	−27.611	1024.2	−27.604	−83
.1052	736.6	9.503	15.310	9.503	−27.371	1024.6	−27.362	−82
.1050	734.7	9.528	15.350	9.528	−27.130	1025.0	−27.122	−81
0.1047	732.7	9.554	15.390	9.554	−26.890	1025.5	−26.881	−80
.1044	730.7	9.579	15.431	9.579	−26.650	1025.9	−26.640	−79
.1041	728.8	9.604	15.472	9.604	−26.410	1026.4	−26.399	−78
.1039	727.0	9.630	15.512	9.630	−26.170	1026.8	−26.159	−77
.1036	725.0	9.655	15.553	9.655	−25.930	1027.3	−25.918	−76
0.1033	723.1	9.680	15.593	9.680	−25.690	1027.7	−25.677	−75
.1030	721.2	9.706	15.634	9.706	−25.450	1028.1	−25.436	−74
.1028	719.3	9.731	15.675	9.731	−25.210	1028.6	−25.195	−73
.1025	717.5	9.756	15.715	9.757	−24.970	1029.0	−24.954	−72
.1022	715.6	9.782	15.756	9.782	−24.730	1029.5	−24.712	−71
0.1020	713.8	9.807	15.796	9.807	−24.490	1029.9	−24.471	−70
.1017	711.9	9.832	15.837	9.833	−24.250	1030.4	−24.230	−69
.1014	710.1	9.858	15.877	9.858	−24.010	1030.8	−23.989	−68
.1012	708.3	9.883	15.917	9.883	−23.770	1031.2	−23.747	−67
.1009	706.4	9.908	15.958	9.909	−23.530	1031.7	−23.506	−66
0.10067	704.7	9.934	15.999	9.934	−23.290	1032.1	−23.264	−65
.10041	702.9	9.959	16.039	9.959	−23.050	1032.6	−23.022	−64
.10016	701.1	9.984	16.080	9.985	−22.810	1033.0	−22.780	−63
.09990	699.3	10.010	16.120	10.010	−22.570	1033.5	−22.538	−62
.09965	697.6	10.035	16.160	10.035	−22.330	1033.9	−22.296	−61
0.09940	695.8	10.060	16.201	10.061	−22.090	1034.3	−22.054	−60
.09915	694.0	10.086	16.242	10.086	−21.850	1034.8	−21.812	−59
.09890	692.3	10.111	16.283	10.112	−21.610	1035.2	−21.569	−58
.09865	690.6	10.136	16.323	10.137	−21.370	1035.7	−21.326	−57
.09841	688.9	10.162	16.364	10.162	−21.130	1036.1	−21.084	−56

Table 4-5. (Continued)

Temp., °F.	Saturation pressure		Saturation humidity, Wt. water vapor /lb. dry air		Saturation moisture content, Wt. water vapor/cu. ft. sat. mixture	
	Lb./sq. in. $\times 10^2$	In. of Hg. $\times 10^2$	Pounds $\times 10^4$	Grains	Pounds $\times 10^6$	Grains
−55	0.11254	0.22913	0.4763	0.3334	4.675	0.03273
−54	.11973	.24377	.5067	.3547	4.962	.03473
−53	.12733	.25926	.5389	.3773	5.264	.03685
−52	.13538	.27564	.5730	.4011	5.583	.03908
−51	.14389	.29296	.6090	.4263	5.919	.04143
−50	0.15288	0.31126	0.6471	0.4529	6.273	0.04391
−49	.16238	.33061	.6873	.4811	6.647	.04653
−48	.17242	.35106	.7298	.5109	7.041	.04929
−47	.18305	.37269	.7748	.5424	7.456	.05220
−46	.19421	.39543	.8221	.5754	7.892	.05525
−45	0.20603	0.41949	0.8721	0.6105	8.352	0.05846
−44	.21850	.44488	0.9249	.6474	8.836	.06185
−43	.23164	.47161	0.9805	.6863	9.345	.06541
−42	.24552	.49988	1.0392	.7275	9.830	.06916
−41	.26014	.52964	1.1011	.7708	10.444	.07311
−40	0.27555	0.56102	1.166	0.8165	11.04	0.07725
−39	.29177	.59405	1.235	.8645	11.66	.08161
−38	.30889	.62890	1.308	.9153	12.31	.08618
−37	.32687	.66551	1.384	.9686	13.00	.09099
−36	.34582	.70409	1.464	1.0247	13.72	.09604
−35	0.36577	0.74472	1.548	1.084	14.48	0.1013
−34	.38676	.78744	1.637	1.146	15.27	.1069
−33	.40881	.83236	1.730	1.211	16.10	.1127
−32	.43204	.87964	1.829	1.280	16.98	.1189
−31	.45643	.92929	1.932	1.353	17.89	.1253
−30	0.48207	0.9815	2.041	1.429	18.86	0.1320
−29	.50900	1.0363	2.155	1.508	19.86	.1390
−28	.53727	1.0939	2.275	1.592	20.92	.1464
−27	.56697	1.1544	2.400	1.680	22.02	.1542
−26	.59816	1.2179	2.533	1.773	23.18	.1623
−25	0.63090	1.2845	2.671	1.870	24.39	0.1707
−24	.66522	1.3544	2.817	1.972	25.66	.1796
−23	.70129	1.4278	2.969	2.079	26.99	.1889
−22	.73899	1.5046	3.129	2.190	28.38	.1986
−21	.77856	1.5852	3.297	2.308	29.83	.2088
−20	0.82009	1.6697	3.473	2.431	31.35	0.2194
−19	.86362	1.7583	3.657	2.560	32.93	.2305
−18	.90924	1.8512	3.850	2.695	34.59	.2422
−17	.95698	1.9484	4.053	2.837	36.33	.2543
−16	1.00696	2.0502	4.265	2.985	38.14	.2670
−15	1.0592	2.1567	4.486	3.140	40.03	0.2802
−14	1.1141	2.2683	4.718	3.303	42.00	.2940
−13	1.1714	2.3849	4.961	3.473	44.07	.3085
−12	1.2313	2.5069	5.215	3.651	46.22	.3235
−11	1.2940	2.6346	5.481	3.837	48.46	.3392

Saturation density, Wt. air plus water vapor/cu. ft. sat. mixture		Volume			Enthalpy			Temp., °F.
Pounds	Grains	Dry air, cu. ft./lb.	Water vapor, cu. ft./lb.	Saturated mixture, cu. ft./lb. dry air	Dry air, B.t.u./lb.	Water vapor, B.t.u./lb.	Saturated mixture, B.t.u./lb. dry air	
0.09816	687.1	10.187	16.404	10.188	−20.890	1036.6	−20.841	−55
.09792	685.4	10.212	16.445	10.213	−20.650	1037.0	−20.597	−54
.09768	683.7	10.238	16.485	10.238	−20.410	1037.5	−20.354	−53
.09744	682.1	10.263	16.526	10.264	−20.170	1037.9	−20.111	−52
.09720	680.4	10.289	16.566	10.289	−19.930	1038.3	−19.867	−51
0.09696	678.7	10.313	16.607	10.315	−19.690	1038.8	−19.623	−50
.09672	677.0	10.339	16.647	10.340	−19.450	1039.2	−19.379	−49
.09648	675.4	10.364	16.688	10.365	−19.210	1039.7	−19.134	−48
.09625	673.8	10.389	16.728	10.391	−18.970	1040.1	−18.889	−47
.09601	672.1	10.415	16.787	10.416	−18.730	1040.6	−18.644	−46
0.09578	670.5	10.440	16.809	10.441	−18.490	1041.0	−18.399	−45
.09555	668.8	10.465	16.850	10.467	−18.250	1041.4	−18.154	−44
.09532	667.2	10.491	16.891	10.492	−18.010	1041.9	−17.908	−43
.09509	665.6	10.516	16.931	10.518	−17.769	1042.3	−17.661	−42
.09486	664.0	10.541	16.971	10.543	−17.529	1042.8	−17.414	−41
0.09463	662.4	10.567	17.012	10.569	−17.289	1043.2	−17.167	−40
.09441	660.8	10.592	17.053	10.594	−17.049	1043.7	−16.920	−39
.09418	659.3	10.617	17.093	10.619	−16.809	1044.1	−16.672	−38
.09396	657.7	10.642	17.134	10.645	−16.569	1044.5	−16.424	−37
.09373	656.1	10.668	17.174	10.670	−16.329	1045.0	−16.176	−36
0.09351	654.6	10.693	17.215	10.696	−16.089	1045.4	−15.927	−35
.09329	653.0	10.718	17.255	10.721	−15.849	1045.9	−15.678	−34
.09307	651.5	10.744	17.296	10.747	−15.609	1046.3	−15.428	−33
.09285	649.9	10.769	17.336	10.772	−15.369	1046.8	−15.178	−32
.09263	648.4	10.794	17.376	10.798	−15.129	1047.2	−14.927	−31
0.09241	646.9	10.820	17.418	10.823	−14.888	1047.6	−14.674	−30
.09220	645.4	10.845	17.458	10.849	−14.648	1048.1	−14.422	−29
.09198	643.9	10.870	17.498	10.874	−14.408	1048.5	−14.170	−28
.09177	642.4	10.896	17.539	10.900	−14.168	1049.0	−13.916	−27
.09155	640.9	10.921	17.580	10.925	−13.928	1049.4	−13.662	−26
0.09134	639.4	10.946	17.620	10.951	−13.688	1049.9	−13.408	−25
.09113	637.9	10.971	17.661	10.977	−13.448	1050.3	−13.152	−24
.09092	636.4	10.997	17.701	11.002	−13.208	1050.7	−12.895	−23
.09071	635.0	11.022	17.742	11.028	−12.968	1051.2	−12.639	−22
.09050	633.5	11.047	17.782	11.053	−12.727	1051.6	−12.380	−21
0.09029	632.0	11.073	17.823	11.079	−12.487	1052.1	−12.122	−20
.09009	630.6	11.098	17.863	11.105	−12.247	1052.5	−11.862	−19
.08988	629.2	11.123	17.904	11.130	−12.007	1053.0	−11.602	−18
.08967	627.7	11.149	17.944	11.156	−11.767	1053.4	−11.340	−17
.08947	626.3	11.174	17.985	11.182	−11.527	1053.8	−11.078	−16
0.08926	624.8	11.199	18.026	11.207	−11.287	1054.3	−10.814	−15
.08906	623.5	11.225	18.066	11.233	−11.047	1054.7	−10.549	−14
.08886	622.0	11.250	18.106	11.259	−10.807	1055.2	−10.284	−13
.08866	620.6	11.275	18.147	11.285	−10.567	1055.6	−10.016	−12
.08846	619.2	11.301	18.188	11.310	−10.326	1056.1	−9.748	−11

Table 4-5. (Continued)

Temp., °F.	Saturation pressure		Saturation humidity, Wt. water vapor/lb. dry air		Saturation moisture content, Wt. water vapor/cu. ft. sat. mixture	
	Lb./sq. in.	In. of Hg	Pounds × 10³	Grains	Pounds × 10⁵	Grains
−10	0.013595	0.027680	0.5759	4.031	5.080	0.3556
−9	.014281	.029076	.6050	4.235	5.324	.3727
−8	.014997	.030534	.6353	4.447	5.579	.3905
−7	.015745	.032056	.6670	4.669	5.844	.4091
−6	.016526	.033648	.7002	4.901	6.121	.4284
−5	0.017343	0.035310	0.7348	5.144	6.409	0.4486
−4	.018196	.037047	.7710	5.397	6.709	.4697
−3	.019085	.038858	.8088	5.661	7.022	.4915
−2	.020014	.040748	.8482	5.937	7.347	.5143
−1	.020982	.042720	.8893	6.225	7.686	.5380
0	0.021994	0.044779	0.9322	6.525	8.039	0.5627
1	.023048	.046925	0.9769	6.839	8.406	.5884
2	.024148	.049165	1.0236	7.166	8.788	.6151
3	.025294	.051500	1.0723	7.506	9.185	.6430
4	.026488	.053929	1.1230	7.861	9.598	.6718
5	0.027733	0.056464	1.176	8.231	10.03	0.7019
6	.029039	.059106	1.231	8.617	10.47	.7331
7	.030384	.061861	1.289	9.020	10.94	.7657
8	.031790	.064724	1.348	9.438	11.42	.7994
9	.033253	.067705	1.411	9.874	11.92	.8344
10	0.034779	0.070810	1.475	10.33	12.44	0.8708
11	.036367	.074043	1.543	10.80	12.98	.9086
12	.038026	.077409	1.613	11.29	13.54	.9479
13	.039735	.080901	1.686	11.80	14.12	.9886
14	.041522	.084539	1.762	12.34	14.73	1.0308
15	0.043384	0.088331	1.842	12.89	15.37	1.076
16	.045316	.092264	1.924	13.47	16.00	1.120
17	.047325	.096355	2.009	14.07	16.68	1.168
18	.049411	.100602	2.098	14.69	17.38	1.216
19	.051583	.105024	2.191	15.34	18.10	1.267
20	0.053838	0.10961	2.287	16.01	18.85	1.320
21	.056179	.11438	2.387	16.71	19.63	1.376
22	.058606	.11932	2.490	17.43	20.44	1.431
23	.061133	.12447	2.598	18.19	21.28	1.489
24	.063751	.12980	2.710	18.97	22.14	1.550
25	0.066474	0.13534	2.826	19.78	23.04	1.613
26	.069295	.14109	2.947	20.63	23.97	1.678
27	.072225	.14705	3.072	21.50	24.93	1.745
28	.075263	.15324	3.202	22.41	25.92	1.815
29	.078417	.15966	3.337	23.36	26.95	1.887
30	0.081684	0.16631	3.476	24.33	28.02	1.962
31	.085072	.17321	3.621	25.35	29.12	2.039
32	.038579	.18035	3.787	26.51	30.38	2.127
33	.092218	.18776	3.943	27.60	31.57	2.210
34	.096000	.19546	4.106	28.74	32.80	2.296

Saturation density, Wt. air plus water vapor/cu. ft. sat. mixture		Volume			Enthalpy			Temp., °F.
Pounds	Grains	Dry air, cu. ft./lb.	Water vapor, cu. ft./lb.	Saturated mixture, cu. ft./lb. dry air	Dry air, B.t.u./lb.	Water vapor, B.t.u./lb.	Saturated mixture, B.t.u./lb. dry air	
0.08826	617.8	11.326	18.228	11.336	−10.086	1056.5	−9.478	−10
.08806	616.5	11.351	18.269	11.362	−9.846	1056.9	−9.207	−9
.08787	615.1	11.376	18.309	11.388	−9.606	1057.4	−8.934	−8
.08767	613.7	11.402	18.350	11.414	−9.366	1057.8	−8.660	−7
.08747	612.3	11.427	18.390	11.440	−9.126	1058.3	−8.385	−6
0.08727	610.9	11.452	18.431	11.466	−8.886	1058.7	−8.108	−5
.08709	609.6	11.478	18.471	11.492	−8.646	1059.2	−7.829	−4
.08689	608.2	11.503	18.512	11.518	−8.406	1059.6	−7.549	−3
.08670	606.9	11.528	18.552	11.544	−8.165	1060.0	−7.266	−2
.08651	605.6	11.553	18.593	11.570	−7.925	1060.5	−6.982	−1
0.08632	604.2	11.579	18.634	11.596	−7.685	1060.9	−6.696	0
.08613	602.9	11.604	18.674	11.622	−7.445	1061.4	−6.408	1
.08594	601.6	11.629	18.715	11.649	−7.205	1061.8	−6.118	2
.08575	600.2	11.655	18.755	11.675	−6.965	1062.3	−5.826	3
.08556	598.9	11.680	18.796	11.701	−6.725	1062.7	−5.531	4
0.08537	597.6	11.705	18.836	11.727	−6.485	1063.1	−5.234	5
.08518	596.3	11.731	18.877	11.754	−6.244	1063.6	−4.935	6
.08500	595.0	11.756	18.917	11.780	−6.004	1064.0	−4.633	7
.08481	593.7	11.781	18.958	11.807	−5.764	1064.5	−4.329	8
.08463	592.4	11.806	18.998	.1.833	−5.524	1064.9	−4.022	9
0.08444	591.1	11.832	19.039	11.860	−5.284	1065.4	−3.712	10
.08426	589.8	11.857	19.079	11.886	−5.044	1065.8	−3.399	11
.08405	588.4	11.882	19.120	11.913	−4.804	1066.2	−3.083	12
.08389	587.3	11.908	19.161	11.940	−4.563	1066.7	−2.765	13
.08371	586.0	11.933	19.201	11.967	−4.323	1067.3	−2.443	14
0.08360	585.2	11.958	19.242	11.984	−4.083	1067.6	−2.1171	15
.08335	583.5	11.983	19.282	12.021	−3.843	1068.0	−1.7882	16
.08317	582.2	12.009	19.322	12.048	−3.603	1068.5	−1.4558	17
.08299	580.9	12.034	19.363	12.075	−3.363	1068.9	−1.1197	18
.08281	579.7	12.059	19.403	12.102	−3.122	1069.3	−0.7797	19
0.08264	578.5	12.085	19.444	12.129	−2.882	1069.8	−0.43572	20
.08246	577.2	12.110	19.485	12.156	−2.642	1070.2	−0.08768	21
.08228	576.0	12.135	19.525	12.184	−2.402	1070.7	+0.26436	22
.08210	574.7	12.160	19.566	12.211	−2.162	1071.1	0.62115	23
.08194	573.6	12.186	19.606	12.239	−1.922	1071.6	0.98221	24
0.08175	572.3	12.211	19.647	12.267	−1.6813	1072.0	1.348	25
.08158	571.1	12.236	19.687	12.294	−1.4411	1072.5	1.719	26
.08140	569.8	12.262	19.728	12.322	−1.2010	1072.9	2.095	27
.08123	568.6	12.287	19.769	12.350	−0.9608	1073.3	2.476	28
.08106	567.4	12.312	19.809	12.378	−0.7206	1073.8	2.862	29
0.08089	566.3	12.337	19.850	12.405	−0.4804	1074.2	3.254	30
.08071	565.0	12.363	19.890	12.435	−0.2402	1074.7	3.651	31
.08055	563.8	12.388	19.918	12.463 ·	0.0000	1075.1	4.071	32
.08037	562.6	12.413	19.959	12.492	0.2402	1075.5	4.481	33
.08020	561.4	12.439	20.000	12.520	0.4804	1076.0	4.898	34

Table 4-5. (Continued)

Temp., °F.	Saturation pressure		Saturation humidity, Wt. water vapor/lb. dry air		Saturation moisture content, Wt. water vapor/cu. ft. sat. mixture	
	Lb./sq. in.	In. of Hg.	Pounds	Grains	Pounds × 10⁴	Grains
35	0.09991	0.20342	0.004274	29.92	3.406	2.384
36	.10396	.21166	.004449	31.14	3.537	2.476
37	.10815	.22019	.004629	32.41	3.672	2.570
38	.11250	.22905	.004817	33.72	3.812	2.668
39	.11699	.23819	.005011	35.08	3.956	2.769
40	0.12164	0.24766	0.005212	36.48	4.105	2.874
41	.12646	.25747	.005420	37.94	4.259	2.981
42	.13144	.26761	.005635	39.45	4.418	3.093
43	.13660	.27812	.005859	41.01	4.582	3.208
44	.14194	.28899	.006090	42.63	4.752	3.327
45	0.14745	0.30021	0.006329	44.30	4.927	3.449
46	.15316	.31183	.006576	46.03	5.107	3.575
47	.15906	.32385	.006833	47.83	5.294	3.706
48	.16516	.33627	.007098	49.68	5.486	3.840
49	.17146	.34909	.007369	51.59	5.682	3.977
50	0.17798	0.36237	0.007655	53.59	5.888	4.121
51	.18471	.37607	.007948	55.64	6.099	4.269
52	.19167	.39024	.008252	57.76	6.316	4.421
53	.19885	.40486	.008565	59.96	6.540	4.578
54	.20627	.41997	.008889	62.23	6.771	4.739
55	0.21394	0.43558	0.009225	64.57	7.009	4.906
56	.22185	.45169	.009571	67.00	7.254	5.078
57	.23002	.46832	.009929	69.50	7.506	5.254
58	.23845	.48548	.010299	72.09	7.766	5.436
59	.24716	.50321	.010681	74.77	8.034	5.624
60	0.25614	0.52150	0.01108	77.53	8.310	5.817
61	.26541	.54038	.01149	80.40	8.595	6.016
62	.27497	.55984	.01191	83.34	8.886	6.220
63	.28483	.57991	.01234	86.39	9.188	6.431
64	.29500	.60062	.01279	89.54	9.498	6.649
65	0.30549	0.62198	0.01326	92.79	9.816	6.871
66	.31630	.64399	.01374	96.15	10.144	7.101
67	.32744	.66667	.01423	99.61	10.479	7.335
68	.33893	.69006	.01474	103.19	10.829	7.580
69	.35077	.71417	.01527	106.88	11.186	7.830
70	0.36297	0.73901	0.01581	110.7	11.55	8.086
71	.37554	.76460	.01638	114.6	11.93	8.352
72	.38848	.79095	.01700	118.7	12.32	8.623
73	.40182	.81811	.01755	122.9	12.72	8.902
74	.41556	.84608	.01817	127.2	13.13	9.189
75	0.42969	0.87485	0.01881	131.7	13.55	9.484
76	.44425	.90449	.01946	136.2	13.98	9.787
77	.45923	.93499	.02014	141.0	14.43	10.098
78	.47467	.96643	.02084	145.9	14.88	10.418
79	.49055	.99876	.02156	150.9	15.35	10.746

Saturation density, Wt. air plus water vapor/cu. ft. sat. mixture		Volume			Enthalpy			Temp., °F.
Pounds	Grains	Dry air, cu. ft./lb.	Water vapor, cu. ft./lb.	Saturated mixture, cu. ft./lb. dry air	Dry air, B.t.u./lb.	Water vapor, B.t.u./lb.	Saturated mixture, B.t.u./lb. dry air	
0.08003	560.2	12.464	20.040	12.549	0.7206	1076.4	5.321	35
.07986	559.0	12.489	20.080	12.578	0.9608	1076.9	5.751	36
.07969	557.8	12.514	20.121	12.607	1.2010	1077.3	6.188	37
.07951	556.6	12.540	20.162	12.637	1.4412	1077.8	6.633	38
.07935	555.4	12.565	20.203	12.666	1.6814	1078.2	7.084	39
0.07918	554.3	12.590	20.243	12.695	1.9216	1078.7	7.544	40
.07901	553.1	12.615	20.284	12.725	2.1618	1079.1	8.010	41
.07884	551.9	12.641	20.324	12.755	2.4020	1079.5	8.485	42
.07867	550.7	12.666	20.365	12.785	2.6422	1080.0	8.969	43
.07851	549.6	12.691	20.404	12.815	2.8824	1080.4	9.462	44
0.07834	548.4	12.717	20.445	12.846	3.1226	1080.9	9.963	45
.07817	547.2	12.742	20.485	12.876	3.3628	1081.3	10.474	46
.07801	546.0	12.767	20.525	12.907	3.6030	1081.8	10.995	47
.07784	544.9	12.792	20.565	12.938	3.8432	1082.2	11.505	48
.07767	543.7	12.818	20.605	12.970	4.0834	1082.7	12.062	49
0.07750	542.5	12.843	20.646	13.002	4.3238	1083.1	12.615	50
.07734	541.4	12.868	20.686	13.033	4.5639	1083.5	13.176	51
.07717	540.2	12.894	20.726	13.065	4.8041	1084.0	13.749	52
.07701	539.0	12.919	20.766	13.097	5.0444	1084.4	14.332	53
.07684	537.9	12.944	20.807	13.129	5.2846	1084.9	14.929	54
0.07668	536.7	12.969	20.847	13.162	5.5248	1085.3	15.536	55
.07651	535.6	12.995	20.887	13.195	5.7650	1085.8	16.157	56
.07635	534.4	13.020	20.927	13.228	6.0053	1086.2	16.790	57
.07618	533.3	13.045	20.967	13.262	6.2456	1086.7	17.438	58
.07602	532.1	13.070	21.006	13.295	6.4858	1087.1	18.097	59
0.07586	531.0	13.096	21.046	13.329	6.7260	1087.5	18.771	60
.07569	529.9	13.121	21.087	13.363	6.9664	1087.9	19.461	61
.07553	528.7	13.146	21.126	13.398	7.2067	1088.3	20.164	62
.07536	527.5	13.172	21.166	13.433	7.4469	1088.8	20.885	63
.07520	526.4	13.197	21.208	13.468	7.6872	1089.2	21.620	64
0.07503	525.2	13.222	21.247	13.504	7.9275	1089.7	22.373	65
.07487	524.1	13.247	21.287	13.540	8.1679	1090.1	23.140	66
.07469	522.8	13.273	21.327	13.580	8.4081	1090.5	23.926	67
.07454	521.8	13.298	21.367	13.613	8.6484	1090.9	24.729	68
.07438	520.7	13.323	21.407	13.650	8.8888	1091.3	25.552	69
0.07421	519.5	13.348	21.447	13.688	9.1290	1091.7	26.392	70
.07405	518.3	13.374	21.486	13.726	9.3694	1092.2	27.255	71
.07389	517.2	13.399	21.527	13.764	9.6098	1092.7	28.137	72
.07372	516.0	13.424	21.566	13.803	9.8500	1093.1	29.037	73
.07356	514.9	13.450	21.606	13.842	10.0904	1093.6	29.962	74
0.07339	513.7	13.475	21.646	13.882	10.331	1094.0	30.907	75
.07323	512.6	13.500	21.685	13.922	10.571	1094.4	31.865	76
.07306	511.4	13.525	21.725	13.963	10.812	1094.9	32.865	77
.07290	510.3	13.551	21.765	14.004	11.052	1095.3	33.880	78
.07273	509.1	13.576	21.804	14.046	11.292	1095.7	34.920	79

Table 4-5. (*Continued*)

Temp., °F.	Saturation pressure		Saturation humidity, Wt. water vapor /lb. dry air		Saturation moisture content, Wt. water vapor/cu. ft. sat. mixture	
	Lb./sq. in.	In. of Hg	Pounds	Grains	Pounds × 10³	Grains
80	0.50689	1.0320	0.02231	156.1	1.583	11.08
81	.52370	1.0663	.02308	161.5	1.645	11.51
82	.54099	1.1015	.02387	167.1	1.684	11.79
83	.55878	1.1377	.02468	172.8	1.736	12.15
84	.57707	1.1749	.02552	178.7	1.789	12.53
85	0.59588	1.2132	0.02639	184.7	1.844	12.91
86	.61522	1.2526	.02728	191.0	1.901	13.30
87	.63510	1.2931	.02821	197.4	1.958	13.71
88	.65555	1.3347	.02916	204.1	2.018	14.12
89	.67656	1.3775	.03014	210.9	2.079	14.55
90	0.69816	1.4215	0.03115	218.0	2.141	14.99
91	.72036	1.4667	.03219	225.3	2.205	15.44
92	.74316	1.5131	.03326	232.8	2.271	15.90
93	.76659	1.5608	.03437	240.6	2.338	16.37
94	.79065	1.6098	.03551	248.5	2.407	16.85
95	0.81537	1.6601	0.03668	256.8	2.478	17.34
96	.84074	1.7118	.03789	265.3	2.550	17.85
97	.86681	1.7648	.03914	274.0	2.624	18.37
98	.89358	1.8193	.04043	283.0	2.701	18.91
99	.92105	1.8753	.04175	292.3	2.779	19.45
100	0.94926	1.9327	0.04312	301.8	2.859	20.01
101	0.97821	1.9916	.04453	311.7	2.941	20.58
102	1.00792	2.0521	.04498	321.9	3.025	21.17
103	1.03842	2.1142	.04748	332.3	3.111	21.77
104	1.06965	2.1788	.04902	343.1	3.198	22.39
105	1.1018	2.2432	0.05061	354.3	3.289	23.02
106	1.1347	2.3103	.05225	365.7	3.381	23.67
107	1.1685	2.3790	.05394	377.6	3.476	24.33
108	1.2031	2.4495	.05568	389.7	3.572	25.00
109	1.2386	2.5218	.05747	402.3	3.671	25.70
110	1.2750	2.5959	0.05932	415.3	3.772	26.41
111	1.3123	2.6719	.06123	428.6	3.876	27.13
112	1.3506	2.7497	.06319	442.4	3.982	27.88
113	1.3897	2.8295	.06522	456.5	4.090	28.63
114	1.4300	2.9114	.06731	471.2	4.202	29.41
115	1.4711	2.9952	0.06946	486.2	4.315	30.20
116	1.5133	3.0811	.07168	501.8	4.431	31.02
117	1.5566	3.1691	.07397	517.8	4.550	31.85
118	1.6008	3.2593	.07633	534.3	4.671	32.69
119	1.6462	3.3517	.07877	551.4	4.795	33.57
120	1.6927	3.4463	0.08128	569.0	4.922	34.45
121	1.7403	3.5432	.08388	587.1	5.052	35.36
122	1.7890	3.6424	.08655	605.9	5.184	36.29
123	1.8389	3.7440	.08931	625.2	5.320	37.24
124	1.8900	3.8480	.09216	645.1	5.458	38.21

Saturation density, Wt. air plus water vapor/cu. ft. sat. mixture		Volume			Enthalpy			Temp., °F.
Pounds	Grains	Dry air, cu. ft./lb.	Water vapor, cu. ft./lb.	Saturated mixture, cu.ft./lb. dry air	Dry air, B.t.u./lb.	Water vapor, B.t.u./lb.	Saturated mixture, B.t.u./lb. dry air	
0.07257	508.0	13.601	21.844	14.088	11.533	1096.2	35.985	80
.07240	506.8	13.626	21.884	14.131	11.773	1096.6	37.077	81
.07223	505.6	13.652	21.923	14.175	12.013	1097.1	38.197	82
.07206	504.4	13.677	21.964	14.219	12.254	1097.5	39.342	83
.07190	503.3	13.702	22.003	14.264	12.494	1097.9	40.515	84
0.07173	502.1	13.728	22.043	14.309	12.735	1098.3	41.718	85
.07156	500.9	13.753	22.082	14.355	12.975	1098.7	42.952	86
.07139	499.8	13.778	22.122	14.402	13.217	1099.1	44.216	87
.07123	498.6	13.803	22.161	14.449	13.457	1099.6	45.515	88
.07106	497.4	13.829	22.200	14.497	13.696	1100.0	46.845	89
0.07088	496.2	13.854	22.240	14.547	13.938	1100.5	48.212	90
.07071	495.0	13.879	22.279	14.597	14.177	1100.9	49.612	91
.07054	493.8	13.904	22.319	14.647	14.418	1101.4	51.050	92
.07037	492.6	13.930	22.358	14.699	14.658	1101.8	52.522	93
.07020	491.4	13.955	22.398	14.751	14.899	1102.3	54.037	94
0.07003	490.2	13.980	22.437	14.804	15.139	1102.7	55.586	95
.06985	489.0	14.006	22.476	14.854	15.380	1103.1	57.179	96
.06968	487.8	14.031	22.515	14.913	15.620	1103.5	58.810	97
.06951	486.6	14.056	22.555	14.968	15.860	1103.9	60.486	98
.06933	485.3	14.081	22 594	15.025	16.101	1104.3	62.209	99
0.06916	484.1	14.107	22.633	15.083	16.341	1104.8	63.980	100
.06898	482.9	14.132	22.673	15.142	16.582	1105.2	65.794	101
.06880	481.6	14.157	22.712	15.202	16.822	1105.6	67.657	102
.06863	480.4	14.182	22.751	15.263	17.063	1106.1	69.577	103
.06845	479.2	14.208	22.790	15.325	17.304	1106.5	71.541	104
0.06827	477.9	14.233	22.829	15.389	17.544	1106.9	73.563	105
.06809	476.7	14.258	22.868	15.453	17.785	1107.3	75.639	106
.06791	475.4	14.283	22.907	15.519	18.025	1107.7	77.771	107
.06773	474.1	14.309	22.946	15.587	18.266	1108.1	79.961	108
.06755	472.8	14.334	22.985	15.655	18.506	1108.5	82.215	109
0.06737	470.6	14.359	23.024	15.725	18.747	1109.0	84.535	110
.06718	470.3	14.384	23.063	15.796	18.987	1109.4	86.915	111
.06700	469.0	14.410	23.101	15.869	19.228	1109.8	89.360	112
.06681	467.7	14.435	23.140	15.944	19.469	1110.2	91.873	113
.06662	466.4	14.460	23.178	16.020	19.709	1110.6	94.461	114
0.06643	465.0	14.486	23.218	16.098	19.950	1111.1	97.128	115
.06624	463.7	14.516	23.256	16.178	20.190	1111.5	99.866	116
.06605	462.4	14.536	23.295	16.259	20.431	1112.0	102.688	117
.06586	461.0	14.561	23.333	16.343	20.672	1112.4	105.586	118
.06567	459.7	14.587	23.372	16.428	20.912	1112.8	108.571	119
0.06547	458.3	14.612	23.408	16.515	21.153	1113.3	111.65	120
.06528	457.0	14.637	23.448	16.603	21.394	1113.7	114.81	121
.06508	455.6	14.662	23.487	16.695	21.634	1114.2	118.07	122
.06488	454.2	14.688	23.526	16.789	21.875	1114.6	121.42	123
.06468	452.8	14.713	23.565	16.885	22.116	1115.0	124.88	124

Table 4-5. (Continued)

Temp., °F.	Saturation pressure		Saturation humidity, Wt. water vapor/lb. dry air		Saturation moisture content, Wt. water vapor/cu. ft. sat. mixture	
	Lb./sq. in.	In. of Hg	Pounds	Grains	Pounds	Grains
125	1.9423	3.9544	0.09511	665.8	0.005600	39.20
126	1.9958	4.0634	.09815	687.1	.005745	40.22
127	2.0506	4.1749	.10129	709.0	.005893	41.25
128	2.1066	4.2891	.10453	731.7	.006045	42.33
129	2.1640	4.4059	.10788	755.2	.006199	43.40
130	2.2227	4.5255	0.1113	779.4	0.006357	44.50
131	2.2828	4.6479	.1149	804.4	.006519	45.63
132	2.3442	4.7729	.1186	830.2	.006683	46.78
133	2.4072	4.9010	.1224	856.9	.006851	47.96
134	2.4715	5.0320	.1264	884.5	.007024	49.16
135	2.5373	5.1659	0.1304	913.0	0.007199	50.39
136	2.6045	5.3028	.1346	942.5	.007377	51.64
137	2.6733	5.4429	.1390	973.1	.007561	52.92
138	2.7436	5.5861	.1435	1,004.5	.007747	54.23
139	2.8155	5.7324	.1482	1,037.3	.007937	55.56
140	2.8890	5.8821	0.1530	1,071	0.008131	56.92
141	2.9641	6.0349	.1580	1,106	.008329	58.31
142	3.0409	6.1912	.1632	1,142	.008532	59.72
143	3.1193	6.3509	.1685	1,180	.008738	61.17
144	3.1915	6.5141	.1741	1,219	.008949	62.64
145	3.2814	6.6809	0.1798	1,259	0.009163	64.14
146	3.3651	6.8512	.1858	1,301	.009383	65.68
147	3.4506	7.0253	.1920	1,344	.009607	67.25
148	3.5379	7.2032	.1984	1,389	.009835	68.84
149	3.6271	7.3847	.2051	1,436	.010066	70.46
150	3.7182	7.5703	0.2120	1,484	0.01030	72.13
151	3.8113	7.7597	.2192	1,534	.01055	73.82
152	3.9063	7.9531	.2267	1,587	.01079	75.54
153	4.0033	8.1506	.2344	1,641	.01104	77.30
154	4.1023	8.3523	.2425	1,698	.01130	79.09
155	4.2034	8.5581	0.2509	1,756	0.01156	80.92
156	4.3066	8.7682	.2596	1,817	.01183	82.78
157	4.4120	8.9828	.2688	1,881	.01210	84.69
158	4.5195	9.2016	.2782	1,948	.01237	86.61
159	4.6292	9.4251	.2881	2,005	.01265	88.58
160	4.7412	9.6531	0.2985	2,089	0.01294	90.59
161	4.8554	9.8856	.3092	2,165	.01323	92.63
162	4.9720	10.1231	.3205	2,244	.01353	94.72
163	5.0909	10.3652	.3323	2,326	.01384	96.85
164	5.2122	10.6162	.3446	2,412	.01414	99.00
165	5.3358	10.864	0.3575	2,502	0.01446	101.2
166	5.4621	11.121	.3710	2,597	.01478	103.5
167	5.5908	11.383	.3851	2,696	.01511	105.8
168	5.7220	11.650	.4000	2,800	.01543	108.1
169	5.8558	11.922	.4156	2,909	.01577	110.4

Saturation density, Wt. air plus water vapor/cu. ft. sat. mixture		Volume			Enthalpy			Temp., °F
Pounds	Grains	Dry air, cu. ft./lb.	Water vapor, cu. ft./lb.	Saturated mixture, cu. ft./lb. dry air	Dry air, B.t.u./lb.	Water vapor, B.t.u./lb.	Saturated mixture, B.t.u./lb. dry air	
0.06448	451.4	14.738	23.602	16.983	22.356	1115.4	128.44	125
.06428	450.0	14.763	23.641	17.084	22.597	1115.8	132.11	126
.06408	448.5	14.789	23.679	17.187	22.838	1116.3	135.91	127
.06387	447.1	14.814	23.718	17.293	23.079	1116.7	139.81	128
.06366	445.6	14.839	23.756	17.402	23.319	1117.1	143.83	129
0.06345	441.2	14.864	23.794	17.514	23.560	1117.6	147.99	130
.06325	442.7	14.890	23.833	17.628	23.801	1118.0	152.27	131
.06303	441.2	14.915	23.871	17.746	24.041	1118.4	156.64	132
.06282	439.7	14.940	23.910	17.867	24.282	1118.8	161.23	133
.06261	438.2	14.965	23.948	17.991	24.523	1119.2	165.95	134
0.06239	436.7	14.991	23.987	18.119	24.764	1119.6	170.79	135
.06217	435.2	15.016	24.025	18.251	25.005	1120.1	175.81	136
.06195	433.7	15.041	24.063	18.386	25.245	1120.5	181.01	137
.06173	432.1	15.067	24.103	18.525	25.486	1120.9	186.35	138
.06150	430.5	15.092	24.139	18.669	25.727	1121.3	191.88	139
0.06128	428.9	15.117	24.178	18.816	25.968	1121.7	197.59	140
.06105	427.3	15.142	24.215	18.969	26.209	1122.1	203.50	141
.06082	425.7	15.168	24.253	19.126	26.449	1122.5	209.62	142
.06058	424.1	15.193	24.290	19.288	26.690	1122.9	215.94	143
.06035	422.5	15.218	24.329	19.454	26.931	1123.3	222.49	144
0.06012	420.8	15.243	24.367	19.626	27.172	1123.7	229.26	145
.05988	419.1	15.269	24.405	19.804	27.413	1124.1	236.29	146
.05964	417.5	15.294	24.443	19.987	27.654	1124.6	243.59	147
.05940	415.8	15.320	24.481	20.176	27.895	1125.0	251.12	148
.05915	414.0	15.344	24.518	20.374	28.135	1125.4	258.94	149
0.05890	412.3	15.370	24.555	20.576	28.376	1125.8	267.06	150
.05866	410.6	15.395	24.593	20.786	28.617	1126.2	275.48	151
.05840	408.8	15.420	24.630	21.004	28.858	1126.6	284.22	152
.05815	407.0	15.445	24.668	21.229	29.099	1127.0	293.30	153
.05789	405.3	15.471	24.705	21.462	29.340	1127.4	302.73	154
0.05763	403.4	15.496	24.742	21.704	29.581	1127.8	312.55	155
.05737	401.6	15.521	24.780	21.955	29.822	1128.2	322.75	156
.05711	399.8	15.546	24.818	22.216	30.063	1128.7	333.41	157
.05684	397.9	15.572	24.855	22.487	30.304	1129.1	344.46	158
.05657	396.0	15.597	24.892	22.769	30.545	1129.5	356.00	159
0.05630	394.1	15.622	24.929	23.063	30.786	1129.9	368.13	160
.05603	392.2	15.647	24.966	23.368	31.027	1130.3	380.56	161
.05575	390.3	15.673	25.003	23.685	31.268	1130.7	393.67	162
.05547	388.3	15.698	25.040	24.017	31.509	1131.1	407.35	163
.05519	386.3	15.723	25.077	24.365	31.750	1131.5	421.66	164
0.05490	384.3	15.748	25.114	24.725	31.991	1131.9	436.61	165
.05462	382.3	15.774	25.151	25.102	32.232	1132.3	452.30	166
.05434	380.3	15.799	25.188	25.492	32.473	1132.7	468.72	167
.05402	378.2	15.824	25.225	25.914	32.714	1133.1	485.95	168
.05373	376.1	15.849	25.261	26.347	32.955	1133.5	504.05	169

Table 4-5. (Continued)

Temp., °F.	Saturation pressure		Saturation humidity, Wt. water vapor/lb. dry air		Saturation moisture content, Wt. water vapor/cu. ft. sat. mixture	
	Lb./sq. in.	In. of Hg	Pounds	Grains	Pounds	Grains
170	5.9923	12.200	0.4320	3,024	0.01612	112.8
171	6.1314	12.484	.4493	3,145	.01647	115.3
172	6.2733	12.772	.4675	3,273	.01682	117.8
173	6.4179	13.067	.4867	3,407	.01719	120.3
174	6.5653	13.367	.5070	3,549	.01756	122.9
175	6.7156	13.673	0.5284	3,699	0.01793	125.5
176	6.8687	13.985	.5511	3,858	.01832	128.2
177	7.0247	14.302	.5752	4,026	.01871	130.9
178	7.1838	14.626	.6008	4,205	.01910	133.7
179	7.3458	14.956	.6279	4,395	.01950	136.5
180	7.5109	15.292	0.6569	4,598	0.01992	139.4
181	7.6791	15.635	.6878	4,815	.02033	142.3
182	7.8504	15.983	.7209	5,046	.02076	145.3
183	8.0247	16.339	.7563	5,294	.02119	148.3
184	8.2027	16.701	.7943	5,560	.02163	151.4
185	8.3836	17.069	0.8352	5,847	0.02207	154.5
186	8.5678	17.444	.8794	6,156	.02253	157.7
187	8.7554	17.826	.9271	6,490	.02299	160.9
188	8.9465	18.215	.9790	6,853	.02346	164.2
189	9.1411	18.611	1.0355	7,249	.02393	167.5
190	9.3392	19.015	1.097	7,681	0.02442	170.9
191	9.5409	19.425	1.165	8,155	.02491	174.4
192	9.7463	19.844	1.240	8,679	.02542	177.9
193	9.9553	20.269	1.322	9,257	.02592	181.5
194	10.1684	20.703	1.414	9,900	.02644	185.1
195	10.385	21.143	1.517	10,629	0.02697	188.8
196	10.605	21.591	1.633	11,432	.02750	192.5
197	10.829	22.048	1.765	12,352	.02805	196.3
198	11.057	22.513	1.915	13,405	.02860	200.2
199	11.289	22.985	2.089	14,620	.02916	204.1
200	11.526	23.466	2.292	16,046	0.02973	208.1
201	11.766	23.955	2.532	17,725	.03031	212.2
202	12.010	24.453	2.820	19,739	.03090	216.3
203	12.259	24.960	3.173	22,212	.03149	220.5
204	12.512	25.474	3.614	25,301	.03210	224.7
205	12.769	25.998	4.181	29,269	0.03272	229.0
206	13.031	26.531	4.939	34,576	.03346	234.2
207	13.297	27.073	6.000	42,000	.03398	237.9
208	13.568	27.624	7.594	53,161	.03463	242.4
209	13.843	28.184	10.248	71,736	.03528	247.0
210	14.122	28.753	15.54	108,773	0.03595	251.6
211	14.407	29.332	31.49	220,451	.03667	256.7

Saturation density, Wt. air plus water vapor/cu. ft. sat. mixture		Volume			Enthalpy			Temp., °F.
Pounds	Grains	Dry air, cu. ft./lb.	Water vapor, cu. ft./lb.	Saturated mixture, cu. ft./lb. dry air	Dry air, B.t.u./lb.	Water vapor, B.t.u./lb.	Saturated mixture, B.t.u./lb. dry air	
0.05343	374.0	15.875	25.298	26.804	33.196	1133.9	523.06	170
.05314	372.0	15.900	25.335	27.272	33.437	1134.3	543.08	171
.05281	369.7	15.925	25.372	27.787	33.678	1134.7	564.15	172
.05251	367.5	15.950	25.408	28.315	33.919	1135.1	586.38	173
.05219	365.3	15.976	25.445	28.876	34.161	1135.5	609.84	174
0.05187	363.1	16.001	25.481	29.465	34.402	1135.9	634.63	175
.05155	360.9	16.026	25.518	30.089	34.643	1136.3	660.88	176
.05123	358.6	16.051	25.554	30.749	34.884	1136.7	688.69	177
.05090	356.3	16.077	25.591	31.449	35.125	1137.1	718.25	178
.05057	356.1	16.102	25.627	32.193	35.366	1137.5	749.63	179
0.05023	351.6	16.127	25.664	32.984	35.607	1137.9	783.08	180
.04989	349.2	16.152	25.700	33.829	35.848	1138.3	818.78	181
.04955	346.8	16.178	25.736	34.731	36.090	1138.7	856.97	182
.04920	344.4	16.203	25.772	35.694	36.331	1139.1	897.79	183
.04885	342.0	16.228	25.809	36.728	36.572	1139.5	941.68	184
0.04850	339.5	16.253	25.844	37.839	36.813	1139.9	988.88	185
.04814	337.0	16.279	25.881	39.037	37.054	1140.3	1,039.79	186
.04778	334.5	16.304	25.916	40.332	37.296	1140.7	1,094.88	187
.04742	331.9	16.329	25.953	41.737	37.537	1141.1	1,154.70	188
.04705	329.3	16.354	25.988	43.265	37.778	1141.5	1,219.80	189
0.04667	326.7	16.380	26.024	44.935	38.019	1141.9	1,291.0	190
.04630	324.1	16.405	26.060	46.764	38.261	1142.3	1,369.5	191
.04592	321.4	16.430	26.093	48.780	38.502	1142.7	1,455.1	192
.04553	318.7	16.455	26.131	51.011	38.743	1143.1	1,550.4	193
.04514	316.0	16.481	26.167	53.488	38.985	1143.5	1,656.2	194
0.04474	313.2	16.506	26.202	56.265	39.226	1143.9	1,775.0	195
.04434	310.4	16.530	26.238	59.381	39.467	1144.3	1,908.3	196
.04394	307.6	16.556	26.273	62.918	39.709	1144.7	2,059.6	197
.04353	304.7	16.582	26.309	66.963	39.950	1145.1	2,232.8	198
.04312	301.8	16.607	26.344	71.630	40.191	1145.5	2,432.7	199
0.04270	298.9	16.632	26.380	77.102	40.433	1145.9	2,667.2	200
.04228	296.0	16.657	26.415	83.543	40.674	1146.2	2,943.0	201
.04185	293.0	16.683	26.450	91.270	40.916	1146.6	3,274.2	202
.04142	290.0	16.708	26.486	100.750	41.158	1146.0	3,679.6	203
.04098	286.9	16.733	26.521	112.590	41.399	1147.4	4,188.6	204
0.04054	283.8	16.758	26.556	127.80	41.640	1147.7	4,840.5	205
.04024	281.7	16.784	26.591	147.60	41.882	1148.1	5,712.8	206
.03965	277.5	16.809	26.626	176.56	42.124	1148.4	6,932.3	207
.03919	274.3	16.834	26.661	219.30	42.366	1148.8	8,766.8	208
.03873	271.1	16.859	26.696	290.44	42.608	1149.2	11,820.3	209
0.03826	267.8	16.885	26.731	432.25	42.849	1149.6	17,906	210
.03779	264.5	16.910	26.765	859.82	43.090	1150.0	36,260	211

*Theoretical volume calculated from the relationship: Vol. at 1 atm. = Vol. at saturation press. $\times \dfrac{\text{saturation press., In. of Hg}}{29.921}$.

Source: Ref. 4-10.

Volume percentage is:

$$Air: \quad 41,800 \div 55,260 = 76\%$$
$$H_2O: \quad 13,460 \div 55,260 = 24\%$$

AIR/MOISTURE MIXTURES

At atmospheric pressure, 14.7 psia, the air can contain a maximum amount of moisture, moisture of saturation, as shown in Fig. 4-1. The saturation time is the 100% curve, to the left of the series of vertical curves. The moisture of saturation, the humidity, is listed on the vertical axis of the chart as pounds of water vapor per pound of dry air. The horizontal axis lists the temperature of the mixture. The saturation moisture can be found as follows:

At 100°F the moisture of saturation, that maximum quantity of moisture that the air can contain, by following the 100% curve, is 0.043 lb H_2O/lb dry air.

For this example if the temperature of the mixture were increased to 150°F, the humidity of the air would drop from 100% to 20%, following the horizontal line representing 0.043 lb H_2O/lb dry air horizontally to the right. If the mixture were to cool to 80°F, the saturation moisture would be 0.022 lb H_2O/lb dry air, reading the 100% saturation line. The difference between the 0.043 lb H_2O/lb dry air contained in the mixture at 150°F and the 0.022 lb H_2O/lb dry air would condense, i.e., would leave the mixture as water droplets.

The properties of the air/moisture mixture including saturation humidity, volume and enthalpy, are listed in Table 4-5.

Chapter 5
Heat Transfer Systems

Incineration is a thermal process and the control of the flow of heat is basic to this process. In this chapter heat transfer calculations will be explained and thermal properties of materials will be presented.

THE NATURE OF HEAT TRANSFER

There are three modes of heat transfer:

- *Conduction.* Heat transfer through a medium by progressive heating of adjacent elements.
- *Convection.* Transfer of heat by physical motion of the heated, or cooled, medium, as motion of a fluid.
- *Radiation.* Heat transfer by electromagnetic radiation between two boundaries not in physical contact with each other and at different temperatures or between one body and another medium, such as air, at different temperatures.

Normally these three heat transfer modes are all present at the same time and must be taken into account in combination. The calculations required to determine heat transfer parameters are complex, but methods have been developed to expedite this calculation process.

CONDUCTION

Referring to Fig. 5-1, the mechanism of conduction transfers heat from a hot surface, T_i, to a cooler surface, T_0, both expressed in degrees Fahrenheit. The conducting material, with width X, in inches, has a property of conduction, or conductivity, K, which has the units of BTU-in./ft^2-hr-$^{\circ}$F. Good insulators, such as fiberglass or perlite, have K values in the range of 0.2–3.0. Moderate insulators, such as lightweight castable refractory or lightweight firebrick, have K

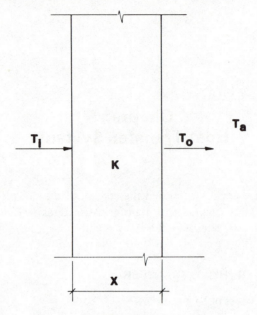

Fig. 5-1. Single wall.

values in the range of 3.0–6.0. Poor insulators, such as standard firebrick, have K values above 6.0, whereas conductors have K values over 100.0.

The basic law of heat transfer by conduction is:

$$Q' = \frac{KA(T_i - T_0)}{X} \tag{5-1}$$

where Q' is the flow of heat in BTU/hr and A is the area, in square feet, perpendicular to the direction of heat flow.

For convenience the quantity resistance R is defined as follows:

$$R = X/K. \tag{5-2}$$

For a wall composed of more than one material the heat transfer noting Fig. 5-2, with

$$Q = Q'/A \tag{5-3}$$

becomes,

$$Q = \frac{K_1}{X_1}(T_i - T_1) = \frac{K_2}{X_2}(T_1 - T_0) \tag{5-4}$$

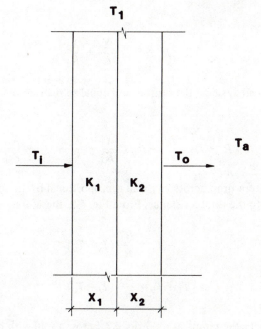

Fig. 5-2. Dual wall.

or

$$Q = \frac{1}{R_1}(T_i - T_1) = \frac{1}{R_2}(T_1 - T_0) \tag{5-5}$$

Solving for the interface temperature, T_1:

$$T_1 = T_i - Q\frac{X_1}{K_1} = T_0 + Q\frac{X_2}{K_2} \tag{5-6}$$

and solving for Q:

$$Q = \frac{(T_i - T_0)}{\dfrac{X_1}{K_1} + \dfrac{X_2}{K_2}} \tag{5-7}$$

or substituting resistance

$$Q = \frac{1}{R}(T_i - T_0) \tag{5-8}$$

where

$$R = R_1 + R_2 = \frac{X_1}{K_1} + \frac{X_2}{K_2} \qquad (5\text{-}9)$$

With multiple wall systems the resistance is equal to the sum of the resistance of each wall:

$$R = R_1 + R_2 + \cdots + R_n = \frac{X_1}{K_1} + \frac{X_2}{K_2} + \cdots + \frac{X_n}{K_n} \qquad (5\text{-}9a)$$

The temperature drop across any wall is proportional to the resistance of that wall compared to the total resistance. From Fig. 5-2, therefore,

$$T_i - T_1 = \frac{R_1}{R} (T_i - T_0) \qquad (5\text{-}10)$$

$$T_1 - T_0 = \frac{R_2}{R} (T_i - T_0) \qquad (5\text{-}11)$$

For multiple walls the temperature drop, ΔT, across a wall with resistance R_x is:

$$\Delta T = \frac{R_x}{R} (T_i - T_0) \qquad (5\text{-}12)$$

EXAMPLES

CONDUCTION ACROSS A SINGLE WALL

What is the cold face temperature with $1200°F$ on the hot face of 3 inches of insulation with a conductivity of 1.33 BTU-in./hr-ft²-°F, allowing a heat loss of 430 BTU/hr-ft²?

From (5-2), $R = X/K = $ 3 in./1.33 BTU-in./hr-ft²-°F = 2.256 hr-ft²-°F/BTU.

From (5-8), $T_0 = T_i - RQ = 1200°F - $ 2.256 hr-ft²-°F/BTU $\times$ 430 BTU/hr-ft² = 230°F, the cold face temperature.

CONDUCTION ACROSS A COMPOSITE WALL

What are the interface temperatures across a composite wall composed of four inches of firebrick ($K = 9.2$ BTU-in./hr-ft²-°F), two inches of insulating block ($K = 0.8$ BTU-in./hr-ft²-°F), and one inch of steel ($K = 380$ BTU-in./hr-ft²-°F). Determine the interface temperatures and heat flow with $1600°F$ hot face (T_i) and $150°F$ (T_0) cold face temperatures.

From (5-2):

$R_1 = X_1/K_1 = 4.0$ in./9.2 BTU-in./hr-ft^2-$^\circ$F = 0.435 hr-ft^2-$^\circ$F/BTU (firebrick)

$R_2 = X_2/K_2 = 2.0$ in./0.8 BTU-in./hr-ft^2-$^\circ$F = 2.500 hr-ft^2-$^\circ$F/BTU (insulation)

$R_3 = X_3/K_3 = 1.0$ in./380 BTU-in./hr-ft^2-$^\circ$F = 0.003 hr-ft^2-$^\circ$F/BTU (steel)

$R = \sum R = 0.435 + 2.500 + 0.003 = 2.938$ hr-ft^2-$^\circ$F/BTU

From (5-10):

$T_i - T_1 = (R_1/R)(T_i - T_0) = (0.435/2.938)(1600 - 150) = 215^\circ$F

$T_1 = T_i - 215 = 1600 - 215 = 1385^\circ$F

Also,

$$T_1 - T_2 = \frac{R_2}{R} (T_i - T_0) = (2.500/2.938)(1600 - 150) = 1234^\circ F$$

$$T_2 = T_1 - 1234 = 1385 - 1234 = 151^\circ F$$

and

$$T_2 - T_0 = \frac{R_3}{R} (T_i - T_0) = (0.003/2.938)(1600 - 150) = 1^\circ F$$

$$T_2 = T_0 + 1 = 150 + 1 = 151^\circ F$$

Therefore, the interface temperature between the brick and the insulation is 1385°F and 151°F between the steel and the insulation.

From (5-8):

$$Q = \frac{1}{R} (T_i - T_0) = \frac{1}{2.938} \times (1600 - 150) = 494 \text{ BTU/hr-ft}^2$$

HEAT CONVECTION

Heat convection is a function of the properties of the convected medium, a fluid, and the geometry of the heated surface. Of specific interest is the flow of heat by convection of air. The basic equation for convection of heat is:

$$Q'_c = hA(T_0 - T_a) \tag{5-13}$$

or

$$Q_c = h(T_0 - T_a);$$
(5-14)

where Q is the flow of heat by conduction in BTU/hr-ft^2, h is the film coefficient, BTU/hr-ft^2-°F, and T_0 and T_a the interface temperature and the ambient temperature respectively, as shown in Fig. 5-1. The film coefficient is difficult to determine, and it is generally calculated from empirical data. One set of such data is as follows:

$$h = 0.25(T_0 - T_a)^{0.25} \quad \text{for vertical plates} \qquad (5\text{-}15)$$

$$h = 0.38(T_0 - T_a)^{0.25} \quad \text{for plates facing up} \qquad (5\text{-}16)$$

$$h = 0.20(T_0 - T_a)^{0.25} \quad \text{for plates facing down} \qquad (5\text{-}17)$$

These values are used for convection in still air. For air at a velocity V, ft/sec:

$$h = (1.0 + 0.225V) \qquad (5\text{-}18)$$

RADIATION

Radiation from a body is governed by the texture and color of that body, expressed as a dimensionless constant, emissivity, ϵ. The equation describing radiation heat loss, Q_R, BTU/hr-ft^2-°F is:

$$Q_R' = Q_R A, \qquad (5\text{-}19)$$

$$Q_R = 0.174\,\epsilon \left[\left(\frac{T_0 + 460}{100} \right)^4 - \left(\frac{T_a + 460}{100} \right)^4 \right] \qquad (5\text{-}20)$$

This equation is derived from the Stephan-Boltzmann equations. Typical values for emissivity ϵ are listed in Table 5-1.

COMBINED HEAT TRANSFER

There will always be flow of heat from a heated to a cooler medium. In practically all situations this flow of heat involves the three modes of heat transfer. Heat is conducted through a medium and is dispersed at the medium boundary through convection and radiation. Expressing this mathematically:

$$Q = Q_R + Q_C \qquad (5\text{-}21)$$

**Table 5-1.   Typical Values
of Emissivity.**

Material	ϵ
Polished aluminum	0.10
Aluminum paint	0.50
White paper	0.70
Rubber, soft gray	0.85
Brick	0.85
Steel surface, oxidized	0.90

Source: Ref. 5-11.

and from (5-8), (5-14), and (5-20),

$$Q = \frac{1}{R}(T_i - T_0) = 0.174\epsilon\left[\left(\frac{T_0 + 460}{100}\right)^4 - \left(\frac{T_a + 460}{100}\right)^4\right] + h(T_0 - T_a) \quad (5\text{-}22)$$

The values for ϵ are as listed in Table 5-1, and can also be found in the heat transfer literature. Values for h are given in Eq. (5-15), (5-16), (5-17), and (5-18).

Typical values of conduction are presented in Tables 5-2 through 5-5. Note that conduction varies as a function of temperature whereas the values for ϵ and h are not, as presented here, temperature dependent. Normally, the internal, external and ambient temperatures are known. The value R must be determined. Once R is known ($R = X/k$), the interface temperatures must be approximated to determine appropriate K values. This is a cumbersome trial-and-error procedure. The first requirement in heat transfer calculations is obtaining the value of R from Eq. (5-22), a complex, unwieldly expression.

THE RESISTANCE GRAPH

A method has been developed to present a complete heat transfer profile on one graph. Referring to Fig. 5-3, the hot face temperature is on the vertical axis. Choosing a hot face temperature and drawing a horizontal line, then choosing a cold face temperature, from the bottom axis, and drawing a vertical line, the R value required is found at the intersection of these two lines. The hot face and cold face temperatures define R. The R, or resistance value, is that resistance required to obtain the desired cold face temperature with a given hot face temperature.

Referring again to Fig. 5-3, as an example, for a hot face temperature of 2100°F, a resistance value of 6.0 is required to obtain a cold face temperature of 200°F. A resistance value of 6.0 defines the type and thickness of refractory wall required. From Eq. (5-2),

$$R = X/K = 6 \text{ ft}^2\text{-}°\text{F-hr/BTU}$$

Table 5-2.   Dense Castables, Typical Materials.

Manufacturer: Material:	Johns Manville Std. Firecrete	Quigley Q-Cast 30-50	JH France Hydricon 186
Maximum temp., °F	2500	3000	2500
Weight, as placed, pcf	130	132	115
Conductivity, BTU-in./ft²-°F-hr			
800°F	4.2	6.3	4.9
1000°F	4.3	6.2	5.0
1200°F	4.6	6.1	5.2
1400°F	4.4	6.1	5.4
1600°F	5.2	6.2	5.5
1800°F	5.5	6.4	5.6

Table 5-3.   Insulating Castables, Typical Materials.

Manufacturer: Material:	Harbison Walker Lightweight Castable 26	Quigley Insulcrete 22	JH France Light Weight Hydricon 2400
Maximum temp., °F	2600	2200	2400
Weight, as placed, pcf	50	43	75
Conductivity, BTU-in./ft²-°F-hr			
800°F	1.6	1.2	1.6
1000°F	1.7	1.3	1.7
1200°F	1.8	1.4	1.7
1400°F	1.9	1.7	1.8
1600°F	2.0	2.0	1.8
1800°F	2.2	2.4	1.9

Table 5-4.   Firebrick, Typical Materials.

Manufacturer: Material:	Harbison Walker Standard Firebrick VARNON	Johns Manville Insulating Firebrick JM-20	JH France Insulating Firebrick FR-20
Maximum temp., °F	3100	2000	2600
Weight, pcf	146	29	33
Conductivity, BTU-in./ft²-°F-hr			
800°F	9.8	1.1	1.9
1000°F	9.8	1.1	2.0
1200°F	9.9	1.2	2.1
1400°F	10.0	1.2	2.2
1600°F	10.2	1.2	2.3
1800°F	10.5	1.3	2.4

Table 5-5. Industrial Insulation, Typical Materials.

Manufacturer: Material:	Celotex Insulation Board IMF-50	Eagle Picher Insulation Board Epitherm 1200	Johns Manville Pipe & Block Insulation Superex-M	Owens Corning Pipe Insulation Fiberglass 25
Maximum temp., °F	1050	1200	1600	450
Density, pcf	8	11	21	4
Conductivity, BTU-in./ft²-°F-hr				
200°F	0.34	0.33	0.58	0.30
400°F	0.50	0.45	0.62	0.45
600°F	0.68	0.59	0.66	–
800°F	1.10	0.78	0.69	–
1000°F	–	1.07	0.73	–

Note: Aluminum covering has a thermal conductivity of 1370 BTU-in./ft²-°F-hr, carbon steel is 380 BTU-in./ft²-°F-hr and stainless steel is 130 BTU-in./ft²-°F-hr, all evaluated at approximately 300°F. These values do not change significantly for metals with changes in temperature.

Using insulating firebrick with K = 1.2 BTU-in./ft²-°F-hr,

$$X = RK = 6 \text{ ft}^2\text{-°F-hr/BTU} \times 1.2 \text{ BTU-in/ft}^2\text{-°F-hr}$$

$$= 7.2 \text{ in.}$$

The calculated thickness, 7.2 inches, is not a standard size. For the next thickest standard size of brick, 9.0 inches, the actual R value is:

$$R = X/K$$

$$= \frac{9.0 \text{ in.}}{1.2 \text{ BTU-in./ft}^2\text{-°F-hr}}$$

$$= 7.5 \text{ ft}^2\text{-°F-hr/BTU}$$

The cold face temperature, again from Fig. 5-3, is read from the line R = 7.5, where intersected by the horizontal line representing the 2100°F hot face temperature. The cold face is approximately 180°F, less than the idealized 200°F by 20°F.

The heat leaving the cold face is found on the top horizontal line and is a function of the cold face temperature. For the face in this example it is 250 BTU/hr-ft² and for 200°F it is 315 BTU/hr-ft².

Four resistance graphs have been included in this chapter. They are each based on an ambient temperature of 80°F. Fig. 5-4 is for an emissivity of 0.9 with the cold face horizontal, facing up. The resistance graph in Fig. 5-5 is for

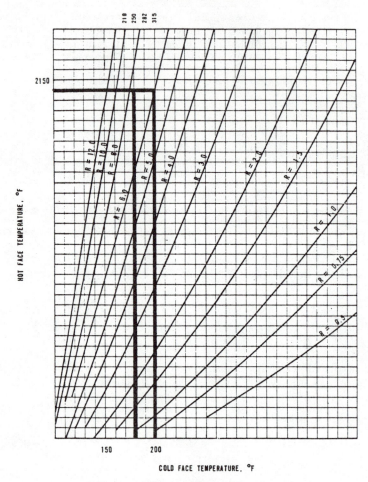

Fig. 5-3. Resistance graph design example.

an emissivity of 0.9 with the cold face vertical. In Fig. 5-6 the cold face is horizontal, looking down. Fig. 5-7 is for an emissivity of 0.1 with the cold face vertical.

The example used in the above discussion, Fig. 5-3, was a representation of the resistance graph with emissivity 0.9, cold face horizontal, facing upward (Fig. 5-4).

The resistance graph is constructed as follows: Values for h, ϵ, and T_a are inserted into Eq. (5-22). Heat transfer Q is calculated for a range of T_0. In Figs. 5-4 through 5-7, values of T_0 from 100°F to 400°F in 10°F increments were

HEAT TRANSFER, BTU/HR-FT^2

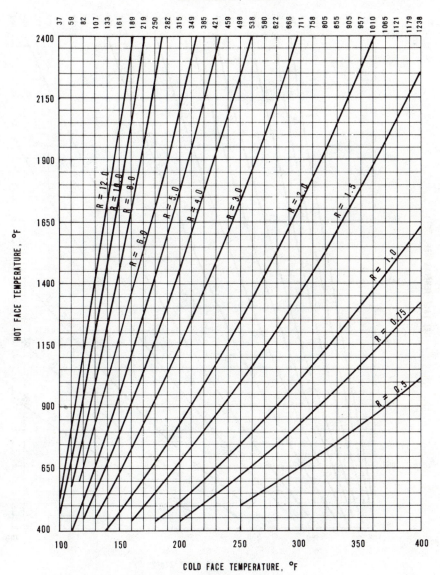

Fig. 5-4. Resistance graph, $E = 0.9$, facing up.

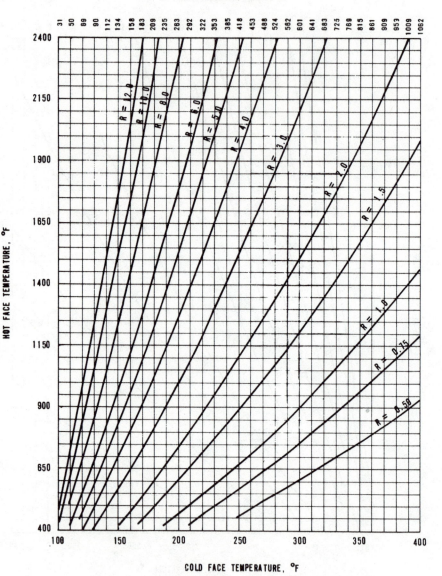

Fig. 5-5. Resistance graph, $E = 0.9$, vertical wall.

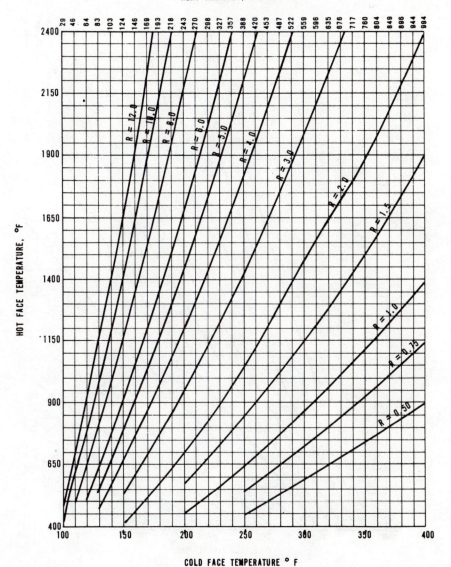

Fig. 5-6. Resistance graph, $E = 0.9$, facing down.

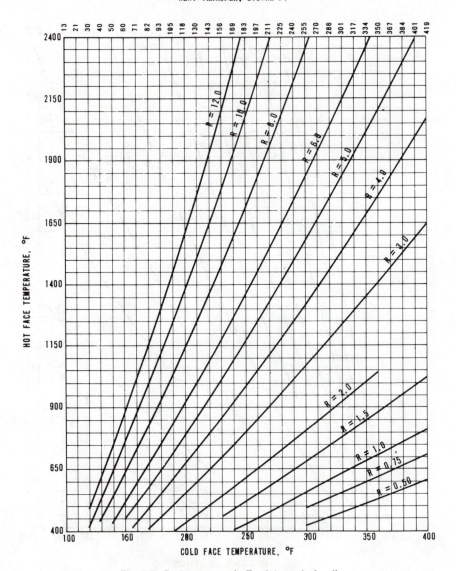

Fig. 5-7. Resistance graph, $E = 0.1$, vertical wall.

calculated. The cold face temperature T_0 is inserted on the bottom horizontal axis and the heat transfer Q is inserted opposite T_0, on the top horizontal axis.

The hot face temperature T_i and the resistance R must be integrated into this graph.

From Eq. (5-8), $Q = (1/R)(T_i - T_0)$ or

$$T_i = T_0 + QR \tag{5-24}$$

With the cold face temperature T_0 and Q the corresponding heat transfer, as calculated above, values of T_i are calculated for various R values using Eq. (5-24). In Figs. 5-4 through 5-7, R values of 0.5–12.0 are used in this calculation.

Plotting resulting hot face temperature T_i against cold face temperatures T_0 for each R value defines a set of curves, the curves drawn in Figs. 5-4 through 5-7. These curves define the resistance graph.

The equations used for the graphs included in this chapter are as follows, using (5-15), (5-16) and (5-17) for convection and (5-19) for radiation.

Fig. 5-4. Ambient temperature of 80°F, emissivity of 0.9 (steel surface), cold face horizontal facing up:

$$Q = 0.174 \times 0.9 \times \left[\left(\frac{T_0 + 460}{100}\right)^4 + \left(\frac{80 + 460}{100}\right)^4\right] + 0.38(T_0 - 80)^{1.25} \tag{5-25}$$

Fig. 5-5. Ambient temperature of 80°F emissivity of 0.9, cold face vertical:

$$Q = 0.174 \times 0.9 \times \left[\left(\frac{T_0 + 460}{100}\right)^4 + \left(\frac{80 + 460}{100}\right)^4\right] + 0.25(T_0 - 80)^{1.25} \tag{5-26}$$

Fig. 5-6. Ambient temperature of 80°F, emissivity of 0.9, cold face horizontal facing down:

$$Q = 0.174 \times 0.9 \times \left[\left(\frac{T_0 + 460}{100}\right)^4 + \left(\frac{80 + 460}{100}\right)^4\right] + 0.20(T_0 - 80)^{1.25} \tag{5-27}$$

Fig. 5-7. Ambient temperature of 80°F, emissivity of 0.1 (polished aluminum), cold face vertical:

$$Q = 0.174 \times 0.1 \times \left[\left(\frac{T_0 + 460}{100}\right)^4 + \left(\frac{80 + 460}{100}\right)^4\right] + 0.25(T_0 - 80)^{1.25} \tag{5-28}$$

These graphs are relatively simple to construct for any set of conditions. For instance, if a polished vertical aluminum surface were exposed to an external air velocity V with 40°F ambient temperature, the heat transfer equation is as follows, using h from (5-18):

$$Q = 0.174 \times 0.1 \times \left[\left(\frac{T_0 + 460}{100}\right)^4 + \left(\frac{40 + 460}{100}\right)^4\right] + (1.0 + 0.225 V)(T_0 - 40)$$

(5-29)

A graph can be drawn for this or for any other condition.

REFRACTORY AND INSULATION

There are a wide variety of refractory and insulation products available. Their significant properties include:

- Maintaining strength under high temperature conditions.
- Cold crushing strength, the ability to withstand handling and shipping without damage and impact strength at low temperature operations.
- Insulating value, the ability to provide resistance to the flow of heat.
- Porosity, the susceptibility to penetration by slags or gases.
- Reheat (ASTM Method C113), a measure of changes in linear dimensions under repeated heating and cooling.
- Modulus of rupture, a standard measure of structural strength provided by load testing (ASTM Methods C16 and C216).
- Erosion, the washing away or physical destruction of material, under forces due to physical contact with the gaseous, liquid, or solid materials on the hot face.
- Corrosion, the chemical process that destroys the refractory bond or the chemical integrity of the insulation.
- Thermal expansion, the reversible change in linear dimensions under heating and cooling. Where refractory is used as a liner in a flue, for instance, the expansion of the refractory must be evaluated with respect to thermal expansion of the flue. If the expansions do not match, provision for material expansion must be considered.
- Operating environment. Of particular interest is the operation of refractory in an oxidizing or reducing atmosphere. A reducing atmosphere, one deficient in oxygen, will tend to degrade refractory material containing iron or silicon components.

MATERIALS DESCRIPTION

Castables (Refractory Concrete)

These materials are supplied dry, to be mixed with water before installation. They are installed by either pouring (casting in place), troweling, pneumatic gunning (as with gunned fireproofing), or ramming. A castable refractory provides a smooth, continuous, monolithic mass. Castable materials are normally placed in an area prepared with pins, mesh, or other anchor devices to hold the refractory in place during placement and curing. Mesh, grid, studs or needles may also be used to enhance the strength of the refractory installation.

Castable refractories are classified as dense or lightweight (insulating). Dense castables have excellent mechanical strength and low permeability. Their insulating properties, however, are relatively poor. As dense materials, over 100 pcf specific weight, they offer good resistance in wet service such as quencher linings.

Lightweight castables are excellent insulators. All castable materials have the advantage of placement in irregular areas such as furnace transitions, burner openings, etc. They are cast in place. Poured castables require a form; however, plastic castable does not. It is rammed in place. Gunned castable is blown in place and formwork is not required. The use of a gunned material, however, is often dictated by the size of the job in question. Gunning requires installation equipment including air compressors and gunning applicators that would be uneconomical to provide when compared to the relatively small amount of equipment needed for poured or plastic castable installation. Normally, gunning is not economical for jobs less than a week in duration.

The quality of a castable installation is a function of mixing and curing as well as application. Mixing is performed at the job site with water added by the installation contractor in accordance with the refractory manufacturers' instructions. If the amount of water added is not within the product specifications, the refractory installation will be deficient. This is related to placement, as well as to the strength developed by the refractory in curing and in service. Too dry a mix will not flow in a form and voids will appear in certain areas. Thickness requirements may not be satisfied in other areas. Too much water in a mix will promote segregation of the mix as it solidifies and as it cures, providing inconsistency in refractory quality as well as poor refractory quality. Note that poured refractory is often vibrated to aid its flow, similar to placement of concrete.

Curing refractory is necessary to set a ceramic bond creating the strength of the refractory matrix. Without proper curing the refractory will not develop mechanical strength or chemical resistance (resistance to corrosion). Refractory is usually either air set or heat cured. Air setting requires only a quiescent period where the refractory is not subject to mechanical changes (motion) or heating effects. Heat curing requires the refractory be heated at a controlled rate to establish the ceramic bond. This bond may require that temperatures of 1000°F to over 2000°F be reached and maintained. Often incinerator temperatures are not high enough to cure heat setting refractories and special operating procedures are necessary to bring the equipment to the curing temperature.

Firebrick

Conventional firebrick is kiln baked to uniform, controlled consistency and quality. The term firebrick refers to dense brick, over 100 pcf, normally placed

Table 5-6. Properties of Refractory Brick.

Type	Class	Weight Percent — Silica (SiO$_2$)	Alumina (Al$_2$O$_3$)	Titania (TiO$_2$)	Temperature Limit (°F)	Density (pcf)	Modulus of Rupture (psi)	Cold Crushing Strength (psi)	Thermal Expansion Ambient to 1800°F (%)
Fireclay	super duty	40-56	40-44	1-3	3185	140/145	730/1975	1830/7325	0.6
	high duty	51-61	40-44	1-3	3175	120/145	610/3660	1830/8550	0.6
	medium duty	57-60	25-38	1-2	3040	120/145	975/3050	2075/7325	0.5-0.6
	low duty	60-70	22-33	1-2	2815	120/145	1220/3050	2440/7325	0.5-0.6
	semi-silica	72-80	18-26	1-2	3040	115/125	365/1100	1220/3660	0.7
High aluminum	45-48%	44-51	45-48	2-3	3245	135/160	1220/1950	3050/7325	0.4-0.5
	60%	31-37	58-62	2-3	3295	140/160	730/2200	2200/8550	0.6
	80%	11-15	78-82	3-4	3390	155/180	1340/3660	4880/10975	0.7
	90%	8-9	89-91	0.4-1	3500	165/190	1465/4275	4880/10975	0.7
	mullite	18-34	60-78	0.5-3	3360	145/165	1220/4275	4275/10975	0.7
	corundum	0.2-1	98-99	trace	3660	170/200	2200/3660	6100/10975	0.8
Silica	super duty	95-97	0.1-0.3 + lime	–	–	105/118	610/1220	1830/4275	1.3
	conventional	94-97	0.4-1.4 + lime	–	–	105/118	730/1465	2200/4880	1.2
	lightweight	94-97	0.4-1.4 + lime	–	–	60	–	–	–
Silicon carbide	bonded	–	–	–	3360	145/165	2440/4880	3050/18300	0.5
Insulating	1600°F	–	–	–	–	36	85/125	110/135	low
	2300°F	–	–	–	–	46	120/200	135/230	low
	2800°F	–	–	–	–	59	200/365	205/370	low
	3000°F	–	–	–	–	60	490/730	975/1220	low
Chrome	fired	Cr$_2$O$_3$ (28-38), MgO (14-49), Al$_2$O$_3$ (15-34), Fe$_2$O$_3$ (11-17)			–	185/205	855/1590	2440/4880	0.8

Source: Selected manufacturers' data.

in direct contact with the hot gas stream. It has relatively poor insulating quality. Insulating firebrick (IFB) is a lightweight, porous brick, normally less than 50 pcf, which can be placed in direct contact with the gas stream and which provides good insulating characteristics. IFB is machined to its final shape providing excellent dimensional control as compared to firebrick which is used as cast.

IFB is lower in strength than firebrick and because of its porosity is a soft material not effective with erosive gas streams, i.e., gas streams with high particulate components. This low abrasion resistance limits the maximum velocities allowed adjacent to it.

Where refractory brick is required and high abrasive resistance is necessary, firebrick will often be provided with insulation block as back-up between the firebrick and the furnace/flue wall.

Anchors are normally provided to hold brick in place. Firebrick and IFB are both self-supporting. However, because of the relative structural strengths of these materials IFB requires a more elaborate anchoring system than firebrick. Brick manufacturers each have their own unique anchor and anchor systems designs.

Table 5-6 lists properties of typical refractory brick materials.

Chapter 6
Combustion Calculations

Incineration is a burning process and as such involves the reaction of combustible components with air. In this chapter the relationship between combustion, air, and products of combustion will be determined.

WASTE CHARACTERISTICS FOR COMBUSTION

Almost by its very definition waste is not able to be classified into a single, or even multiple, known chemical structure. Often the chemical nature of a particular waste is unknown. Calculations involving the combustion of a waste material, therefore, in most cases must necessarily be based on a number of assumptions. These assumptions include the following:

- All hydrogen present converts to water vapor, H_2O, unless otherwise noted below.
- All chloride (or fluoride) converts to hydrogen chloride, HCl (or hydrogen fluorde, HF).
- All carbon converts to carbon dioxide, CO_2.
- All sulfur converts to sulfur dioxide, SO_2.
- Alkali metals convert to hydroxides: sodium to sodium hydroxide ($2Na + O_2 + H_2 \rightarrow 2NaOH$) and potassium to potassium hydroxide ($2K + O_2 + H_2 \rightarrow 2KOH$).
- Non-alkali metals convert to oxides: copper to copper oxide ($2Cu + O_2 \rightarrow 2CuO$), iron to iron oxide ($4Fe + 3O_2 \rightarrow 2Fe_2O_3$).
- All nitrogen from the waste, the fuel, or air, will take the form of a diatomic molecule, i.e., nitrogen is present as N_2.

EQUILIBRIUM EQUATIONS

Chemical equilibrium equations are developed to indicate conservation of matter, i.e., the molecular weight on the left side of the equation equals the molecular weight on the right. For carbon, for instance:

$$C + O_2 \longrightarrow CO_2 \qquad (6\text{-}1)$$

From Table 6-1, the atomic weight of carbon is 12.01, and of oxygen, 16.00. For this simple example, therefore, the atomic weight on the left-hand side is $1 \times 12.01 + 2 \times 16.00 = 44.01$, which is equal to the weight of CO_2 on the right-hand side of the equation.

In the wastes and fuels normally encountered, the major constituents include carbon and hydrogen. For hydrogen:

$$2H_2 + O_2 \rightarrow 2H_2O \qquad (6\text{-}2)$$

Note that oxygen or hydrogen exist as diatomic molecules (O_2 and H_2), as does nitrogen (N_2).

Table 6-1. Chemical Elements.

Element	Symbol	Atomic No.	Atomic Weight
Actinium	Ac	89	227.
Aluminum	Al	13	26.98
Americium	Am	95	243.
Antimony	Sb	51	121.75
Argon	Ar	18	39.95
Arsenic	As	33	74.92
Astatine	At	85	210.
Barium	Ba	56	137.34
Berkelium	Bk	97	247.
Beryllium	Be	4	9.01
Bismuth	Bi	83	208.98
Boron	B	5	10.81
Bromine	Br	35	79.90
Cadmium	Cd	48	112.40
Calcium	Ca	20	40.08
Californium	Cf	98	251.
Carbon	C	6	12.01
Cerium	Ce	58	140.12
Cesium	Cs	55	132.91
Chlorine	Cl	17	35.45
Chromium	Cr	24	52.00
Cobalt	Co	27	58.93
Columbium (*see* Niobium)			
Copper	Cu	29	63.55
Curium	Cm	96	247.
Dysprosium	Dy	66	162.50
Einsteinium	Es	99	254.
Erbium	Er	68	167.26
Europium	Eu	63	151.96
Fermium	Fm	100	253.
Fluorine	F	9	19.00

Table 6-1.   (Continued)

Element	Symbol	Atomic No.	Atomic Weight
Francium	Fr	87	223.
Gadolinium	Gd	64	157.25
Gallium	Ga	31	69.72
Germanium	Ge	32	72.59
Gold	Au	79	196.97
Hafnium	Hf	72	178.49
Helium	He	2	4.00
Holmium	Ho	67	164.93
Hydrogen	H	1	1.01
Indium	In	49	114.82
Iodine	I	53	126.90
Iridium	Ir	77	192.2
Iron	Fe	26	55.85
Krypton	Kr	36	83.80
Lanthanum	La	57	138.91
Lead	Pb	82	207.19
Lithium	Li	3	6.94
Lutetium	Lu	71	174.97
Magnesium	Mg	12	24.31
Manganese	Mn	25	54.94
Mendelevium	Md	101	256.
Mercury	Hg	80	200.59
Molybdenum	Mo	42	95.94
Neodymium	Nd	60	144.24
Neon	Ne	10	20.18
Neptunium	Np	93	237.
Nickel	Ni	28	58.71
Niobium	Nb	41	92.91
Nitrogen	N	7	14.01
Nobelium	No	102	254.
Osmium	Os	76	190.2
Oxygen	O	8	16.00
Palladium	Pd	46	106.4
Phosphorus	P	15	30.97
Platinum	Pt	78	195.09
Plutonium	Pu	94	242.
Polonium	Po	84	210.
Potassium	K	19	39.10
Praseodymium	Pr	59	140.91
Promethium	Pm	61	147.
Protactinium	Pa	91	231.
Radium	Ra	88	226.
Radon	Rn	86	222.
Rhenium	Re	75	186.2
Rhodium	Rh	45	102.91
Rubidium	Rb	37	85.47

Table 6-1. *(Continued)*

Element	Symbol	Atomic No.	Atomic Weight
Ruthenium	Ru	44	101.07
Samarium	Sm	62	150.53
Scandium	Sc	21	44.96
Selenium	Se	34	78.96
Silicon	Si	14	28.09
Silver	Ag	47	107.87
Sodium	Na	11	22.99
Strontium	Sr	38	87.62
Sulphur	S	16	32.06
Tantalum	Ta	73	180.95
Technetium	Tc	43	99.
Tellurium	Te	52	127.60
Terbium	Tb	65	158.92
Thallium	Tl	81	204.37
Thorium	Th	90	232.04
Thulium	Tm	69	168.93
Tin	Sn	50	118.69
Titanium	Ti	22	47.90
Tungsten	W	74	183.85
Uranium	U	92	238.03
Vanadium	V	23	50.94
Xenon	Xe	54	131.30
Ytterbium	Yb	70	173.04
Yttrium	Y	39	88.91
Zinc	Zn	30	65.37
Zirconium	Zr	40	91.22

Source: Refs, 6-11 and 6-13.

The equilibrium equation for methane, CH_4, is:

$$CH_4 + 2O_2 \rightarrow CO_2 + 2H_2O \qquad (6\text{-}3)$$

Note that the total molecules of each element on the left is equal to the total molecules of each element in the right (one C, four H, four O).

AIR PROPERTIES

Air is a mixture of nitrogen, oxygen, water vapor, helium, carbon dioxide and other gases. Its composition varies in different parts of the world, at different elevations, and at different times of the year. For purposes of combustion calculations, however, air is considered to be a mixture of only nitrogen and oxygen, as noted in Table 6-2.

Table 6-2.   Air Composition

	By Weight	By Volume
Oxygen in air	0.2315	0.21
Nitrogen in air	0.7685	0.79
Air to oxygen	4.3197	4.7619
Air to nitrogen	1.3012	1.2658
Oxygen to nitrogen	0.3012	0.2658
Nitrogen o oxygen	3.3197	3.7619
Molecular weight, average	28.9414	

BASIC COMBUSTION CALCULATIONS

In most situations waste, or fuel, is combusted with air. Air, as shown in Table 6-2, is composed of nitrogen and oxygen. For a burning reaction to proceed oxygen must be introduced and nitrogen will also be present along with the oxygen, 3.76 molecules N_2 per molecule O_2 from Table 6-2. For carbon, from equation (6-1):

$$C + O_2 + 3.76N_2 \rightarrow CO_2 + 3.76N_2 \qquad (6\text{-}4)$$

The numbers of molecules of each substance present (moles) is proportional to the volume of that substance.

The weight of each constituent is proportional to its atomic weight. Inserting the atomic weights as listed in Table 1 into Eq. (6-3):

$$
\begin{array}{ccccc}
12.01 & 32.00 & 28.02 & 44.01 & 28.02 \\
C & + \; O_2 \; + & 3.76N_2 \rightarrow & CO_2 \; + & 3.76N_2 \\
12.01 & 32.00 & 105.36 & 44.01 & 105.36 \\
1.00 & 2.66 & 8.77 & 3.66 & 8.77
\end{array}
\qquad (6\text{-}5)
$$

The first line beneath (6-5), the equilibrium equation, represents the total weights of each element present as calculated from its molecular weight above the equation. The second line relates each element to one pound of carbon and is obtained by dividing the constituent weights by the weight of carbon, 12.01.

Note first that the sum of the constituent weights on one side, 12.01 + 32.00 + 105.36 = 149.37, equals that on the other side of the equation, 44.01 + 105.36 = 149.37. Likewise, normalized to one pound of carbon the constituent weights are equal, one side to the other: 1.00 + 2.66 + 6.77 = 12.43, 3.66 + 8.77 = 12.43.

By normalizing this reaction to one pound of carbon the following determination is apparent:

For burning one pound of carbon, 2.66 pounds O_2 is required, $2.66 + 8.77 = 11.43$ pounds of air is required, 3.66 pounds of CO_2 is generated, and $3.66 + 8.77 = 12.43$ pounds of combustion products are produced.

Applying similar calculations to the combustion of hydrogen, Eq. (6-2), gives

$$
\begin{array}{ccccc}
2.02 & 32.00 & 28.02 & 18.02 & 28.02 \\
2H_2 + & O_2 + & 3.76N_2 \rightarrow & 2H_2O + & 3.76N_2 \\
4.04 & 32.00 & 105.36 & 35.04 & 105.36 \\
1.00 & 7.92 & 26.08 & 8.92 & 26.08
\end{array}
\tag{6-6}
$$

Therefore:

For burning one pound of hydrogen 7.92 pounds O_2 is required, $7.92 + 26.08 = 34.00$ pounds of air is required, 8.92 pounds of H_2O is generated, $8.92 + 26.08 = 35.00$ pounds of combustion products are produced.

STOICHIOMETRIC BURNING

The examples of combustion illustrated in Eqs. (6-5) and (6-6) utilized only that amount of air required for complete combustion to CO_2 or to H_2O. All of the oxygen provided was used up in the generation of CO_2 and H_2O and no free oxygen was left in the products of combustion. This condition is known as *complete combustion*, provision of 100% of total air, zero excess air, and defines the *stoichiometric* condition. When stoichiometric air or stoichiometric oxygen is referred to it defines the condition where only that amount of air, or oxygen, is provided to insure complete combustion in accordance with the chemical equilibrium equation. For carbon the stoichiometric air requirement is 11.43 pounds per pound carbon, and for hydrogen, 34.00 pounds stoichiometric air per pound of hydrogen, as previously calculated.

EXCESS AIR

Combustion calculations must relate to actual burning equipment requirements if they are to have any validity. In practice the air provided for complete burning must be greater than the stoichiometric requirement. The stoichiometric air value implies a burning process with 100% efficiency, where air is in contact with 100% of the waste, or fuel, surface, no air is wasted, and the burning reaction occurs instantaneously. In actuality, in order to achieve complete burning an amount of air greater than the stoichiometric requirement must be provided.

For burning gaseous fuel a total air requirement of 5% above stoichiometric may be sufficient, but for burning solid wastes or sludges in a multiple chamber furnace excess air of 100–200% of stoichiometric may be required.

EXCESS AIR CALCULATIONS

Burning a simple fuel, benzene, C_6H_6, at stoichiometric conditions:

$$78.12 \quad 32.00 \qquad 28.02 \qquad\quad 44.01 \quad 18.02 \quad 28.02$$

$$C_6H_6 + 7.5O_2 + (7.5 \times 3.76)N_2 \longrightarrow 6CO_2 + 3H_2O + 28.20N_2 \quad (6\text{-}7)$$

$$
\begin{array}{lllll}
78.12 & 240.00 & 790.16 & 264.06 & 54.06 & 790.16 \\
1.00 & 3.07 & 10.11 & 3.38 & 0.69 & 10.11
\end{array}
$$

The stoichiometric oxygen is 3.07 lb per lb C_6H_6, 7.5 moles O_2 per mole of C_6H_6. For 20% excess air an additional amount of oxygen, 20% of 7.5 or 1.5 moles of oxygen must be added to this reaction. With 7.5 moles of oxygen required for combustion of fuel, therefore, the additional 1.5 moles of oxygen added in the air will appear in the flue gas, as follows:

$$78.12 \qquad 32.00 \qquad\quad 28.02 \qquad\quad 44.01 \quad 18.02 \quad 32.00 \qquad 28.02$$

$$C_6H_6 + (7.5 + 1.5)O_2 + (9 \times 3.76)N_2 \longrightarrow 6CO_2 + 3H_2O + 1.5O_2 + 33.64N_2$$

$$
\begin{array}{llllll}
78.12 & 288.00 & 948.20 & 264.06 & 54.06 & 48.00 & 948.20 \\
1.00 & 3.69 & 12.14 & 3.38 & 0.69 & 0.62 & 12.14
\end{array}
$$

$$(6\text{-}8)$$

When burning a waste or fuel containing oxygen, such as cellulose ($C_6H_{10}O_5$), the main constituent of paper, calculations of excess air must consider the oxygen component of the fuel.

For stoichiometric conditions:

$$162.16 \qquad 32.00 \qquad\quad 28.02 \qquad\quad 44.01 \quad 18.02 \qquad 28.02$$

$$C_6H_{10}O_5 + 6O_2 + (6 \times 3.76)N_2 \longrightarrow 6CO_2 + 5H_2O + 22.56N_2 \quad (6\text{-}9)$$

$$
\begin{array}{llllll}
162.16 & 192.00 & 632.13 & 264.06 & 90.10 & 632.13 \\
1.00 & 1.19 & 3.90 & 1.63 & 0.56 & 3.90
\end{array}
$$

When calculating excess air requirements for burning, consider that 6 moles of O_2 define the stoichiometric requirements. (This is not true when calculating heating value or NO_x and CO emissions which will be covered in a later chapter.)

Therefore, for 150% excess air, an additional 6 × 1.5 = 9.0 moles of O_2, must be introduced, as follows:

$$
\begin{array}{ccccccc}
162.16 & 32.00 & 28.02 & 44.01 & 18.02 & 32.00 & 28.02 \\
\end{array}
$$

$$C_6H_{10}O_5 + (6+9)O_2 + (15 \times 3.76)N_2 \longrightarrow 6CO_2 + 5H_2O + 9O_2 + 56.40N_2$$

$$
\begin{array}{ccccccc}
162.16 & 480.00 & 1580.33 & 264.06 & 90.10 & 288.00 & 1580.33 \\
1.00 & 2.69 & 9.75 & 1.63 & 0.56 & 1.77 & 9.75 \\
\end{array}
$$

$$(6\text{-}10)$$

MEASUREMENT OF EXCESS AIR

Burning equipment is usually rated to be utilized within a specific range of excess air. If air below this range is provided, the waste or fuel will not combust completely. Above this range the excess air is greater than that required to assure complete combustion and this air flow will reduce the combustion temperature, increasing the supplemental fuel requirement. A measurement of excess air, therefore, is an important parameter of operation of burning equipment or systems.

Normally an oxygen or carbon dioxide gas analyzer is installed in the furnace exhaust stream and the oxygen or carbon dioxide present is related to excess air. From the burning of benzene and cellulose, as previously discussed, this relationship can be illustrated. Of note is that there are four modes of measurement that are used:

1. Fraction O_2, or CO_2, in total flue gas, by weight
2. Fraction O_2, or CO_2, in dry flue gas, by weight
3. Fraction O_2, or CO_2, in total flue gas, by volume
4. Fraction O_2 or CO_2, in dry flue gas, by volume

Flue gas is the gaseous product of combustion exiting a furnace.

For benzene, at stoichiometric conditions (0% excess air), from Eq. (6-7):

Total flue gas weight = 3.38 lb CO_2 + 0.69 lb H_2O + 10.11 lb N_2 = 14.18 lb

Dry flue gas weight = 3.38 lb CO_2 + 10.11 lb N_2 = 13.49 lb

Total flue gas volume = $6CO_2 + 3H_2O + 28.20N_2$ = 37.20 moles

Dry flue gas volume = $6CO_2 + 28.20N_2$ = 34.20 moles

For oxygen in the flue gas, at stoichiometric conditions, the O_2 content in the flue gas is zero.

For CO_2 in the flue gas (3.38 lb or 6 moles):

$$CO_2 \text{ by weight, total gas:} \qquad \frac{3.38}{14.18} = 23.84\%$$

$$\text{dry gas:} \qquad \frac{3.38}{13.49} = 25.06\%$$

$$CO_2 \text{ by volume, total gas:} \qquad \frac{6.00}{37.20} = 16.13\%$$

$$\text{dry gas:} \qquad \frac{6.00}{34.20} = 17.54\%$$

For 20% excess air, Eq. (6-8):

Total flue gas, by weight = 3.38 lb CO_2 + 0.69 lb H_2O + 0.62 lb O_2
$$+ 12.14 \text{ lb } N_2 = 16.83 \text{ lb}$$

Dry flue gas, by weight = 3.36 lb CO_2 + 0.62 lb O_2 + 12.14 lb N_2
$$= 16.14 \text{ lb}$$

Total flue gas, by volume = $6CO_2 + 3H_2O + 1.5O_2 + 33.64N_2$
$$= 44.34 \text{ moles}$$

Dry flue gas, by volume = $6CO_2 + 1.5O_2 + 33.84N_2 = 41.34$ moles

Oxygen fraction (0.62 lb, 1.5 moles in flue gas):

$$\text{Oxygen, by weight, total gas:} \qquad \frac{0.62}{16.83} = 3.68\%$$

$$\text{dry gas:} \qquad \frac{0.62}{16.14} = 3.84\%$$

$$\text{Oxygen, by volume, total gas:} \qquad \frac{1.5}{44.34} = 3.38\%$$

$$\text{dry gas:} \qquad \frac{1.5}{41.34} = 3.63\%$$

CO_2 fraction (3.38 lb, 6 moles in flue gas)

$$CO_2, \text{ by weight, total gas:} \qquad \frac{3.38}{16.83} = 20.08\%$$

$$\text{dry gas:} \qquad \frac{3.38}{16.14} = 20.94\%$$

$$CO_2, \text{ by volume, total gas:} \qquad \frac{6}{44.34} = 13.53\%$$

$$\text{dry gas:} \qquad \frac{6}{41.34} = 14.51\%$$

Similarly, for cellulose, at stoichiometric or 100% total air, Eq. (6-9):

O_2, by weight or volume: zero

$$CO_2, \text{ by weight, total gas:} \qquad \frac{1.63}{6.9} = 26.77\%$$

$$\text{dry gas:} \qquad \frac{1.63}{5.53} = 29.48\%$$

$$CO_2, \text{ by volume, total gas:} \qquad \frac{6}{33.56} = 17.88\%$$

$$\text{dry gas:} \qquad \frac{6}{28.56} = 21.01\%$$

At 150% excess air, or 250% total air, Eq. (6-10):

$$O_2, \text{ by weight, total gas:} \qquad \frac{1.77}{13.71} = 12.91\%$$

$$\text{dry gas:} \qquad \frac{1.77}{13.15} = 13.46\%$$

$$O_2, \text{ by volume, total gas:} \qquad \frac{9}{76.40} = 11.78\%$$

$$\text{dry gas:} \qquad \frac{9}{71.40} = 12.61\%$$

$$O_2, \text{ by weight, total gas:} \qquad \frac{1.63}{13.71} = 11.89\%$$

$$\text{dry gas:} \qquad \frac{1.63}{13.15} = 12.40\%$$

$$O_2, \text{ by volume, total gas:} \qquad \frac{6}{76.40} = 7.85\%$$

$$\text{dry gas:} \qquad \frac{6}{71.40} = 8.40\%$$

A more desirable method of measurement is based on volumetric percentage, with moisture eliminated from the calculations, i.e., based on dry flue gas by volume. Equipment is available to dry a sample prior to measurement. This eliminates concern for the moisture content of the fuel, as will be discussed in a later chapter. Summarizing the fuel gas constituents dry flue gas by volume, in Table 6-3, note the following:

- As excess air increases, the oxygen component of the flue gas increases.
- As excess air increases, the CO_2 component of the flue gas decreases.

The oxygen fraction will increase, by definition, as the excess air quantity increases. The excess, that which is not used in the reaction, must necessarily appear in the products of combustion. The reason for a decrease in CO_2 level with an increase in excess air is that a fixed amount of carbon produces a fixed amount of CO_2 regardless of the amount of excess air introduced. As the total gas flow increases, however, and excess air is increased, the proportion of CO_2 will decrease, i.e., the amount of CO_2 is fixed in an increased volume of flue gas.

PYROLYSIS

Up to now in this chapter burning processes have been discussed. Another thermal process of interest is *pyrolysis*, the degradation of carbonaceous material (a material that contains carbon) in the absence of oxygen, or air, upon application of heat. For instance, pyrolysis of cellulose is:

$$C_6H_{10}O_5 \longrightarrow 2CO + CH_4 + 3H_2O + 3C \qquad (6\text{-}11)$$

Note that in the absence of oxygen the cellulose will degrade, or fractionate, to carbon monoxide (CO), methane, (CH_4) and water vapor (H_2O), which will be in a gaseous phase, and carbon (C), a char, which will be in a solid or liquid

Table 6-3.

	Benzene		Cellulose	
	0% excess air	20% excess air	0% excess air	150% excess air
O_2	0	3.63%	0	12.61%
CO_2	17.54%	14.51%	21.01%	8.4%

phase. The off gas has a heating value, since two of its constituents, CO and CH_4, will burn in oxygen (or air), providing a heat release.

In practice true pyrolysis does not occur because it is difficult to reduce the oxygen infiltration to zero. A condition between the pyrolysis and stoichiometric burning is defined as a *starved air* mode. There is equipment designed to operate in a starved air mode, as will be discussed in a subsequent chapter. Often the terms *pyrolysis* and *starved air* are used interchangeably.

FLUE GAS PROPERTIES

A good approximation of the properties of flue gas can be made by using the properties of air as the dry flue gas component. Flue gas will contain carbon dioxide, nitrogen, oxygen, and water vapor. The proportions of these elements will vary with changes in excess air and other conditions of operation. If calculations of carbon dioxide, oxygen, and nitrogen were performed for each variation in operation, analysis of flue gas properties would be exceedingly complex and time consuming. Using air as dry gas is a convenient simplification which allows quick calculations, accurate calculations, use of air/humidity tables, etc. Incineration normally involves large amounts of excess air for proper combustion, and this excess air supply passes to the exiting flue gas. A large component of the flue gas is, therefore, air from the excess air supply.

To illustrate the properties of air with regard to the individual gas components of the flue gas, consider the flue gas produced when burning cellulose with 150% excess air [Eq. (6-10)], assuming 8500 BTU produced. The reaction temperature will be determined.

Flue gas components:

$$1.63 \text{ lb } CO_2$$
$$1.77 \text{ lb } O_2$$
$$9.75 \text{ lb } N_2$$
$$0.56 \text{ lb } H_2O$$

Enthalpy, from Chapter 4:

	CO_2	O_2	N_2	H_2O	Total
	1.63 lb	1.77 lb	9.75 lb	0.56 lb	
h at $t = 2100°F$, BTU/lb	562.8	511.4	552.7	2128.70	8403
h at $t = 2200°F$, BTU/lb	594.3	538.6	582.0	2189.92	8823

The reaction temperature for generation of 8500 BTU is interpolated from the above data as follows:

$$t = 2100 + (2200 - 2100)\frac{8500 - 8403}{8823 - 8403} = 2123°F$$

Considering the dry flue gas as air:

$$1.63 + 1.77 + 9.75 = 13.15 \text{ lb Air}$$
$$0.56 \text{ lb } H_2O$$

	Air	H_2O	Total
h at t = 2100, BTU/lb	538.72	2128.70	8276
h at t = 2200, BTU/lb	567.52	2189.92	8689

The reaction temperature for generation of 8500 BTU is interpolated as follows:

$$t = 2100 + (2200 - 2100)\frac{8500 - 8276}{8689 - 8276} = 2154°F$$

Comparing the calculated reaction temperature using the actual gas components so that temperature calculating with air as the dry gas component:

$$100\% \times \frac{2154 - 2123}{2123} = 1.46\%$$

This error will vary with the amount of excess air introduced and with the nature of the combusted waste. But an error on the order of $1\frac{1}{2}\%$ is truly insignificant when considering the reliability of other data used in incinerator design, particularly heating value. Heating value is never known with such a degree of accuracy. At best, the determination of heating value is rarely repeatable to an accuracy better than 5% even with the best of instruments. The variability of waste quality from one lot to another will often exceed 10% or 15% in heating value and other parameters.

Chapter 7
Air Emissions Calculations

As more attention is focused upon incineration as a method of waste disposal, air pollution control is becoming increasingly rigorous. The identification and control of emissions from incinerators is often the subject of intense public scrutiny. In this chapter methods of predicting the generation of air emissions will be presented.

THE AIR

Beside oxygen, nitrogen and water vapor, clean air at sea level will typically contain the components listed in Table 7-1. These compounds are the result of natural processes occurring on the earth, from plant photosynthesis to forest fires, lightning, and the eruption of volcanoes.

The concentration of human life in centralized locations around the globe necessarily results in discharges into the earth, water, and air environments. Of interest in the evaluation of incineration and other burning processes are discharges into the air environment. Such emissions include inorganic gas and organic gas discharges and the release of particulate matter. Some of these discharges appear harmless; however, many of them have been found to be injurious to life: plant, animal and/or human. The map in Fig. 7-1 is an indication of the visual effects of air pollution within the United States.

INORGANIC GAS DISCHARGES

The burning process can produce carbon dioxide, carbon monoxide, oxides of nitrogen and, where sulfur is present, oxides of sulfur. Carbon dioxide is not considered a pollutant; however, there is concern that excessive quantities of this gas within the atmosphere might produce a "greenhouse effect." This is the mechanism whereby carbon dioxide molecules absorb heat energy as does glass in a greenhouse, preventing the normal radiation of heat from the earth. An indication of this effect can be that the first half of this century saw a rise in the mean temperature of the earth of approximately $1°F$. At the same time the level of carbon dioxide in the atmosphere rose over 10%.

Table 7-1.   Minor Components of Clean, Dry
Air at Sea Level.

	% by Volume	ppm
Argon	0.93	9300
Carbon dioxide	0.0318	318
Neon	0.0018	18
Helium	0.00052	512
Krypton	0.0001	1
Xenon	0.000008	0.08
Nitrous oxide	0.000025	0.25
Hydrogen	0.00005	0.5
Methane	0.00015	1.5
Nitrogen dioxide	0.0000001	0.001
Ozone	0.000002	2.02
Sulfur dioxide	0.00000002	0.0002
Ammonia	0.000001	0.01
Carbon monoxide	0.00001	0.1

Source: Ref. 7-17.

Carbon monoxide is a danger to human health. It has the ability to pass through the lungs, directly into the bloodstream of an organism, where it destroys the ability of red blood cells to carry oxygen. At an exposure of only 0.10% carbon monoxide in air by volume (1000 ppm), a human being will be comatose in less than two hours. The Federal Government has established maximum exposure standards for carbon monoxide of 9.0 ppm for an eight hour average, and 13.0 ppm for any one hour.

Nitrogen is an extremely active substance forming a wide range of compounds with oxygen, as listed in Table 7-2. As seen in this listing, nitrogen dioxide is most significant as a pollutant of all of the nitrogen oxides. Federal standards for NO_x, which normally includes NO and NO_2, expressed as NO_2, are 0.05 ppm maximum on a yearly average and 0.13 ppm maximum for any twenty four hour average.

Note that where NO and NO_2 are present, the calculation for NO_x is as follows:

NO_x is expressed as NO_2. The molecular weight of NO_2 is 46.01 whereas the molecular weight of NO is 30.01. If 10 lbs of NO_2 and 150 lbs of NO are present, the quantity of NO_x, expressed as NO_2 is the sum of:

NO present $\times$ ratio of weight of NO_2 to NO, i.e.,

$$150 \text{ lb NO} \times \frac{46.01 \text{ lb } NO_2}{30.01 \text{ lb NO}} = 299.97 \text{ lb } NO_2 \text{ equivalent}$$

plus NO_2 present,   10.00 lb
Total NO_x 309.97 lb, expressed as NO_2

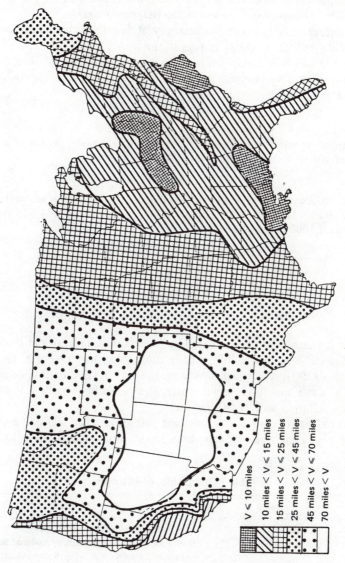

Fig. 7-1. Visual range in suburban and nonurban airports of the United States, 1974–1976. *Source:* Ref. 7-21.

V ≤ 10 miles

10 miles < V ≤ 15 miles

15 miles < V ≤ 25 miles

25 miles < V ≤ 45 miles

45 miles < V ≤ 70 miles

70 miles < V

It is important to note that knowing only the quantity of NO_x present it is not possible to estimate the NO and/or the NO_2 components.

The concept of NO_x is a convenient way of describing the magnitude of nitrogen oxide pollutants existing in a gas stream.

If the NO and NO_2 components of a gas stream are required, and the NO_x quantity is known, by assuming a ratio of NO to NO_2, these components can be estimated. For example:

A gas stream contains 1000 lbs of NO_x, expressed as the NO_2 with a ratio of NO to NO_2 of 100:1. Determine the NO and NO_2 present.

With the NO_2 quantity as X and the NO quantity as $100X$, equivalent to $46.01/30.01 \times 100X = 153.32X$ NO_2, the total NO_2 equivalent is $154.32X$ lb. For a total flow of 1000 lb $= 154.32X$,

$$X = 6.48 \text{ lb } NO_2$$

$$100X = 648.00 \text{ lb NO}$$

Sulfur is released into the atmosphere from burning processes in the form of sulfur dioxide, SO_2, and sulfur trioxide, SO_3. More than 95% of sulfur oxides generated are SO_2, which will slowly oxidize to SO_3. Sulfur trioxide is highly soluble in water and forms sulfuric acid, H_2SO_4. Sulfur dioxide is less soluble and forms sulfurous acid, H_2SO_3, in water.

Sulfur oxides are considered significant pollutants. They will spot and bleach leaves of plants and trees and have been found, in concentrations as low as 0.5

Table 7-2.   Oxides of Nitrogen.

Formula	Name	Effects
N_2O	nitrous oxide	Inert, not a pollutant (laughing gas)
NO	nitric oxide	Main product of combustion, considered harmless by itself. Converts to NO_2. Some indication that it may disintegrate the ability of red blood cells to carry oxygen.
N_2O_3	dinitrogen trioxide	Unstable, rare, not a significant pollutant
NO_2	nitrogen dioxide	Causes significant effects in the atmosphere, e.g., smog, yellows white fabric, creates plant leaf injury, reduces plant yields.
N_2O_5	dinitrogen pentoxide	Unstable, rare, not a significant pollutant.

ppm, to cause damage to fruit trees such as apple and pear. Alfalfa, barley, and various species of pine and other conifers are also sensitive to the presence of sulfur oxides.

Rain will wash sulfur oxides from the atmosphere; however, the rain will turn acidic. Acid rain is increasingly of concern because of its detrimental effect on plant life, fabrics, metals and other structural materials. It is a major cause of deterioration of statuary, building façades, and other structures throughout the industrial (and semi-industrial) world.

In human life, SO_2 is an eye irritant, causes and aggravates respiratory diseases such as emphysema and bronchitis, and studies have found a link between sulfur oxides and the occurrence of lung cancer.

Federal standards have been established for SO_x, measured as SO_2, (see the previous discussion of NO_x to relate SO_2 and SO_3 emissions to SO_x) as 0.14 ppm for a twenty-four-hour period and 0.03 ppm on a yearly basis.

Beside specific direct pollutant effects of sulfur oxides, they are contributors in gross atmospheric effects such as smog and haze.

ORGANIC GAS DISCHARGES

The majority of gaseous organic discharges into the atmosphere occur from natural sources. Of the many organic discharges from industrial sources the more significant ones are as follows:

- Oxygenated hydrocarbons such as aldehydes, ketones, alcohols, and acids. In sufficient quantity they will produce eye irritation, reduce visibility in the atmosphere, and react with other components of the atmosphere to form additional pollutants.
- Halogenated hydrocarbons, such as carbon tetrachloride, perchloroethylene, etc. These will act as the oxygenated compounds, above, and may also generate odor.
- Hydrocarbons, such as paraffins, olefins, and aromatics. These compounds may cause plant damage as well as eye irritation, odor, and a reduction in atmospheric visibility. Such compounds may also be highly reactive, readily combining with other elements of the atmospheric environment to create additional significant danger to life and comfort.

In general of the above organic gases the incineration process will produce reactive hydrocarbons. These compounds are significant in that they will form complex and harmful components in a series of reactions with nitrogen oxides and sulfur oxides.

PARTICULATE DISCHARGE

The most obvious particulate discharge from a burning process is smoke, which will be discussed in a later section of this chapter.

Of note is that in 1980 there were five major eruptions of Mount St. Helens, the Washington State volcano. Just one of those eruptions discharged over 325,000,000 tons of particulate into the atmosphere. If all the municipal refuse and sewage sludge generated in the United States were incinerated (assuming 5 lbs per capita per day for municipal refuse and 0.2 lbs per capita per day for sewage solids), at current emissions standards, it would take over 8000 years to equal a single day's eruption of the volcano.

This calculation was made as follows:

220,000,000 people × 5 lb × 365 days = 200,000,000 tons/yr refuse

Assuming 7 lb/ton emissions × 0.05 discharge (95% efficiency)

200,000,000 tons × 0.35 lb/ton = 35,000 tons/yr particulate from refuse

220,000,000 people × 0.2 lb × 365 days = 8,000,000 tons/yr sewage solids

With 1.3 lb/dry ton allowable emissions,

8,000,000 tons/yr × 1.3 lb/ton = 5200 tons/yr particulate from sludge

Total sludge and refuse = 40,200 tons/yr particulate

Volcano: 325,000,000 tons

Allowable years: 325,000,000 tons/40,200 tons/yr = 8085 years

Metallic particulate discharges into the atmosphere account for less than 1% of the total particulate discharge. However, metals can have severe toxic effects. It has been found that inhalation of metals has a greater effect on the human body than receiving the same metals damage by digestion.

There are four metals which have been shown to have a significant negative effect on human health:

1. *Lead.* This metal accumulates within the human body. It causes symptoms such as anemia, headaches, sterility, miscarriages, or the birth of handicapped children. Such handicaps, known as lead encephalopathy, include convulsions, coma, blindness, mental retardation, and/or death.

2. *Nickel.* Nickel carbonyl, NiCO, is considered to be a form of nickel which

is hazardous to human health. It causes changes in the lung structure, which causes respiratory diseases, including lung cancer. It has been found to be present in tobacco smoke.

3. *Cadmium.* This element is associated with the incineration of cadmium-containing products such as automobile tires and certain plastics. In the human body it has been found to interfere with the natural processes of zinc and copper metabolism. Cadmium has also been found to cause cardiovascular disease and hypertension.

4. *Mercury.* Mercury has long been recognized as a severe pollutant in the water environment. Because of its low boiling point it is also released into the atmosphere when it is present within a combustible, or heated, material. Mercurial poisoning in humans is characterized by blindness, progressive weakening of the muscles, numbness, paralysis, coma, and death. It also leads to severe birth deformities.

SMOKE

Smoke is a suspension of solid or liquid particulate matter within a gaseous discharge. The particles range from fractions of a micrometer (micron) to over 50 microns. The visibility of smoke is related to the quantity of particles present, rather than the weight of the particulate matter. The weight of particulate emissions is therefore not necessarily indicative of the density of a smoke discharge. The color of smoke is not necessarily indicative of smoke density either.

White Smoke

The formation of white or other opaque, non-black smoke is usually due to insufficient furnace temperatures when burning carbonaceous materials. Hydrocarbons will be heated to a level where evaporation and/or cracking will occur within the furnace when white smoke is produced. The temperatures will not be high enough to produce complete combustion of these hydrocarbons. With a stack temperature in the range of 300–500°F, many of these hydrocarbons will condense to liquid particulate and with the solid hydrocarbons these will appear as non-black smoke. A method of control of white smoke is, therefore, an increase in the furnace/stack temperatures and increased turbulence, to help insure uniformity of this higher temperature within the off-gas flow. Excessive air flow may provide excessive cooling, and an evaluation of reducing white smoke discharges would include investigating the air quantity introduced into the furnace. Inorganics in the exit gas may also produce a non-black smoke discharge. For instance, sulfur and sulfur compounds will appear yellow in a discharge, calcium and silicon oxides in the discharge will appear light to dark brown.

Black Smoke

When burned in an oxygen deficient atmosphere, hydrocarbons will not be completely destroyed, and carbon particles will be found in the off-gas. Related to oxygen deficiency is poor atomization, inadequate turbulence (or mixing), and poor air distribution within a furnace chamber. These factors will all produce carbon particles which produce dark, black smoke in the off-gas. Black smoke is also present where hydrocarbons are heated with insufficient oxygen and undergo a pyrolysis reaction. This generates stable, less complex hydrocarbon compounds that form a dark, minute particulate, generating black smoke.

One common method of reducing, or eliminating, black smoke is steam injection into the furnace. The carbon present is converted to methane and carbon monoxide as follows:

$$3C \text{ (smoke)} + 2H_2O \longrightarrow CH_4 + 2CO$$

Similar reactions occur with other hydrocarbons present. The methane and carbon monoxide produced burn clean in the heat of the furnace, eliminating the black carbonaceous smoke that would have been produced without steam injection:

$$CH_4 + 2O_2 \longrightarrow CO_2 + 2H_2O \text{ (smokeless)}$$

$$2CO + O_2 \longrightarrow 2CO_2 \text{ (smokeless)}$$

Steam injection normally requires from 20 to 80 pounds of steam per hundred pounds of flue gas.

GENERATION OF CARBON MONOXIDE

Carbon monoxide, CO, is produced where insufficient oxygen is provided to completely combust a fuel. It is, therefore, indicative of burning, or combustion efficiency. The greater the amount of air present, and the greater the degree of turbulence, the less carbon monoxide will be formed.

Turbulence as a combustion parameter cannot be easily quantified. The amount of air present and the combustion temperature affect the equilibrium constant, the relationship of CO to CO_2 produced for a given reaction. Table 7-3 is derived from equilibrium analyses and lists the formation of CO as a function of excess air, temperatures, and the ratio of carbon to hydrogen in the fuel.

Note that the generation of CO decreases with an increase in the air supply, and increases with an increase in temperature. As expected, with increased carbon in the fuel, there will be an increased rate of CO formed.

Table 7-3. Generation of Carbon Monoxide (CO) Pounds CO per Pound Stoichiometric Air.

Temp. °F	Excess Air							
	0	10%	20%	30%	50%	100%	150%	200%
For C_1H_0 (carbon)								
1000	—	—	—	—	—	—	—	—
1500	2.060E-07	1.407E-09	1.039E-09	8.827E-10	7.344E-10	5.997E-10	5.474E-10	5.193E-00
1832	9.544E-06	1.480E-07	1.093E-07	9.287E-08	7.725E-08	6.303E-08	5.758E-08	5.474E-00
2192	1.170E-04	6.110E-06	4.506E-06	3.827E-06	3.183E-06	2.597E-06	2.370E-06	2.249E-00
2500	4.830E-04	5.128E-05	3.778E-05	3.207E-05	2.666E-05	2.176E-05	1.985E-05	1.883E-00
3000	3.418E-03	9.796E-04	7.248E-04	6.159E-04	5.118E-04	4.174E-04	3.807E-04	3.610E-00
For C_3H_4 ($C_1H_{1.33}$)								
1000	—	—	—	—	—	—	—	—
1500	2.591E-08	1.080E-09	7.968E-10	6.755E-10	5.606E-10	4.554E-10	4.149E-10	3.928E-00
1832	8.131E-06	1.137E-07	8.375E-08	7.105E-08	5.896E-08	4.791E-08	4.362E-08	4.133E-00
2192	8.858E-05	4.692E-06	3.454E-06	2.930E-06	2.429E-06	1.975E-06	1.796E-06	1.702E-00
2500	3.870E-04	5.468E-05	2.897E-05	2.456E-05	2.036E-05	1.653E-05	1.505E-05	1.425E-00
3000	2.792E-03	7.567E-04	5.574E-04	4.723E-04	3.912E-04	3.175E-04	2.889E-04	2.735E-00
For CH_4								
1000	2.919E-10	—	—	—	—	—	—	—
1500	8.122E-08	7.970E-10	5.418E-10	4.589E-10	3.799E-10	3.075E-10	2.794E-10	2.642E-00
1832	5.572E-08	7.749E-08	5.699E-08	4.828E-08	3.997E-08	3.236E-08	2.940E-08	2.778E-00
2192	6.161E-05	3.199E-06	2.350E-06	1.991E-06	1.647E-06	1.333E-06	1.204E-06	1.144E-00
2500	2.743E-04	2.686E-05	1.972E-05	1.669E-05	1.380E-05	1.116E-05	1.014E-05	9.581E-00
3000	2.022E-03	5.174E-04	3.801E-04	3.213E-04	2.654E-04	2.146E-04	1.946E-04	1.839E-00

Source: Derived from Ref. 7-2.
Note: 15E-03 equals .0015.

Burning methane, CH_4, the quantity of carbon monoxide formed can be estimated as follows:

$$16.05 \quad 32.00 \qquad 44.01 \quad 18.02 \qquad 28.01$$

$$CH_4 + 2O_2 \longrightarrow CO_2 + 2H_2O + CO \text{ (trace)}$$

$$16.05 \quad 64.00 \qquad 44.01 \quad 36.04$$
$$1.00 \quad\;\; 3.99 \qquad\;\; 2.74 \quad\;\; 2.25$$

The equilibrium equation defines the stoichiometric oxygen required for burning methane, 3.99 lb oxygen per pound of methane. Assuming that this reaction will take place at 2500°F, with 20% excess air supplied, the amount of CO formed per pound of stoichiometric air is 1.972×10^{-5}, from Table 7-3. For one pound of methane, therefore:

3.99 lb O_2 required

3.99 lb $O_2 \times 4.3197$ lb air/lb $O_2 = 17.24$ lb air

17.24 lb air $\times 1.972 \times 10^{-5}$ lb CO/lb air = 0.00034 lb CO

Therefore 0.00034 lb CO is produced when burning one pound of methane at 2500°F with 20% excess air.

To determine the volume of CO produced in parts per million (ppm), relative to the volume of methane, the volumes of each component must be calculated. Using the perfect gas law (Chapter 4):

$$W = \frac{144P}{RT}$$

where

$$W = m/V \text{ lb/ft}^3$$

where M is weight, pounds, and V is volume, cubic feet.

Therefore

$$V = \frac{mRT}{144P}$$

Substituting $R = 1545/M$, where M is the molecular weight,

$$V = \frac{1545mT}{144PM}$$

For the case in question, the volume of CO vs. the volume of methane, both components are at the same temperature and pressure. Therefore the volume ration is simplified to the following:

$$\frac{V_{CO}}{V_{CH_4}} = \frac{m_{CO}}{m_{CH_4}} \times \frac{M_{CH_4}}{M_{CO}}$$

$$= \frac{0.00034 \text{ lb}}{1.0 \text{ lb}} \times \frac{16.05}{28.01} \times 1{,}000{,}000$$

$$= \underline{195 \text{ ppm}} \text{ by volume, CO to } CH_4$$

To relate the CO present to the total flue gas flow, at 20% excess air, the flue gas quantity must be determined, noting 20% excess air corresponds to $1.2 \times 2 = 24$ molecules of oxygen, which carries with it $2.4 \times 3.7619 = 9.03$ molecules nitrogen.

16.05	(2×16.00)	(2×14.01)		44.01	18.02	(2×16.00)	(2×14.01)
$CH_4 +$	$2.4O_2 +$	$9.03N_2$	$\rightarrow$	$CO_2 +$	$2H_2O +$	$0.4O_2 +$	$9.03N_2$
16.05	76.80	253.02		44.01	36.04	12.80	253.02
1.00	4.79	15.76		2.74	2.25	0.80	15.76

The outlet flue gas volume is:

$$V_{FG} = \frac{1545}{144}\frac{T}{P}\left[\frac{m_{CO_2}}{M_{CO_2}} + \frac{m_{H_2O}}{M_{H_2O}} + \frac{m_{O_2}}{M_{O_2}} + \frac{m_{N_2}}{M_{N_2}}\right]$$

The ratio of CO to flue gas volume is therefore:

$$\frac{V_{CO}}{V_{FG}} = \frac{m_{CO}/M_{CO}}{m_{FG}/M_{FG}}$$

$$= \frac{\dfrac{0.00034 \text{ lb}}{28.01 \text{ lb}}}{\dfrac{2.74 \text{ lb}}{24.01 \text{ lb}} + \dfrac{2.25 \text{ lb}}{18.02 \text{ lb}} + \dfrac{0.80 \text{ lb}}{16.00 \text{ lb}} + \dfrac{15.76 \text{ lb}}{14.01 \text{ lb}}} \times 1{,}000{,}000$$

$$= 8.91 \text{ ppm} \text{ by volume CO in exiting flue gas}$$

GENERATION OF NITROGEN OXIDES

The amount of nitrogen oxides generated is a function of temperature, excess air, and fuel composition. The greater the amount of excess air, the greater the amount of nitrogen present, and higher quantities of NO_x will be expected. The series of graphs in Fig. 7-2 show this relationship. The scale lb NO_x per million BTU relates to fuel type or composition and is directly related to mol, or molecules of oxygen required for stoichiometric combustion, which is the second of the vertical scales.

Tables 7-4 and 7-5 list generation rates for NO and NO_2, respectively. To illus-

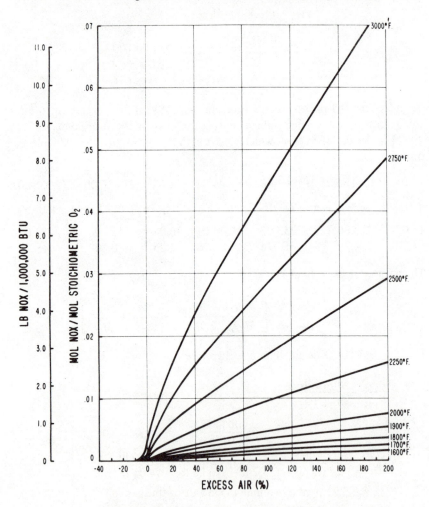

Fig. 7-2. Generation of NO_x. *Source:* Ref. 7-2.

Table 7-4. Generation of Nitrogen Oxide (NO) Pounds NO per Pound Stoichiometric Air.

Temp. °F	0	10%	20%	30%	50%	100%	150%	200%
					Excess Air			
For C_1H_0 (carbon)								
1000	—	9.587E-07	1.416E-06	1.805E-06	2.503E-06	4.088E-06	5.599E-06	7.082E-06
1500	—	2.931E-05	4.331E-05	5.523E-05	7.657E-05	1.251E-04	1.712E-04	2.166E-04
1832	1.712E-06	1.260E-04	1.862E-04	2.375E-04	3.293E-04	5.380E-04	7.366E-04	9.316E-04
2192	2.022E-05	4.081E-04	6.033E-04	7.696E-04	1.074E-03	1.744E-03	2.388E-03	3.022E-03
2500	8.198E-05	8.510E-04	1.260E-03	1.607E-03	2.232E-03	3.645E-03	4.993E-03	6.318E-03
3000	5.677E-04	2.204E-03	3.242E-03	4.194E-03	5.760E-03	9.420E-03	1.291E-02	1.633E-02
For C_3H_4 ($C_1H_{1.33}$)								
1000	—	9.587E-07	1.416E-06	1.805E-06	2.503E-06	4.088E-06	5.599E-06	7.082E-06
1500	2.575E-07	2.933E-05	4.331E-05	5.523E-05	7.657E-05	1.251E-04	1.712E-04	2.166E-04
1832	1.932E-06	1.260E-04	1.862E-04	2.375E-04	3.293E-04	5.380E-04	7.366E-04	9.316E-04
2192	1.899E-05	4.079E-04	6.033E-04	7.694E-04	1.068E-03	1.744E-03	2.388E-03	3.022E-03
2500	7.870E-05	8.502E-04	1.259E-03	1.670E-03	2.230E-03	3.645E-03	4.993E-03	6.318E-03
3000	5.341E-04	2.191E-03	3.241E-03	4.138E-03	5.751E-03	9.411E-03	1.290E-02	1.632E-02
For C_0H_2 (hydrogen)								
1000	—	9.587E-07	1.416E-06	1.805E-06	3.372E-06	4.088E-06	5.599E-06	7.082E-00
1500	2.835E-07	2.933E-05	4.331E-05	5.523E-05	7.657E-05	1.251E-04	1.712E-04	2.166E-00
1832	1.849E-06	1.260E-04	1.862E-04	2.375E-04	3.293E-04	5.380E-04	7.366E-04	9.316E-00
2192	1.792E-05	4.079E-04	6.031E-04	7.694E-04	1.068E-03	1.744E-03	2.388E-03	3.018E-00
2500	6.949E-05	8.493E-04	1.258E-03	1.606E-03	2.230E-03	3.645E-03	4.993E-03	6.315E-00
3000	4.281E-04	2.169E-03	3.222E-03	4.125E-03	5.738E-03	9.398E-03	1.288E-02	1.632E-00

Source: Derived from Ref. 7-2.
Note: 1.5E-03 equals .0015.

Table 7-5.   Generation of Nitrogen Dioxide (NO_2) Pounds NO_2 per Pound Stoichiometric Air.

Temp. °F	Excess Air							
	0	10%	20%	30%	50%	100%	150%	200%
For C_1H_0 (carbon)								
1000	—	1.208E-07	2.416E-07	3.622E-07	6.039E-07	1.208E-06	1.812E-06	2.416E-00
1500	—	4.045E-07	8.086E-07	1.213E-06	2.022E-06	4.045E-06	6.065E-06	8.089E-00
1832	—	6.528E-07	1.352E-06	2.029E-06	3.386E-06	6.768E-06	1.015E-05	1.354E-00
2192	6.538E-10	1.018E-06	2.008E-06	3.066E-06	5.113E-06	1.024E-05	1.536E-05	2.049E-00
2500	1.357E-08	1.330E-06	2.671E-06	4.015E-06	6.701E-06	1.343E-05	2.015E-05	2.668E-00
3000	1.359E-07	1.863E-06	3.718E-06	5.589E-06	9.341E-06	1.875E-05	2.816E-05	3.758E-00
For C_3H_4 ($C_1H_{1.33}$)								
1000	—	1.180E-07	2.364E-07	3.552E-07	5.936E-07	1.192E-06	1.793E-06	2.395E-00
1500	—	3.948E-07	7.916E-07	1.189E-06	1.988E-06	3.991E-06	6.006E-06	8.020E-00
1832	—	6.598E-07	1.324E-06	1.989E-06	3.326E-06	6.681E-06	1.005E-05	1.342E-00
2192	6.961E-10	9.944E-07	1.998E-06	3.004E-06	5.027E-06	1.010E-05	1.262E-05	2.031E-00
2500	1.219E-08	1.296E-06	2.611E-06	3.932E-06	6.585E-06	1.325E-05	1.994E-05	2.664E-00
3000	1.172E-07	1.798E-06	3.615E-06	5.453E-06	9.151E-06	1.846E-05	2.782E-05	3.718E-00
For C_0H_2 (hydrogen)								
1000	—	1.107E-07	2.228E-07	3.362E-07	5.656E-07	1.149E-06	1.740E-06	2.335E-00
1500	—	3.705E-07	7.460E-07	1.126E-06	1.894E-06	3.848E-06	5.826E-06	7.820E-00
1832	—	6.189E-07	1.247E-06	1.883E-06	3.168E-06	6.438E-06	9.751E-06	1.309E-00
2192	—	9.318E-07	1.882E-06	2.843E-06	4.787E-06	9.734E-06	1.475E-05	1.980E-00
2500	1.471E-08	1.213E-06	2.457E-06	3.718E-06	6.269E-06	1.276E-05	1.934E-05	2.597E-00
3000	7.021E-08	1.654E-06	3.376E-06	5.123E-06	8.679E-06	1.775E-05	2.694E-05	3.655E-00

Source: Derived from Ref. 7-2.
Note: 1.5E-03 equals .0015.

trate the use of these charts, from the example for the burning of methane, it was found that 17.24 lb air was the stoichiometric demand for burning one pound of methane, CH_4. Also, methane has the formula CH_4, which is between the values for C_3H_4 and C_0H_2 listed in Tables 7-3 and 7-4. A closer examination of the tables, however, indicates that NO and NO_2 production is not significantly affected by fuel composition at the 20% excess air and 2500°F parameters used in this example. Therefore, for NO use 1.258×10^{-3}, and for NO_2 use 2.5×10^{-6} lbs generated per pound of stoichiometric air.

For NO (atomic weight of 30.01)

$$17.24 \text{ lb air} \times \frac{1.258 \times 10^{-3} \text{ lb NO}}{\text{lb air}} = 0.022 \text{ lb NO}$$

$$\frac{V_{NO}}{V_{CH_4}} = \frac{m_{NO}}{m_{CH_4}} \times \frac{M_{CH_4}}{M_{NO}} \times 1{,}000{,}000 = \frac{0.022 \text{ lb}}{1 \text{ lb}} \times \frac{16.05}{30.01} \times 1{,}000{,}000$$

$$= 11766 \text{ ppm by volume, NO to } CH_4$$

For NO_2 (atomic weight of 46.01):

$$17.24 \text{ lb air} \times \frac{2.5 \times 10^{-6} \text{ lb } NO_2}{1 \text{ lb air}} = 0.0000431 \text{ lb } NO_2$$

$$\frac{V_{NO_4}}{V_{CH_4}} = \frac{m_{NO_2}}{m_{CH_4}} \times \frac{M_{CH_4}}{M_{NO_2}} \times 1{,}000{,}000 = \frac{0.000431 \text{ lb}}{1 \text{ lb}} \times \frac{16.05}{46.01} \times 1{,}000{,}000$$

$$= 15 \text{ ppm by volume, } NO_2 \text{ to } CH_4$$

To obtain the total weight of NO_x expressed as NO_2:

NO: $\quad 0.022 \text{ lb NO} \times \dfrac{46.01 \text{ lb } NO_2}{30.01 \text{ lb NO}} = 0.034 \text{ lb } NO_2 \text{ equivalent}$

NO_2:

$$\frac{0.0000431 \text{ lb}}{\text{Total } NO_x \quad 0.034 \text{ lb per lb } CH_4}$$

Similarly, the volumetric presence of NO_x would be calculated as:

$$11766 \text{ ppm NO} \times \frac{46.01 \text{ lb } NO_2}{30.01 \text{ lb NO}} = 18{,}034 \text{ ppm NO}$$

NO_2: $\qquad\qquad\qquad$ 15 ppm NO_2

Total: $\qquad\qquad\qquad$ 18,054 ppm NO_x by volume related to CH_4

To relate the NO, NO_2 and NO_x generated to the exiting flue gas flue use the methods in the previous discussion of CO generation.

SULFUR OXIDE EMISSIONS

It is difficult to predict the quantities of SO_2 and SO_3 generated in incineration. It has been found, however, that many wastes, municipal refuse, and sewage sludge for insistance, have low sulfur content and the release of sulfur oxides is not significant.

Once determining the sulfur component in the waste, assume that all of the sulfur exists the stack and that this airborne sulfur is in the form of SO_2.

Table 7-6 lists expected emissions from refuse burning incinerators which can be used as a guide when no other data is available.

PARTICULATE GENERATION

The generation of particulate matter is a function of the waste burned, furnace design, charging method, combustion air supply, and other furnace parameters. Table 7-6 lists particulate emission which can be expected when burning refuse in a number of different types of incinerators. There are additional tables in other chapters of this book which provide additional estimates of particulate emission.

METALS DISCHARGES

There is little data available on the release of heavy metals into the gas stream from the incineration process. Of the four metals of interest, lead, nickel, cadmium and mercury, a worst case situation is that they are released to the stack as their oxide and that no portion of the elements remains in the ash. Table 7-7 lists a typical heavy metals mass balance for the burning of sewage sludge in a fluid

Table 7-6. Emission Factors for Refuse Incinerators Without Controls.

Incinerator Type	Particulates lb/ton	Sulfur Oxides[a] lb/ton	Carbon Monoxide lb/ton	Hydrocarbons[b] lb/ton	Nitrogen Oxides[c] lb/ton
Industrial/commercial					
Multiple chamber	7	2.5	10	3	3
Single chamber	15	2.5	20	15	2
Conical (wood waste)	7	0.1	130	11	1

[a]Expressed as sulfur dioxide.
[b]Expressed as methane.
[c]Expressed as nitrogen dioxide.
Source: Ref. 7-18.

Table 7-7.   Heavy Metal Mass Balance Fluid Bed
Incinerator Burning Sewage Sludge.

Metal	Ash	Scrubber Water Percent by Weight	Stack
Cadmium	80	20	0
Chromium	95	4	1
Copper	78	21	1
Lead	87	12	1
Mercury	0.4	2	97.6
Nickel	80	20	0
Zinc	79	20	1

Source: Ref. 7-19.

bed incinerator. Note that the ash is removed from the gas stream with the scrubber water.

ORGANIC GAS GENERATION

It has not been possible to reliably predict the quantities or even the nature of organic components discharged into the atmosphere. These discharges should be determined by measurement at the incinerator or at the stack discharge of an operating unit. Estimates of these discharges are, in general, unreliable.

Chapter 8
Waste Characteristics

Although a waste may be heterogeneous, defying rigorous analysis, assumptions about waste composition and quality must be made in order to determine incinerator parameters. In this chapter waste characteristics with respect to combustion will be presented.

WASTE CHARACTERIZATION

In Chapter 1 general characteristics of waste streams were presented. Design of incineration systems and equipment requires identification of many additional parameters of waste generation and composition, as follows:

1. *General parameters.* Compositional weight fractions including component percentages of paper, metals, glass, special wastes, etc.

Process weight fractions such as percentages of combustables, non-combustables, salvageable materials, etc.

2. *Physical parameters.* The total waste, its unit size and unit shape, weight, density, age, odor, void fraction, generation/storage temperatures, compactability, angle of repose, physical state, etc.

If a solid waste, its solubility, combustibility, specific heat, heat conduction, volatile and ash fractions, shape, melting point, etc.

For a liquid waste, temperature/viscosity relationship, turbidity, specific gravity, specific heat, vapor pressure, total, settleable and suspended solids, aqueous fraction, etc.

If a gas waste its temperature/pressure quality, viscosity, density, odor, color, etc.

3. *Chemical parameters.* In general, its chemical constituents, heating value, products of combustion, organic/inorganic component fractions, etc.

4. *Caveats.* Is the waste a hazardous waste in accordance with the definition of a hazardous waste as it is generated, stored, transported, or disposed of? Can it be hazardous under any condition of handling and/or operation? Is the waste dangerous in any other way with regard to storage or disposal?

PROXIMATE ANALYSIS

The proximate analysis is a relatively quick and inexpensive laboratory determination of the percentages of moisture, volatile matter, fixed carbon and ash. The analytic procedure is as follows:

1. Heat a sample for one hour at 105-110°C (221-230°F). Report the weight loss fraction as percent moisture.
2. Raise the temperature of the dried sample, in a covered crucible, to 725°C (1337°F) and hold it at this temperature for 7 minutes. Report the sample weight loss fraction as volatile matter percentage.
3. Ignite the remaining sample in an open crucible at 950°C (1742°F) and allow it to burn to a constant weight. Report the sample weight loss as percent fixed carbon.
4. The sample residual is to be reported as percent ash. The sum of moisture, volatiles, fixed carbon, and ash should equal 100%.

ULTIMATE ANALYSIS

An ultimate analysis is a standard procedure used for a determination of the quantities of elemental components present in a sample. It is required in order to determine the products of combustion of a material, its combustion air requirement, and the nature of the off-gas or combustion products.

In this procedure the following element percentages are normally determined:

- Carbon.
- Hydrogen.
- Sulfur.
- Oxygen.
- Nitrogen.
- Halogens (chlorine, fluorine, etc.).
- Heavy metals (mercury, lead, etc.).
- Other elements that can effect the combustion process.

In addition to these components, analyses may be performed under the heading of ultimate analysis for the presence of certain compounds that may be in the waste such as benzene, PCB (polychlorinated biphenyls), dioxins, etc.

Ultimate analyses are performed by specialty laboratories with specialty equipment developed specifically for elemental analyses such as gas chromatographs, infrared scanners, or mass spectrometers.

HEATING VALUE MEASUREMENT

The most reliable method of determining heating value quantities is by testing. The most common equipment for testing for heating value is the oxygen bomb calorimeter. In this instrument a measured sample, usually one gram, is ignited in an atmosphere of pure oxygen by an electric wire. The sample heat of combustion heats a water bath surrounding the bomb. The temperature rise of the water is measured and the heat of combustion is calculated from this temperature increase.

When dealing with waste this method of analysis is often impractical. A sample as small as one gram will not be an accurate representation of a heterogeneous waste such as refuse or wastewater skimmings. Attempts are being made by the USEPA to design and construct a calorimeter for analysis of a 100 gram to one kilogram sample. Even when it is operational, this calorimeter will still not be able to handle a representative sample of a material as diffuse as common waste materials encountered.

With heterogeneous waste streams heating value measurement is made on a case by case procedure.

TABULAR VALUES

Heating values and other combustion characteristics of common chemical substances are listed in Table 8-1. Note that the high (or gross) heating value assumes that the moisture of combustion condenses within the combustion system, i.e., the exit temperature of the flue gas is below 212°F at atmospheric pressure. The lower heating value (net) is based on the water present in the vapor phase, i.e., if the exit temperature from the combustion system is greater than 212°F.

Table 8-2 lists heating values of common wastes, their density, and typical ash and moisture fractions.

HEAT OF FORMATION

The heat of formation of a compound is that amount of heat absorbed when it is formed from its prime elements. If the formation reaction generates heat the heat of formation (ΔH_f) has a negative sign and the reaction is determined as exothermic. If the reaction absorbs heat it is termed endothermic and ΔH_f is positive.

Using values for the heat of formation the heat of combustion can be calculated. The heat of combustion of a substance is equal to that amount of heat released when the substance is completely burned.

Table 8-3 lists the heat of formation of selected organic compounds, Table 8-4 that of the solid phase of inorganic oxides and Table 8-5, the heat of formation of miscellaneous materials. The units used are those in which this parameter is

Table 8-1. Combustion Constants.

Columns grouped as: **Heat of Combustion** — Btu per Cu Ft (Gross High / Net Low), Btu per Lb (Gross High / Net Low); **For 100% Total Air — Moles per mole of Combustible or Cu Ft per Cu Ft of Combustible** — Required for Combustion (O₂, N₂, Air), Flue Products (CO₂, H₂O, N₂); **For 100% Total Air — Lb per Lb of Combustible** — Required for Combustion (O₂, N₂, Air), Flue Products (CO₂, H₂O, N₂).

No.	Substance	Formula	Molecular Weight	Lb per Cu Ft	Cu Ft per Lb	Sp Gr Air 1.0000	Btu/CuFt Gross (High)	Btu/CuFt Net (Low)	Btu/Lb Gross (High)	Btu/Lb Net (Low)	Moles O₂ (Req)	Moles N₂ (Req)	Moles Air (Req)	Moles CO₂ (Flue)	Moles H₂O (Flue)	Moles N₂ (Flue)	Lb O₂ (Req)	Lb N₂ (Req)	Lb Air (Req)	Lb CO₂ (Flue)	Lb H₂O (Flue)	Lb N₂ (Flue)
1	Carbon*	C	12.01	...	...	...	...	...	14,093	14,093	1.0	3.76	4.76	1.0	...	3.76	2.66	8.86	11.53	3.66	...	8.86
2	Hydrogen	H₂	2.016	0.0053	187.723	0.0696	325	275	61,100	51,623	0.5	1.88	2.38	...	1.0	1.88	7.94	26.41	34.34	...	8.94	26.41
3	Oxygen	O₂	32.000	0.0846	11.819	1.1053	...	...	...	...	...	...	...	...	...	...	...	...	...	...	...	...
4	Nitrogen (atm)	N₂	28.016	0.0744	13.443	0.9718	...	...	...	...	...	...	...	...	...	...	...	...	...	...	...	...
5	Carbon monoxide	CO	28.01	0.0740	13.506	0.9672	322	322	4,347	4,347	0.5	1.88	2.38	1.0	...	1.88	0.57	1.90	2.47	1.57	...	1.90
6	Carbon dioxide	CO₂	44.01	0.1170	8.548	1.5282	...	...	...	...	...	...	...	...	...	...	...	...	...	...	...	...
	Paraffin series																					
7	Methane	CH₄	16.041	0.0424	23.565	0.5543	1013	913	23,879	21,520	2.0	7.53	9.53	1.0	2.0	7.53	3.99	13.28	17.27	2.74	2.25	13.28
8	Ethane	C₂H₆	30.067	0.0803	12.455	1.0488	1792	1641	22,320	20,432	3.5	13.18	16.68	2.0	3.0	13.18	3.73	12.39	16.12	2.93	1.80	12.39
9	Propane	C₃H₈	44.092	0.1196	8.365	1.5617	2590	2385	21,661	19,944	5.0	18.82	23.82	3.0	4.0	18.82	3.63	12.07	15.70	2.99	1.63	12.07
10	n-Butane	C₄H₁₀	58.118	0.1582	6.321	2.0665	3370	3113	21,308	19,680	6.5	24.47	30.97	4.0	5.0	24.47	3.58	11.91	15.49	3.03	1.55	11.91
11	Isobutane	C₄H₁₀	58.118	0.1582	6.321	2.0665	3363	3105	21,257	19,629	6.5	24.47	30.97	4.0	5.0	24.47	3.58	11.91	15.49	3.03	1.55	11.91
12	n-Pentane	C₅H₁₂	72.144	0.1904	5.252	2.4872	4016	3709	21,091	19,517	8.0	30.11	38.11	5.0	6.0	30.11	3.55	11.81	15.35	3.05	1.50	11.81
13	Isopentane	C₅H₁₂	72.144	0.1904	5.252	2.4872	4008	3716	21,052	19,478	8.0	30.11	38.11	5.0	6.0	30.11	3.55	11.81	15.35	3.05	1.50	11.81
14	Neopentane	C₅H₁₂	72.144	0.1904	5.252	2.4872	3993	3693	20,970	19,396	8.0	30.11	38.11	5.0	6.0	30.11	3.55	11.81	15.35	3.05	1.50	11.81
15	n-Hexane	C₆H₁₄	86.169	0.2274	4.398	2.9704	4762	4412	20,940	19,403	9.5	35.76	45.26	6.0	7.0	35.76	3.53	11.74	15.27	3.06	1.46	11.74
	Olefin series																					
16	Ethylene	C₂H₄	28.051	0.0746	13.412	0.9740	1614	1513	21,644	20,295	3.0	11.29	14.29	2.0	2.0	11.29	3.42	11.39	14.81	3.14	1.29	11.39
17	Propylene	C₃H₆	42.077	0.1110	9.007	1.4504	2336	2186	21,041	19,691	4.5	16.94	21.44	3.0	3.0	16.94	3.42	11.39	14.81	3.14	1.29	11.39
18	n-Butene	C₄H₈	56.102	0.1480	6.756	1.9336	3084	2885	20,840	19,496	6.0	22.59	28.59	4.0	4.0	22.59	3.42	11.39	14.81	3.14	1.29	11.39
19	Isobutene	C₄H₈	56.102	0.1480	6.756	1.9336	3068	2869	20,730	19,382	6.0	22.59	28.59	4.0	4.0	22.59	3.42	11.39	14.81	3.14	1.29	11.39
20	n-Pentene	C₅H₁₀	70.128	0.1852	5.400	2.4190	3836	3586	20,712	19,363	7.5	28.23	35.73	5.0	5.0	28.23	3.42	11.39	14.81	3.14	1.29	11.39
	Aromatic series																					
21	Benzene	C₆H₆	78.107	0.2060	4.852	2.6920	3751	3601	18,210	17,480	7.5	28.23	35.73	6.0	3.0	28.23	3.07	10.22	13.30	3.38	0.69	10.22
22	Toluene	C₇H₈	92.132	0.2431	4.113	3.1760	4484	4284	18,440	17,620	9.0	33.88	42.88	7.0	4.0	33.88	3.13	10.40	13.53	3.34	0.78	10.40
23	Xylene	C₈H₁₀	106.158	0.2803	3.567	3.6618	5230	4980	18,650	17,760	10.5	39.52	50.02	8.0	5.0	39.52	3.17	10.53	13.70	3.32	0.85	10.53
	Miscellaneous gases																					
24	Acetylene	C₂H₂	26.036	0.0697	14.344	0.9107	1499	1448	21,500	20,776	2.5	9.41	11.91	2.0	1.0	9.41	3.07	10.22	13.30	3.38	0.69	10.22
25	Naphthalene	C₁₀H₈	128.162	0.3384	2.955	4.4208	5854	5654	17,298	16,708	12.0	45.17	57.17	10.0	4.0	45.17	3.00	9.97	12.96	3.43	0.56	9.97
26	Methyl alcohol	CH₃OH	32.041	0.0846	11.820	1.1052	868	768	10,259	9,078	1.5	5.65	7.15	1.0	2.0	5.65	1.50	4.98	6.48	1.37	1.13	4.98
27	Ethyl alcohol	C₂H₅OH	46.067	0.1216	8.221	1.5890	1600	1451	13,161	11,929	3.0	11.29	14.29	2.0	3.0	11.29	2.08	6.93	9.02	1.92	1.17	6.93
28	Ammonia	NH₃	17.031	0.0456	21.914	0.5961	441	365	9,668	8,001	0.75	2.82	3.57	...	1.5	3.32	1.41	4.69	6.10	...	1.59	5.51
29	Sulfur*	S	32.06	...	...	...	...	...	3,983	3,983	1.0	3.76	4.76	SO₂ 1.0	...	3.76	1.00	3.29	4.29	SO₂ 2.00	...	3.29
30	Hydrogen sulfide	H₂S	34.076	0.0911	10.979	1.1898	647	596	7,100	6,545	1.5	5.65	7.15	1.0	1.0	5.65	1.41	4.69	6.10	1.88	0.53	4.69
31	Sulfur dioxide	SO₂	64.06	0.1733	5.770	2.2640	...	...	...	...	...	...	...	...	...	...	...	...	...	...	...	...
32	Water vapor	H₂O	18.016	0.0476	21.017	0.6215	...	...	...	...	...	...	...	...	...	...	...	...	...	...	...	...
33	Air		28.9	0.0766	13.063	1.0000	...	...	...	...	...	...	...	...	...	...	...	...	...	...	...	...

*Carbon and sulfur are considered as gases for molal calculations only.

Note: This table is reprinted from *Fuel Flue Gases*, 1941 Edition, courtesy of American Gas Association. All gas volumes corrected to 60° F and 30 in Hg dry.

Table 8-2. Characteristics of Selected Materials.

Waste	B.T.U. value/lb. as fired	Wt. in lbs. per cu. ft. (loose)	Wt. in lbs. per cu. ft.	Content by weight in percentage ASH	MOISTURE
Type 0 Waste	8,500	10		5	10
Type 1 Waste	6,500	10		10	25
Type 2 Waste	4,300	20		7	50
Type 3 Waste	2,500	35		5	70
Type 4 Waste	1,000	55		5	85
Kerosene	18,900		50	.5	0
Benzene	18,210		55	.5	0
Toluene	18,440		52	.5	0
Hydrogen	61,000		.0053	0	0
Acetic acid	6,280		65.8	.5	0
Methyl alcohol	10,250		49.6	0	0
Ethyl alcohol	13,325		49.3	0	0
Turpentine	17,000		53.6	0	0
Naphtha	15,000		41.6	0	0
Newspaper	7,975	7		1.5	6
Brown paper	7,250	7		1.0	6
Magazines	5,250	35		22.5	5
Corrugated paper	7,040	7		5.0	5
Plastic coated paper	7,340	7		2.6	5
Coated milk cartons	11,330	5		1.0	3.5
Citrus rinds	1,700	40		.75	75
Shoe Leather	7,240	20		21.0	7.5
Butyl sole composition	10,900	25		30.0	1
Polyethylene	20,000	40-60	60	0	0
Polyurethane (foamed)	13,000	2	2	0	0
Latex	10,000	45	45	0	0
Rubber waste	9,000-11,000	62-125		20-30	
Carbon	14,093		138	0	0
Wax paraffin	18,621		54-57	0	0
1/3 wax-2/3 paper	11,500	7-10		3	1
Tar or asphalt	17,000	60		1	0
1/3 tar-2/3 paper	11,000	10-20		2	1
Wood sawdust (pine)	9,600	10-12		3	10
Wood sawdust	7,800-8,500	10-12		3	10
Wood bark (fir)	9,500	12-20		3	10
Wood bark	8,000-9,000	12-20		3	10
Corn cobs	8,000	10-15		3	5
Rags (silk or wool)	8,400-8,900	10-15		2	5
Rags (linen or cotton)	7,200	10-15		2	5
Animal fats	17,000	50-60			0
Cotton seed hulls	8,600	25-30		2	10
Coffee grounds	10,000	25-30		2	20
Linoleum scrap	11,000	70-100		20-30	1

The above chart shows the various BTU values of materials commonly encountered in incinerator designs. The values given are approximate and may vary based on their exact characteristics or moisture content. The BTU value is the higher heating value.

Source: Ref. 8-4.

Table 8-3. Heat of Formation of Selected Organic Compounds.

Formula	Weight, g/mole	Name	ΔH_f, kcal/mole
CBr_3	251.71	Carbon tribromide (g)	+ 42.0
CBr_4	331.61	Carbon tetrabromide (g)	+ 19.0
CBr_4	331.61	Carbon tetrabromide (c)	+ 4.5
CCl_3	118.36	Carbon trichloride (g)	+ 14.0
CCl_4	153.81	Carbon tetrachloride (g)	− 24.6
CCl_4	153.81	Carbon tetrachloride (liq)	− 32.4
CF_2O	66.01	Carbonyl difluoride (g)	− 151.7
CF_3	69.01	Carbon trifluoride (g)	− 114.0
CF_4	88.01	Carbon tetrafluoride (g)	− 223.0
CO_3Fe	115.86	Iron carbonate (c)	− 177.0
$CHBr_3$	252.72	Bromoform (g)	+ 4.0
$CHBr_3$	252.72	Bromoform (liq)	− 6.8
$CHCl_3$	119.37	Chloroform (g)	− 24.7
$CHCl_3$	119.37	Chloroform (liq)	32.1
CHF_3	70.02	Fluoroform (g)	− 164.5
CHI_3	393.72	Iodoform (c)	33.7
CH_2N_4	70.07	Tetrazole (c)	+ 56.7
CH_2N_4O	86.07	5-hydroxytetrazole (c)	+ 1.5
CH_2O	30.03	Formaldehyde (g)	− 26.0
CH_2O_2	46.03	Formic acid (liq)	− 101.5
CH_3Br	94.94	Methyl bromide (g)	− 8.4
CH_3Cl	50.49	Methyl chloride (g)	− 19.3
CH_3Hg	215.63	Methyl mercury (g)	+ 40.0
CH_3I	141.94	Methyl iodide (g)	+ 3.1
CH_3I	141.94	Methyl iodide (liq)	− 3.7
CH_4	16.05	Methane (g)	− 17.9
CH_4N_2O	60.07	Urea (c)	− 79.7
CH_4O	32.05	Methanol (g)	− 48.0
CH_4O	32.05	Methanol (liq)	− 57.0
$CMnO_3$	114.95	Manganese carbonate (c)	− 213.7
CO_3Zn	125.38	Zinc carbonate (c)	− 194.3
C_2F_4	100.02	Tetrafluoroethylene (g)	− 155.5
C_2F_6	138.02	Hexaflouroethane (g)	− 310.0
$C_2H_2Cl_4$	167.84	1,1,2,2-tetrachloroethane (g)	− 35.7
$C_2H_2CoO_4$	148.97	Cobaltous formate (c)	− 208.7
$C_2H_2CuO_4$	153.59	Copper formate (c)	− 186.7
$C_2H_2MnO_4$	144.98	Manganese formate (c)	− 249.7
$C_2H_2NiO_4$	148.75	Nickel formate (c)	− 208.4
$C_2H_2O_2$	58.04	Glyoxal (g)	− 50.7
$C_2H_2O_4$	90.04	Oxalic acid (c)	− 197.7
$C_2H_2O_4Pb$	297.23	Lead formate (c)	− 210.0
$C_2H_2O_4Zn$	155.41	Zinc formate (c)	− 235.8
$C_2H_3AgO_2$	166.92	Silver acetate (c)	− 95.3
C_2H_3Br	106.95	Vinyl bromide (g)	+ 18.7
C_2H_3BrO	122.95	Acetyl bromide (liq)	− 53.4
$C_2H_3Br_3O_2$	298.75	Bromal hydrate (c)	− 112.0

Table 8.3.   (*Continued*)

Formula	Weight, g/mole	Name	ΔH_f, kcal/mole
C_2H_3Cl	62.50	Vinyl chloride (g)	+ 8.5
C_2H_3Cl	62.50	Vinyl chloride (liq)	+ 3.5
C_2H_3ClO	78.50	Acetyl chloride (g)	− 58.2
C_2H_3ClO	78.50	Acetyl chloride (liq)	− 65.4
$C_2H_3ClO_2$	94.50	Chloral hydrate (c)	− 137.7
$C_2H_3ClO_2$	94.50	Chloral hydrate (g)	− 107.2
C_2H_4	28.06	Ethylene (g)	+ 12.5
$C_2H_4O_2$	60.06	Acetic acid (liq)	− 115.7
$C_2H_4O_2$	60.06	Methyl formate (g)	− 83.7
$C_2H_4O_2$	60.06	Methyl formate (liq)	− 90.6
$C_2H_4O_3$	76.06	Gylcolic acid (c)	− 158.6
$C_2H_4O_4$	92.06	Glyoxylic acid (c)	− 199.7
C_2H_5Br	108.97	Ethyl bromide (g)	− 15.4
C_2H_5Br	108.97	Ethyl bromide (liq)	− 22.0
C_2H_5Cl	64.52	Ethyl chloride (g)	− 26.8
C_2H_5Cl	64.52	Ethyl chloride (liq)	− 32.6
C_2H_5ClO	80.52	Ethylene chlorohydrin (liq)	− 70.6
C_2H_5I	143.89	Ethyl iodide (g)	− 1.8
C_2H_5I	143.89	Ethyl iodide (liq)	− 9.6
C_2H_5N	43.08	Ethylenimine (liq)	+ 21.9
C_2H_6	30.08	Ethane (g)	− 20.2
C_2H_6O	46.08	Ethanol (g)	− 56.2
C_2H_6O	46.08	Ethanol (liq)	− 66.4
$C_3H_4O_3$	78.07	Ethylene carbonate (c)	− 138.9
C_3H_6	42.09	Propylene (g)	+ 4.9
C_3H_6O	58.09	Propanone (g)	− 51.8
C_3H_6O	58.09	Propanone (liq)	− 59.2
$C_3H_6O_2$	74.09	Methane acetate (liq)	− 106.4
C_3H_8	44.11	Propane (g)	− 24.8
$C_3H_9N_5O_4$	179.17	Acetamideguanidine nitrate (c)	− 119.1
C_3H_9P	76.09	Trimethylphomphine (liq)	− 30.0
$C_3H_{10}N_2$	74.15	1,2-propanediamine (liq)	− 23.4
$C_3H_{10}N_2O_3$	122.15	Trimethylamine nitrate (c)	− 74.1
$C_3H_{10}O_3Si$	122.22	Trimethexymilane (liq)	− 199.0
$C_3H_{10}Sn$	164.82	Trimethyl tin (g)	+ 5.0
$C_3H_{10}Sn$	164.82	Trimethyl tin (liq)	− 2.1
$C_3H_{12}N_6O_3$	180.21	Guanidine carbonate (c)	− 232.1
$C_4H_2N_2S$	110.14	4-cyanothiazole (c)	+ 52.6
$C_4H_4N_2$	80.10	Pyraxine (c)	+ 33.4
$C_4H_4N_2$	80.10	Pyraxine (g)	+ 46.9
C_4H_4NS	99.16	4-Methylthiazole (liq)	+ 16.3
C_4H_6	54.1	Butadiene (g)	+ 38.8
C_4H_6	54.0	1-butyne (g)	+ 39.5
$C_4H_6O_4Pb$	325.29	Lead acetate (c)	− 230.4
$C_4H_6O_4Zn$	183.47	Zinc acetate (c)	− 257.8
$C_4H_6O_6$	150.10	Tartaric acid (c)	− 308.5

Table 8.3.   (Continued)

Formula	Weight, g/mole	Name	ΔH_f, kcal/mole
C_4H_8	56.12	Cyclobutane (liq)	+ 0.8
C_4H_8	56.12	Butene (g)	− 2.7
C_4H_8O	72.12	Butanone (g)	− 59.3
C_4H_8O	72.12	Butanone (liq)	− 65.3
C_4H_9N	71.14	Pyrrolidine (g)	− 0.9
C_4H_9N	71.14	Pyrrolidine (liq)	− 9.9
C_4H_{10}	58.14	Butane (g)	− 30.1
$C_4H_{10}O$	74.14	Butanol (liq)	− 78.2
$C_4H_{10}O$	74.14	Diethyl ether (g)	− 60.3
$C_4H_{11}N$	73.16	Butylamine (liq)	− 30.5
$C_5H_4N_4O$	136.13	Hypoxanthine (c)	− 26.2
$C_5H_4N_4O_2$	152.13	Xanthine (c)	− 90.5
$C_5H_4N_4O_3$	168.13	Uric acid (c)	− 147.7
$C_5H_4O_2$	96.09	Furfural (liq)	− 47.8
C_5H_5N	79.11	Pyridine (g)	+ 33.6
C_5H_5N	79.11	Pyridine (liq)	+ 24.0
$C_5H_6N_4O_4$	186.15	Pseudouric acid (c)	− 221.7
$C_5H_6O_2$	98.11	Furfuryl alcohol (liq)	− 66.0
C_5H_8	68.13	Pentadine (g)	+ 33.6
$C_5H_9NO_2$	115.15	DL-proline (c)	− 125.7
$C_5H_9NO_3$	131.15	L-hydroxyproline (c)	− 158.1
C_5H_{10}	70.15	Cyclopentane (liq)	− 25.3
C_5H_{10}	70.15	Pentene (g)	− 5.0
$C_5H_{10}O$	86.15	2-pentanone (liq)	− 71.1
C_5H_{12}	72.17	Pentane (g)	− 35.0
C_5H_{12}	88.17	Pentane (liq)	− 41.4
$C_5H_{12}O$	88.17	Pentanol (liq)	− 64.3
$C_5H_{12}O_5$	152.17	Xylitol (c)	− 267.3
C_6H_6	78.12	Benzene (g)	+ 19.6
C_6H_6	78.12	Benzene (liq)	+ 11.7
C_6H_6O	94.12	Phenol (c)	− 39.5
C_6H_7N	93.14	Aniline (liq)	+ 7.6
C_6H_7N	93.14	Methylpyridine (g)	+ 24.1
C_6H_7N	93.14	Methylpyridine (liq)	+ 13.8
$C_6H_8O_2$	112.14	Sorbic acid (c)	− 93.4
$C_6H_8O_6$	176.14	Ascorbic acid (c)	− 278.3
$C_6H_8O_7$	192.14	Citric acid, anhydrous (c)	− 269.0
C_6H_{10}	82.16	Hexyne (g)	+ 29.6
$C_6H_{10}O_2$	114.16	Hydrosorbic acid (liq)	− 110.2
$C_6H_{10}O_5$	162.16	Saccharinic acid lactone (c)	− 249.6
$C_6H_{10}O_8$	210.16	Citric acid monohydrate (c)	− 439.4
$C_6H_{10}O_8$	210.16	Mucic acid (c)	− 423.0
C_6H_{12}	84.18	Cyclohexane (liq)	− 37.3
C_6H_{12}	84.18	Hexene (g)	− 9.9
$C_6H_{12}O_2$	116.18	Caproic acid (liq)	− 139.7
$C_6H_{12}O_3$	132.18	Acetone glycerol (liq)	− 163.0

Table 8.3. (Continued)

Formula	Weight, g/mole	Name	ΔH_f, kcal/mole
$C_6H_{12}O_5$	164.18	Fucose (c)	− 262.7
$C_6H_{12}O_6$	180.18	Fructose (c)	− 302.2
$C_6H_{12}O_6$	180.18	Glucose (c)	− 203.8
$C_6H_{12}O_7$	196.18	Gluconic acid (c)	− 379.3
C_6H_{14}	86.20	Dimethylbutane (liq)	− 51.0
C_6H_{14}	86.20	Hexane (liq)	− 47.5
$C_6H_{14}O$	102.20	Hexanol (liq)	− 90.7
$C_6H_{14}O_6$	182.20	Dulcitol (c)	− 321.8
$C_6H_{14}O_7$	198.20	Glucose hydrate (c)	− 375.0
$C_6H_{15}N$	101.22	Triethylamine (liq)	− 32.1
C_7H_6O	106.13	Benzaldehyde (liq)	− 20.1
$C_7H_6O_2$	122.13	Benzoic acid (c)	− 92.0
$C_7H_6O_3$	138.13	Salicylic acid (c)	− 140.9
C_7H_7N	105.15	Vinylpyridine (liq)	+ 37.2
C_7H_7NO	121.15	Benzamide (c)	− 48.4
C_7H_8	92.15	Toluene (liq)	+ 2.9
C_7H_8O	108.15	Cresol (c)	− 48.9
C_7H_8S	124.21	Benzyl mercaptan (liq)	+ 10.5
C_7H_9N	107.17	Ethylpyridine (liq)	− 1.2
C_7H_9N	107.17	Methylaniline (liq)	+ 7.7
C_7H_{12}	96.19	Heptyne (g)	+ 24.6
C_7H_{14}	98.21	Cycloheptane (liq)	− 37.5
C_7H_{14}	98.21	Heptene (g)	− 14.9
C_7H_{14}	98.21	Methylcyclohexane (liq)	− 45.4
C_7H_{16}	100.23	Dimethylpentane (liq)	− 57.0
C_7H_{16}	100.23	Heptane (liq)	− 53.6
$C_8H_6O_4$	166.14	Phthalic acid (c)	− 191.9
C_8H_7N	117.16	Indole (c)	+ 29.8
C_8H_7NO	133.16	Oxindole (c)	− 41.2
$C_8H_7NO_2$	149.16	Dioxindole (c)	− 76.9
C_8H_8	104.16	Ethenylbenzene (liq)	+ 24.8
$C_8H_8N_2O_2$	164.18	Phthalamide (c)	− 104.4
$C_8H_8O_2$	136.16	Methyl benzoate (liq)	− 79.8
C_8H_9NO	135.18	Acetanilide (c)	− 50.3
C_8H_{10}	106.18	Ethylbenzene (liq)	− 3.0
$C_8H_{10}N_4O_2$	194.22	Caffeine (c)	− 76.2
$C_8H_{10}O$	122.18	Ethylphenol (c)	− 49.9
$C_8H_{11}N$	121.20	Dimethylaniline (liq)	+ 8.2
C_8H_{14}	110.22	Octyne (g)	+ 19.7
C_8H_{16}	112.24	Cyclooctane (liq)	− 40.6
C_8H_{16}	112.24	Dimethylcyclohexane (liq)	− 53.3
C_8H_{16}	112.24	Dimethylcyclohexane (liq)	− 51.5
C_8H_{16}	112.24	Dimethylcyclohexane (liq)	− 51.5
C_8H_{16}	112.24	Dimethylcyclohexane (liq)	− 53.1
C_8H_{16}	112.24	Ethylcyclohexane (liq)	− 50.7
C_8H_{16}	112.24	Propylcyclopentane (liq)	− 45.2

Table 8.3.   (Continued)

Formula	Weight, g/mole	Name	ΔH_f, kcal/mole
$C_8H_{16}N_2O_3$	188.26	DL-leucylglycine (c)	− 205.1
$C_8H_{16}O_2$	144.24	Caprylic acid (liq)	− 151.9
$C_8H_{17}N$	127.26	Couline (liq)	− 57.6
C_8H_{18}	114.26	Dimethylhexane (liq)	− 62.6
C_8H_{18}	114.26	Ethylhexane (liq)	− 59.9
C_8H_{18}	114.26	Octane (liq)	− 59.7
$C_8H_{20}Pb$	323.47	Tetraethyl lead (g)	− 26.2
$C_8H_{20}Pb$	323.47	Tetraethyl lead (liq)	+ 12.6
C_9H_9N	131.19	Skatole (c)	+ 16.3
$C_9H_{10}N_2$	146.21	Dipyrrylmethane (c)	+ 31.4
C_9H_{12}	120.21	Isopropylbenzene (liq)	− 9.8
$C_9H_{14}O_6$	218.23	Glyceryl triacetate (liq)	− 318.3
$C_9H_{14}O_6$	218.23	Mannitol triformal (c)	− 242.0
$C_9H_{15}N$	137.25	Phyllopyrrole (c)	− 20.4
C_9H_{16}	124.25	Nonyne (g)	+ 14.8
C_9H_{18}	126.27	Nonene (g)	− 24.7
C_9H_{18}	126.27	Propylcyclohexane (liq)	− 57.0
$C_9H_{18}O_2$	158.27	Methyl caprylate (liq)	− 141.1
C_9H_{20}	128.29	Nonane (liq)	− 65.8
$C_9H_{21}N$	143.31	Tri-n-propylamine (liq)	− 49.5
$C_{10}H_8$	128.18	Naphthalene (c)	+ 18.0
$C_{10}H_8$	128.18	Naphthalene (g)	+ 35.6
$C_{10}H_9N$	143.20	Phenylpyrrole (c)	+ 34.5
$C_{10}H_9N$	143.20	Quinaldene (c)	+ 39.3
$C_{10}H_{10}O_4$	194.20	Dimethyl phthalate (c)	− 171.0
$C_{10}H_{11}NO_3$	193.22	Benzoyl sarcosine (c)	− 135.7
$C_{10}H_{11}NO_4$	209.22	Animoyl glycine (c)	− 180.9
$C_{10}H_{12}O_4$	196.22	Glyceryl benzoate (c)	− 185.8
$C_{10}H_{13}NO_2$	179.24	Phenacetin (c)	− 101.1
$C_{10}H_{14}N_2$	162.26	Nicotine (liq)	+ 9.4
$C_{10}H_{16}O_2$	168.26	Dehydrocampholenolactone (c)	− 130.0
$C_{10}H_{16}O_3$	184.26	Methyl ethyl heptane lactone (c)	− 183.5
$C_{10}H_{18}$	136.26	Decyne (g)	+ 9.9
$C_{10}H_{20}$	140.30	Decene (g)	− 29.6
$C_{10}H_{20}O_2$	172.30	Capric acid (c)	− 170.6
$C_{10}H_{20}O_2$	172.30	Capric acid (liq)	− 163.6
$C_{10}H_{20}O_2$	172.30	Methyl pelargonate (liq)	− 147.3
$C_{10}H_{22}$	142.32	Decane (liq)	− 71.9
$C_{10}H_{22}O_7$	254.32	Dipentaerythritot (c)	− 376.9
$C_{11}H_{12}N_2O_2$	204.25	Tryptophane (c)	− 99.8
$C_{11}H_{14}N_2O_3$	222.27	Glycylphenylalanine (c)	− 163.9
$C_{11}H_{20}$	152.31	1-undecyne (g)	+ 5.0
$C_{11}H_{20}N_2O_2$	212.33	Valylleucyl anhydride (c)	− 150.1
$C_{11}H_{22}$	154.33	1-undecene (g)	− 34.6
$C_{11}H_{22}O_2$	186.33	Methyl caprate (liq)	− 153.1
$C_{11}H_{24}$	156.35	Undecane (liq)	− 78.0

Table 8.3.   (*Continued*)

Formula	Weight, g/mole	Name	ΔH_f, kcal/mole
$C_{12}H_8N_2$	180.22	Phenazine (c)	+ 56.4
$C_{12}H_9N$	167.22	Carbazole (c)	+ 30.3
$C_{12}H_{10}S_2$	218.34	Diphenyl disulfide (c)	+ 35.8
$C_{12}H_{14}N_2O_2$	218.28	Alanylphenylalanyl anhydride (c)	− 89.3
$C_{12}H_{14}H_4O_6$	310.30	Desoxyamalic acid (c)	− 285.7
$C_{12}H_{14}N_4O_8$	342.30	Amalic acid (c)	− 367.0
$C_{12}H_{15}NO_4$	237.28	DL-phenylalanine-*N*-carboxylic acid dimethyl ester (c)	− 184.3
$C_{12}H_{16}N_2O_3$	236.30	Alanylphenylalanine (c)	− 170.2
$C_{12}H_{16}O_8$	288.28	Levoglucosan triacetate (c)	− 371.3
$C_{12}H_{22}$	166.34	Dodecyne (g)	+ 0.1
$C_{12}H_{22}N_2O_2$	226.36	Leucine anhydride (c)	− 160.0
$C_{12}H_{22}O_6$	262.34	Diacetonemannitol (c)	− 350.0
$C_{12}H_{22}O_{11}$	342.34	Cellobiose (c)	− 532.5
$C_{12}H_{22}O_{11}$	342.34	Lactose (c)	− 530.1
$C_{12}H_{22}O_{11}$	342.34	Maltose (c)	− 530.8
$C_{12}H_{22}O_{11}$	342.34	Sucrose (c)	− 531.9
$C_{12}H_{24}$	168.36	Dodecene (g)	− 39.5
$C_{12}H_{24}O_2$	200.36	Lauric acid (c)	− 185.1
$C_{12}H_{24}O_2$	200.36	Lauric acid (liq)	− 176.4
$C_{12}H_{24}O_2$	200.36	Methyl undecylate (liq)	− 159.0
$C_{12}H_{24}O_{12}$	360.36	Lactose monohydrate (c)	− 602.0
$C_{12}H_{26}$	170.38	Dodecane (liq)	− 84.1
$C_{12}H_{26}O_{13}$	378.38	Trehalose dihydrate (c)	− 676.1
$C_{13}H_9N$	179.23	Acridine (c)	+ 44.8
$C_{13}H_{10}O$	182.23	Benzophenone (c)	− 8.0
$C_{13}H_{11}NO$	197.25	Benzanilide (c)	− 22.3
$C_{13}H_{24}$	180.37	Tridecyne (g)	− 4.9
$C_{13}H_{26}$	182.39	Tridecene (g)	− 44.4
$C_{13}H_{26}O_2$	214.39	Methyl laurate (liq)	− 165.6
$C_{13}H_{28}$	184.41	Tridecane (liq)	− 90.2
$C_{14}H_{10}$	178.24	Anthracene (c)	+ 29.0
$C_{14}H_{10}$	178.24	Anthracene (g)	+ 53.7
$C_{14}H_{10}$	178.24	Phenanthrene (c)	+ 27.3
$C_{14}H_{10}$	178.24	Phenanthrene (g)	+ 48.4
$C_{14}H_{18}N_2O_2$	246.34	Valylphenylalanyl anhydride (c)	− 94.3
$C_{14}H_{19}N_3O_4$	293.36	Glycylalanyl phenylalanine (c)	− 222.0
$C_{14}H_{10}N_2O_3$	254.26	Valylphenylalanine (c)	− 183.5
$C_{14}H_{20}O_9$	332.34	Rhamnose triacetate (c)	− 455.4
$C_{14}H_{23}N_3O_{10}$	393.40	Diethylenetriaminepentaacetic acid (c)	− 531.8
$C_{14}H_{26}$	194.40	Tetradecyne (g)	− 9.8
$C_{14}H_{28}$	196.42	Tetradecene (g)	− 49.3
$C_{14}H_{28}O_2$	228.42	Myristic acid (c)	− 199.2
$C_{14}H_{28}O_2$	228.42	Myristic acid (liq)	− 188.5
$C_{14}H_{30}$	198.44	Tetradecane (liq)	− 96.3
$C_{15}H_{20}O_6$	296.35	Tricyclobutyrin (liq)	− 247.0

Table 8.3. (Continued)

Formula	Weight, g/mole	Name	ΔH_f, kcal/mole
$C_{15}H_{28}$	208.43	Pentadecyne (g)	− 14.7
$C_{15}H_{30}$	210.45	Pentadecene (g)	− 54.3
$C_{15}H_{30}O_2$	242.45	Methyl myristate (liq)	− 177.8
$C_{15}H_{32}$	212.47	Pentadecane (liq)	− 102.4
$C_{16}H_{10}$	202.26	Fluoranthene (c)	+ 45.8
$C_{16}H_{10}$	202.26	Pyrene (c)	+ 27.4
$C_{16}H_{10}N_2O_2$	262.28	Indigotin (c)	− 32.0
$C_{16}H_{12}N_2O_4$	296.30	Insatide (c)	− 139.0
$C_{16}H_{15}NO_3$	269.32	Benzoylphenylalanine (c)	− 129.6
$C_{16}H_{22}O_{11}$	390.38	Galactose pentaacetate (c)	− 532.8
$C_{16}H_{22}O_{11}$	390.38	Glucose pentaacetate (c)	− 532.1
$C_{16}H_{30}$	222.46	Hexadecyne (g)	− 19.6
$C_{16}H_{32}$	224.48	Hexadecene (g)	− 59.1
$C_{16}H_{32}O_2$	256.48	Palmitic acid (c)	− 213.1
$C_{16}H_{32}O_2$	256.48	Palmitic acid (liq)	− 200.4
$C_{16}H_{34}$	226.50	Hexadecene (liq)	− 108.5
$C_{17}H_{21}NO_4$	303.39	Morphine monohydrate (c)	− 170.1
$C_{17}H_{32}$	236.49	Heptadecyne (g)	− 24.6
$C_{17}H_{34}$	238.51	Heptadecene (g)	− 64.1
$C_{17}H_{36}$	240.53	Heptadecane (liq)	− 114.6
$C_{18}H_{12}$	228.30	Naphthacene (c)	+ 38.3
$C_{18}H_{12}$	228.30	Triphenylene (c)	+ 33.7
$C_{18}H_{18}N_2O_2$	294.38	Phenylalanyl anhydride (c)	− 69.3
$C_{18}H_{23}NO_4$	317.42	Codeine monohydrate (c)	− 151.2
$C_{18}H_{26}O_6$	338.44	Tricyclovalerin (liq)	− 270.0
$C_{18}H_{32}O_{16}$	504.50	Melezitose (c)	− 815.0
$C_{18}H_{32}O_{16}$	504.50	Raffinose (c)	− 761.0
$C_{18}H_{34}$	250.52	Octadecyne (g)	− 29.5
$C_{18}H_{34}O_2$	282.52	Oleic acid (c)	− 187.2
$C_{18}H_{34}O_2$	282.52	Oleic acid (liq)	− 178.9
$C_{18}H_{36}$	252.54	Octadecene (g)	− 69.0
$C_{18}H_{36}O_2$	284.54	Stearic acid (c)	− 226.5
$C_{18}H_{36}O_2$	284.54	Stearic acid (liq)	− 212.5
$C_{18}H_{38}$	254.56	n-octadeacne (liq)	− 120.7
$C_{18}H_{42}O_{21}$	594.60	Raffinose penthydrate (c)	−1122.0
$C_{19}H_{21}NO_3$	311.41	Thebaine (c)	+ 63.0
$C_{19}H_{22}N_2O$	294.43	Cinchonine (c)	+ 7.4
$C_{19}H_{24}N_2O$	296.45	Cinchonamine (c)	− 10.4
$C_{19}H_{25}N_3O_4$	359.47	Cinchonamine nitrate (c)	− 79.9
$C_{19}H_{32}$	260.51	Androstane (c)	− 75.0
$C_{19}H_{36}$	264.55	Nonadecyne (g)	− 34.4
$C_{19}H_{36}O_2$	296.55	Methyl elaidate (liq)	− 175.8
$C_{19}H_{36}O_2$	296.55	Methyl oleate (liq)	− 174.2
$C_{19}H_{38}$	266.57	1-nonadecene (g)	− 73.9
$C_{19}H_{40}$	268.59	Nonadecane (liq)	− 126.8
$C_{20}H_{12}$	252.32	Perylene (c)	+ 43.7

Table 8.3. (*Continued*)

Formula	Weight, g/mole	Name	ΔH_f, kcal/mole
$C_{20}H_{21}NO_4$	339.42	Papaverine (c)	$-$ 120.2
$C_{20}H_{24}N_2O_2$	324.26	Quinidine (c)	$-$ 38.3
$C_{20}H_{24}N_2O_2$	324.26	Quinine (c)	$-$ 37.1
$C_{20}H_{27}NO_{11}$	457.48	Amygdalin (c)	$-$ 455.0
$C_{20}H_{38}$	278.58	Elcosyne (g)	$-$ 39.4
$C_{20}H_{40}$	280.60	Elcosene (g)	$-$ 78.9
$C_{20}H_{40}O_2$	312.60	Arachidic acid (c)	$-$ 241.8
$C_{20}H_{40}O_2$	312.60	Arachidic acid (liq)	$-$ 224.6
$C_{20}H_{42}$	282.62	Eicosane (liq)	$-$ 132.9
$C_{21}H_{16}N_2$	296.39	Lophine (c)	$+$ 65.0
$C_{21}H_{18}N_2$	298.41	Amarine (c)	$+$ 63.0
$C_{21}H_{19}N_2O_5$	379.42	Amarine heminhydrate (c)	$+$ 29.0
$C_{21}H_{22}N_2O_2$	334.45	Strychnine (c)	$-$ 41.0
$C_{21}H_{42}O_4$	358.63	Glyceryl ntearate (c)	$-$ 315.8
$C_{22}H_{23}NO_7$	413.46	Narcotine (c)	$-$ 210.9
$C_{22}H_{42}O_2$	338.64	Erucic acid (c)	$-$ 207.0
$C_{22}H_{44}O_2$	340.66	Behenic acid (c)	$-$ 235.0
$C_{22}H_{44}O_4$	372.66	Dihydroxybehenic acid (c)	$-$ 337.0
$C_{23}H_{26}N_2O_4$	394.51	Brucine (c)	$-$ 118.6
$C_{23}H_{31}NO_{10}$	481.55	Narceine dihydrate (c)	$-$ 421.2
$C_{24}H_{20}O_6$	404.44	Glyceryl tribenzoate (c)	$-$ 214.0
$C_{24}H_{24}N_2O_3$	388.50	Anisine (c)	$-$ 51.0
$C_{24}H_{40}O_{20}$	648.64	Diamylose (c)	$-$ 850.0
$C_{24}H_{42}O_{21}$	666.66	Stachyose (c)	$-$ 987.0
$C_{32}H_{36}N_4O_2$	508.72	Pyrroporphyrin monomethyl ester (c)	$+$ 88.6
$C_{32}N_{38}N_4$	478.74	Aetioporphyrin (c)	$-$ 1.8
$C_{34}H_{34}N_4O_4$	562.72	Protoporphyrin (c)	$-$ 120.6
$C_{34}H_{36}N_4O_3$	548.74	Phylloerythrin monomethyl ester (c)	$-$ 83.2
$C_{36}H_{36}N_4O_6$	620.76	Methyl pheophorbide b (c)	$-$ 200.7
$C_{36}H_{38}N_4O_4$	590.78	Protoporphyrin dimethyl ester (c)	$-$ 122.1
$C_{36}H_{38}N_4O_5$	606.78	Methyl pheophorbide a (c)	$-$ 156.1
$C_{36}H_{38}N_4O_5$	606.78	Pheoporphyrin a_5 dimethyl ester (c)	$-$ 164.3
$C_{36}H_{40}N_4O_6$	624.80	Chlorin p_6 trimethyl ester (c)	$-$ 292.0
$C_{36}H_{42}N_4O_4$	594.82	Mesoporphyrin (1x) dimethyl ester (c)	$-$ 196.4
$C_{36}H_{46}N_4$	534.86	Octaethylporphyrin (c)	$-$ 39.9
$C_{36}H_{60}O_{30}$	972.96	Tetamylose (c)	$-$1360.0
$C_{37}H_{40}N_4O_7$	652.81	Dimethyl pheopurpurin 7 (c)	$-$ 245.8
$C_{39}H_{74}O_6$	639.13	Glyceryl trilaurate (c)	$-$ 489.0
$C_{40}H_{46}N_4O_8$	710.90	Coproporphyrin (I) tetramethyl ester (c)	$-$ 348.3
$C_{45}H_{86}O_6$	723.31	Glyceryl trimyristate (c)	$-$ 520.3
$C_{47}H_{88}O_5$	733.35	Glyceryl dibrassidate (c)	$-$ 472.0
$C_{47}H_{88}O_5$	733.35	Glyceryl dierucate (c)	$-$ 447.0
$C_{48}H_{54}N_4O_{16}$	943.06	Isouroporphyrian octamethyl ester (c)	$-$ 620.1
$C_{48}H_{80}O_{40}$	1297.28	Hexamylose (c)	$-$1853.0
$C_{69}N_{128}O_6$	1053.97	Glyceryl tribrassidate (c)	$-$ 625.0
$C_{69}H_{128}O_6$	1053.97	Glyceryl trierucate (c)	$-$ 596.0

(g) Gaseous.
(c) Crystalline.
(liq) Liquid.
Source: Compiled from Refs. 8-1, 8-2, and 8-3.

Table 8-4.   Heat of Formation of Inorganic Oxides, Solid State.

Formula	Weight, g/mole	Name	ΔH_f, kcal/mole
Al_2O_3	101.96	Aluminum oxide	−390.0
B_2O_3	69.62	Boric oxide	−299.8
BaO	153.34	Barium oxide	−132.3
BeO	25.01	Beryllium oxide	−145.7
BrO_2	111.90	Bromine dioxide	11.6
CaO	56.08	Calcium oxide	−151.8
CdO	128.40	Cadmium oxide	−61.7
Co_3O_4	240.79	Cobalt oxide	−213.0
CrO_2	84.00	Chromium dioxide	−143.0
CrO_3	100.00	Chromium trioxide	−140.9
Cr_3O_4	220.00	Chromium oxide	−366.0
CuO	79.55	Cupric oxide	−37.6
Cu_2O	143.10	Cuprous oxide	−40.3
FeO	71.85	Ferrous oxide	−65.0
Fe_2O_3	159.70	Ferric oxide	−197.0
HgO	216.59	Mercuric oxide	−21.4
K_2O	94.20	Potassium oxide	−86.8
Li_2O	29.88	Lithium oxide	−143.1
MgO	40.31	Magnesium oxide	−142.9
MnO	70.94	Manganese oxide	−92.1
MnO_2	86.94	Manganese dioxide	−124.3
MoO_2	127.94	Molybdenium dioxide	−140.8
Na_2O	61.98	Sodium oxide	−99.9
NiO	74.71	Nickel monoxide	−57.3
P_4O_6	219.88	Phosphorus trioxide	−392.0
PbO	223.19	Lead monoxide	−52.3
PbO_2	239.19	Lead dioxide	−66.3
Pt_3O_4	649.27	Platinum oxide	−39.0
PuO_2	274.00	Plutonium dioxide	−252.9
RaO	242.00	Radium oxide	−125.0
RhO	118.91	Rhodium monoxide	92.0
RhO_2	134.91	Rhodium dioxide	44.0
SO_3	80.06	Sulfur trioxide	−108.6
SiO_2	60.09	Silicon dioxide	−217.3
SnO	134.69	Tin monoxide	−68.3
SnO_2	150.69	Tin dioxide	−138.8
SrO	103.62	Strontium monoxide	−141.5
SrO_2	119.62	Strontium peroxide	−151.4
TeO_2	159.60	Tellurium dioxide	−77.1
ThO_2	264.04	Thorium dioxide	−293.2
TiO_2	79.90	Titanium dioxide	−224.6
UO_2	270.03	Uranium dioxide	−259.2
UO_3	286.03	Uranium trioxide	−292.0
VO	66.94	Vanadium monoxide	−103.2
V_2O_5	181.88	Vanadium pentoxide	−370.6
WO_2	215.85	Tungsten dioxide	−140.9
ZnO	81.37	Zinc oxide	−83.2
ZrO_2	123.22	Zirconium dioxide	−263.0

Source: Compiled from Refs. 8-1, 8-2, and 8-3.

Table 8-5. Heat of Formation of Miscellaneous Materials

Formula	Weight, g/mole	Name	ΔH_f, kcal/mole
CO	28.01	Carbon monoxide, gas	− 26.4
CO_2	44.01	Carbon dioxide, gas	− 94.1
$CaCO_3$	100.09	Calcium carbonate, solid	−289.5
$Ca(OH)_2$	74.10	Calcium hydroxide, solid	−235.6
$Ca(OH)_2$	74.10	Calcium hydroxide, liquid	−239.7
ClO	51.45	Chlorine monoxide, gas	24.3
ClO_2	67.45	Chlorine dioxide, gas	24.5
ClO_3	83.45	Chlorine trioxide, gas	37.0
Cl_2O	86.90	Dichlorine monoxide, gas	19.2
FO	35.00	Fluorine monoxide, gas	− 5.2
F_2O	54.00	Oxygen difluoride, gas	4.3
HCl	36.46	Hydrogen chloride, gas	− 22.1
HCl	36.46	Hydrogen chloride, liquid	− 40.0
H_2O	18.02	Water, liquid	− 68.3
H_2O	18.02	Water, steam	− 57.8
H_2O_2	34.02	Hydrogen peroxide, liquid	− 44.9
H_2O_2	34.02	Hydrogen peroxide, gas	− 32.6
H_2SO_3	82.02	Sulfurous acid, liquid	−146.8
H_2SO_4	98.08	Sulfuric acid, liquid	−193.9
IO	142.90	Iodine monoxide, gas	41.8
MgS	56.37	Magnesium sulfide, solid	− 84.2
$Mg(OH)_2$	58.33	Magnesium hydroxide, solid	−221.9
$MgCO_3$	84.32	Magnesium carbonate, solid	−261.7
$MgSO_4$	120.37	Magnesium sulfate, solid	−304.9
NO	30.01	Nitric oxide, gas	21.6
NO_2	46.01	Nitrogen dioxide, gas	7.9
N_2O	44.02	Nitrous oxide, gas	19.6
Na_2CO_3	105.99	Sodium carbonate, solid	−269.5
Na_2CO_3	105.99	Sodium carbonate, liquid	−275.1
N_2O_5	108.02	Nitrogen pentoxide, gas	2.7
NaCl	58.44	Sodium chloride, solid	− 98.2
NaCl	58.44	Sodium chloride, aqueous	− 97.3
$NaHCO_3$	84.01	Sodium bicarbonate, solid	−226.0
$NaHCO_3$	84.01	Sodium bicarbonate, liquid	−222.1
NaF	41.99	Sodium flouride, solid	−135.9
NaI	149.89	Sodium iodide, solid	− 69.3
NaOH	40.00	Sodium hydroxide, liquid	−112.2
O_3	48.00	Ozone, gas	34.1
SO	48.06	Sulfur monoxide, gas	1.5
SO_2	64.06	Sulfur dioxide, gas	− 70.9
SO_3	80.06	Sulfur trioxide, gas	− 94.6
SbO	137.75	Antimony monoxide, gas	47.7
VO	66.94	Vanadium monoxide, gas	25.0
VO_2	82.94	Vanadium dioxide, gas	− 57.1

Source: Compiled from Refs. 8-1, 8-2, and 8-3.

normally stated, kcal/mole and gram/mole. The conversion to English units is as follows:

$$\left[\frac{kcal}{mole} \times \frac{BTU}{0.252\ kcal}\right] \div \left[\frac{gram}{mole} \times \frac{lb}{454\ gram}\right]$$

$$1\ kcal/gram = 1802\ BTU/lb$$

The calculation for heating value using the heat of combustion can be illustrated by the burning of benzene, C_6H_6, as follows:

$$-57.8\ (g)$$

$$+11.7 \qquad\qquad -94.1 \quad -68.3\ (l)$$

$$C_6H_6 + 7.5\ O_2 \longrightarrow 6CO_2 + 3H_2O$$

$$78.12$$

78.12 and 11.7 are the molecular weight and heat of formation of liquid benzene, from Table 8-3, and -57.8 and -68.3 the heat of formation of gaseous water (steam) and liquid water respectively, and -94.1 the heat of formation of carbon dioxide from Table 8-5.

Assuming all water that is produced is maintained as steam, and bringing the values of heat of formation of all compounds multiplied by the moles of each compound present to the right hand side of the equation, the sum of the heat of formation ΔH_f, is calculated:

$$\Delta H_f\ H_2O(g):\quad -57.8 \times 3 = -173.4$$

$$\Delta H_f\ CO_2:\quad -94.1 \times 6 = -564.6$$

$$\Delta H_f\ C_6H_6:\ (-)11.7 \times 1 = \underline{\ -11.7\ }$$

$$\sum \Delta H_f = -749.7\ kcal/mole$$

Relating $\sum \Delta H_f$ to the weight of benzene present:

$$\frac{1}{78.12\ gram/mole} \times (-749.7)\ kcal/mole \times 1802 = -17293\ BTU/lb$$

If the benzene were converted to liquid water instead of steam (the higher heating value of substance) the heat produced would be calculated as follows (note the use of -68.3 as the heat of formation of liquid water in lieu of -57.8, that of steam):

$$\Delta H_f \; H_2O(l): \quad -68.3 \times 3 = -204.9$$

$$\Delta H_f \; CO_2: \quad -94.1 \times 6 = -564.6$$

$$\Delta H_f \; C_6H_6: (-)11.7 \times 1 = \underline{\;-11.7\;}$$

$$\sum \Delta H_f = -781.2 \; kcal/mole$$

Relating $\sum \Delta H_f$ to the weight of benzene present:

$$\frac{1}{78.12 \; gram/mole} \times (-781.2) \; kcal/mole \times 1802 = -18020 \; BTU/lb$$

Note that in the above examples the heating value is negative, indicative of an exothermic or heat releasing reaction. These values, 18020 BTU/lb and 17293 BTU/lb for the high and low heating value of benzene, correlate with the published values as found in Table 1, 18210 BTU/lb and 17480 BTU/lb respectively.

This method of analysis can be used to determine the loss of heat in a furnace resulting from ash formations. For instance, if calcium is present as a liquid and, due to furnace heating, loses moisture as steam and forms lime, CaO, the heat loss is calculated as follows:

$$-239.6 \qquad -151.8 -57.8$$

$$Ca(OH)_2 \longrightarrow CaO + H_2O$$

$$74.28$$

$$\Delta H_f \; H_2O(g) = \quad -57.8$$

$$\Delta H_f \; CaO = \quad -151.8$$

$$\Delta H_f \; Ca(OH)_2 = (+)\underline{239.7}$$

$$\sum \Delta H_f = \quad +30.1 \; kcal/mole$$

$$\frac{30.1 \; kcal/mole}{74.28 \; gram/mole} \times 1802 = 730 \; BTU/lb$$

The heat of formation for calcium hydroxide and steam were found in Table 8-5 and for lime in Table 8-4. The resultant "heat of combustion," 730 BTU/lb, is a positive quantity, indicating an endothermic reaction, one in which heat must be added for the reaction to occur. Therefore when calcium hydroxide is present in a combustion reaction, 730 BTU must be subtracted from the furnace fuel heating value for each pound of calcium hydroxide.

HEATING VALUE APPROXIMATION, HYDROCARBONS

As an approximation for the heating value of a hydrocarbon when the heat of formation or other thermodynamic data are not available, the figure of 184,000 BTU released for every required pound mole of stoichiometric oxygen is used. (The figures above the chemical formula are the elemental atomic weights. The lower figure is the molecular weight.)

For benzene:

$$12.01$$
$$C_6H_6 + 7.5\,O_2 \rightarrow 6CO_2 + 3H_2O$$
$$78.12$$

$$\frac{7.5\text{ lb mole }O_2}{\text{mole }C_6H_6} \times \frac{\text{mole }C_6H_6}{78.12\text{ lb}} \times \frac{184{,}000\text{ BTU}}{1\text{lb mole }O_2} = 17{,}665\text{ BTU/lb}$$

For another substance, isobutane:

$$12.01$$
$$C_4H_{10} + 6.5\,O_2 \rightarrow 4CO_2 + 5H_2O$$
$$58.14$$

$$\frac{6.5\text{ lb mole }O_2}{\text{mole }C_4H_{10}} \times \frac{\text{mole }C_4H_{10}}{58.14\text{ lb}} \times \frac{184{,}000\text{ BTU}}{1\text{ lb mole }O_2} = 20{,}571\text{ BTU/lb}$$

Table 8-6.   Typical Combustion Parameters for Stoichiometric Burning

BTU	lb air/ 10 kB	lb dry gas/ 10 kB	lb H_2O/ 10 kB	C	H_2	S	O_2	N_2
3998	6.880	8.711	0.671	0.15	0.030	0.01	0.011	0.799
5004	6.955	8.328	0.625	0.20	0.035	0.01	0.015	0.740
6003	7.006	8.076	0.596	0.25	0.040	0.01	0.020	0.680
7001	7.043	7.897	0.574	0.30	0.045	0.01	0.025	0.620
8000	7.070	7.762	0.559	0.35	0.050	0.01	0.030	0.560
8998	7.092	7.657	0.546	0.40	0.055	0.01	0.035	0.500
10005	7.108	7.571	0.536	0.45	0.060	0.01	0.039	0.441
11003	7.122	7.503	0.528	0.50	0.065	0.01	0.044	0.381
12002	7.134	7.445	0.521	0.55	0.070	0.01	0.049	0.321
13000	7.143	7.397	0.516	0.60	0.075	0.01	0.054	0.261
13999	7.152	7.356	0.511	0.65	0.080	0.01	0.059	0.201
14997	7.159	7.320	0.506	0.70	0.085	0.01	0.064	0.141
16004	7.165	7.287	0.503	0.75	0.090	0.01	0.068	0.082
17002	7.171	7.260	0.499	0.80	0.095	0.01	0.073	0.022

Note: 10 kB ≡ 10000 BTU.

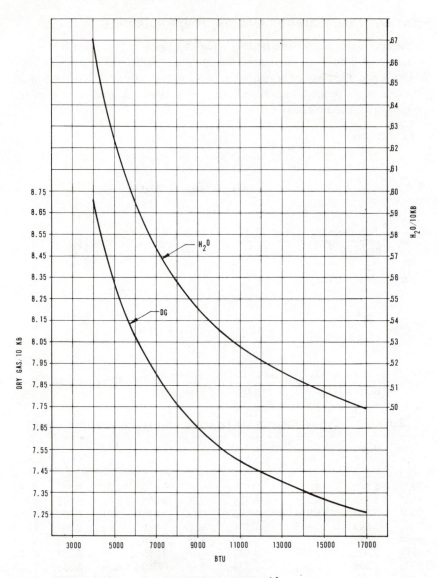

COMBUSTION PARAMETERS

Fig. 8-1. Combustion parameters.

The actual heating value of these substances, from the chart, Table 8-1, is 17,480 BTU/lb for benzene and 19,629 BTU/lb for isobutane.

This method of heating value approximation is valid only for hydrocarbons. The presence of oxygen, nitrogen, halogens, or other compounds may lead to erroneous values by use of this calculation.

DU LONG'S APPROXIMATION

An empirical method of determining the heating value of coal, the Du Long formula, can be used as a rough approximation for the heating value of other carbonaceous materials. This is only an approximation and its validity is questionable:

$$Q = 14544C + 62028(H_2 - 0.125O_2) + 4050S$$

where Q is in BTU/lb and C, H, O, and S are the weight fractions of carbon, hydrogen, oxygen, and sulfur present. The weight fractions should add to 100% unless an inert material is present, such as nitrogen. With the presence of nitrogen, for instance, the weight fraction (percent) of C, H_2, O_2, and S will total 100% less the fraction of nitrogen present.

Table 8-6 lists combustion parameters including heating value and products of combustion for an organic waste, based on Du Long's approximation. These values are shown in graphic form in Fig. 8-1.

Chapter 9
Incinerator Analysis

All incinerators utilize air and heat to produce the conditions which will destroy the combustible portion of the waste. Incinerator calculations, therefore, are basically the same for any incineration system or condition.

COMBUSTION PROPERTIES

The efficiency of combustion is related to the combination of three factors, the "three - T's": Temperature in the furnace, Time of residence of the combustion products at the furnace temperature, and Turbulence within the furnace.

In general a solid or liquid must be converted to a gaseous phase before burning will occur. (Examine a lit match or a burning log. The flame does not rise directly from the solid. There is a zone immediately above the match, or log, where the gaseous fuel phase has been generated and is mixing with combustion air prior to burning). The three T's are factors which control the rapidity of conversion of solid and liquid fuel to the gaseous phase.

Flame properties include the following:

Flame

Flame is the envelope or zone where combustion reactions occur and visible radiation is produced.

Flame Front

This is the plane along which combustion starts. It is the dividing line between fuel/air mixture and products of combustion.

Flame Speed

This is also called flame velocity, ignition velocity or rate of propagation.

1. If the flame is stable the flame front appears stationary. The flame moves toward the burner at the same speed that the fuel/air mixture is injected into the burner.
2. If the fuel/air mixture is too fast the flame may blow out.
3. If the fuel/air mixture is too slow, a flashback to the burner will occur (consideration of safety factors are necessary).

Flame Stability

This must be maintained after ignition or combustion may be extinguished. Severe furnace pulsations or explosions could result from instability of the location of the flame front or variations in flame speed.

Flame Temperature

The adiabatic or theoretical temperature assumes no heat losses. Therefore, the enthalpy of products of combustion equals the enthalpy of the reactants. The actual flame temperature (non-adiabatic) differs from the adiabatic temperature. This difference is caused by:

1. Radiation losses—radiation from a bunsen burner flame may be 12-18%.
2. Convection losses—heat to the products of combustion and incoming fuel mixture.
3. Excess air carries heat away.
4. Conduction losses—direct contact with surroundings such as impingement on a furnace wall.
5. All energy in the fuel is not instantaneously released. Therefore the full heating value is not necessarily generated within the flame front.
6. Dissociation of gases will absorb energy (heat) generated during combustion.

The furnace flame temperature can be increased in practice by liberating the heat of combustion as fast as possible using minimum excess air, intimate mixing, preheating air or using oxygen or oxygen enriched air.

Note that organics will not burn in a stable manner with a flame temperature below about 500°F (cigarettes burn steadily at approximately 800°F). Solid waste incinerators operate at internal temperatures of 1,200-2,000°F. At lower temperatures incomplete combustion may occur.

Carbon and hydrogen containing organic compounds typically have flame temperatures ranging from 3,000-4,000°F at stoichiometric conditions. (With halogenated hydrocarbons the range would be 3,000°F down to nonflammable).

Temperatures in excess of 4,000°F are achieved by blends of hydrocarbon gas and by using more pure oxygen. For instance the flame temperature of 9% acetylene in air is 4,200°F, and the flame temperature of 18% acetylene in oxygen is 5,200°F (stoichiometric conditions).

Solid propellants used in the space program have flame temperatures in the 5,000–6,000°F range. Flame temperatures of materials normally disposed of by incineration are usually less than 3,000°F.

Furnace Temperature

Furnace temperature is a function of fuel heating value, furnace design, air admission, and combustion control. The minimum temperature must be higher than the ignition temperature of the waste. The upper temperature limit is normally a function of the enclosure materials. Over 2,400°F operation requires use of special refractory materials. The rates of combustion reactions increase rapidly with increased temperatures. Of the three T's (temperature, time, and turbulence), only temperature can be significantly controlled after a furnace is constructed. Time and turbulence are fixed by furnace design and air flow rate and can normally be controlled only over a limited range.

Furnace Temperature Control

This can be achieved as follows:

Excess air control, i.e., control of the air/fuel ratio. Temperature produced is a direct function of the fuel properties and excess air introduced. Excess air control requires either automatic control or close manual supervision.

Direct heat transfer by the addition of heat absorbing material within the furnace, such as water-cooled furnace walls. The addition of water sprays in the combustion zone (1000 Btu/lb water evaporated is equivalent to $\frac{1}{4}$ Btu/°F sensible heat in the flue gas) will reduce the furnace temperature. Use of water sprays must be carefully controlled to avoid thermal shock to the furnace refractory.

Furnace Gas Turbulence

Turbulence is an expression relating the physical relationship of fuel and combustion air in a furnace. A high degree of turbulence, intimate mixing of air and fuel, is desirable. Burning efficiency is enhanced with increased surface area of fuel particles exposed to the air. Fuel atomization maximizes the exposed particle surface. Turbulence helps to increase particle surface area by promoting fuel vaporization. In addition, good turbulence exposes the fuel to air in a rapid manner helping to promote rapid combustion, maximizing fuel release.

A burner requiring no excess air and producing no smoke is said to have perfect turbulence (a *turbulence factor* of 100%). If, for instance, 15% excess air is required to achieve a no-smoke condition the turbulence factor is calculated as follows:

$$\text{turbulence factor} = \frac{\text{stoichiometric air}}{\text{total air}} \times 100\%$$

$$= \frac{1.00}{1.15} \times 100\% = 87\%$$

Fuel gas burners can be designed to produce a turbulence factor close to 100%.

Retention Time

Combustion does not occur instantaneously. Sufficient space must be provided within a furnace chamber to allow fuel and combustible gases the time required to fully burn. This factor, termed dwell time, residence time, or retention time, is a function of furnace temperature, degree of turbulence, and fuel particle size.

Retention time required may be a fraction of a second, as when gaseous waste is burned, or many minutes, as when solid granular waste such as powdered carbon is burned.

THE MASS BALANCE

The flow into an incinerator must equal the flow of products leaving the incinerator. Input includes waste fuel, air (including humidity entrained within the air), and supplementary fuel. The flow exiting the incinerator includes moisture and dry gas of the exhaust as well as ash, both in the exhaust as fly ash and exiting as bottom ash.

Table 9-1, the mass flow table, provides an orderly method in establishing a mass balance surrounding a combustion system. Of initial interest is waste quality: its moisture content, ash content (non-combustible fraction), and heat content. Of prime importance is the generation of moisture and dry flue gas from the combustion process as described in Chapter 6.

For the purpose of this example the following waste will be assumed, at the indicated firing rate and other combustion parameters:

8000 pounds per hour of paper waste.
5250 BTU per pound as fired.
22.5% moisture as fired.
5% ash as fired.
7.50 pounds of dry flue gas generated per 10000 BTU released.

0.51 pounds of moisture generated per 10000 BTU released.
Fired with 125% excess air.

Following Table 9-1:

Wet Feed, as received charging rate, is 8000 pounds per hour.

Moisture. Percent moisture, by weight, of the wet feed, is 22.5%. The moisture rate is 22.5% of the wet feed rate, 0.225 X 8000 lb/hr = 1800 lb/hr.

Dry Feed equals total wet feed less moisture, 8000 lb/hr - 1800 lb/hr = 6200 lb/hr.

Ash is percent of total dry feed that remains after combustion. From the data provided, 5% of the total wet feed, 0.05 X 8000 lb/hr = 400 lb/hr ash. The ash fraction of the dry feed is 400 lb/hr ÷ 6200 lb/hr = 0.0645.

Volatile is that portion of dry feed that is combusted. It is found by subtracting ash from dry feed: 6200 lb/hr - 400 lb/hr = 5800 lb/hr.

Volatile Heating Value is the BTU value of the waste per pound of volatile matter. The total heating value as charged is 8000 lb/hr X 5250 BTU/lb = 42,000,000 BTU/hr. With M representing one million the total waste heat value is 42.00 MBH. On a unit volatile basis, with 5800 lb/hr volatile charged, the heating value per pound volatile is 42 MBH ÷ 5800 lb/hr = 7241 BTU/lb.

Table 9-1.   Mass Flow.

	Example
Wet Feed, lb/hr	8,000
Moisture, %	22.5
lb/hr	1,800
Dry Feed, lb/hr	6,200
Ash, %	6.45
lb/hr	400
Volatile	5,800
Vol. Htg. Value, BTU/lb	7,241
MBH	42.00
Dry Gas, lb/10 kB	7.5
lb/hr	31,500
Comb. H_2O, lb/10 kB	.51
lb/hr	2142
DG + Comb. H_2O, lb/hr	33,642
100% Air, lb/hr	27,842
Total Air Fract.	2.25
Total Air, lb/hr	62,645
Excess Air, lb/hr	34,803
Humid/Dry Gas (Air), lb/lb	0.01
Humidity, lb/hr	626
Total H_2O, lb/hr	4,568
Total Dry Gas, lb/hr	66,303

Dry Gas produced from combustion of the waste is given for this example as 7.5 lb/10kB, where k represents the number 1000. The dry gas flow is this figure multiplied by the BTU released or (7.5 ÷ 10000) lb/BTU X 42MBH = 31500 lb/hr dry gas.

Combustion H_2O is the moisture generated from burning the waste, in this example 0.51 lb/10 kB. Therefore, the combustion moisture flow rate is (0.51 ÷ 10,000) (lb/BTU) X 42MBH = 2142 lb/hr moisture.

DG + COMB H_2O is the sum of the dry gas and the moisture products of combustion, 31,500 lb/hr + 2,142 lb/hr = 33,642 lb/hr. This figure is convenient in order to obtain the amount of air required for combustion.

100% Air is the dry gas and moisture weights that are produced by combustion of the volatile component, which equals the weight of the volatile component plus the weight of the air provided. Likewise, the air requirement is equal to the sum of the dry gas and moisture of combustion less the volatile component. This air requirement is the stoichiometric air requirement, that amount of air necessary for complete combustion of the volatile component of the waste. Using the above figures, the value for 100% air is as follows:

$$33,642 \text{ lb/hr (DG} + H_2O) - 5,800 \text{ lb/hr (volatile)} = 27,842 \text{ lb/hr}$$

Total Air Fraction is the air required for effective combustion. This is basically a function of the physical state of the fuel (gas, liquid, or solid) and the nature of the burning equipment. In this example, 125% air is required, providing 1.25 + 1.00 = 2.25 total air fraction.

Total Air is the stoichiometric requirement multiplied by the total air fraction, 27,842 lb/hr X 2.25 = 62,645 lb/hr.

Excess Air provided to the system is the total air less the stoichiometric air requirement, 62,645 lb/hr - 27,842 lb/hr = 34,803 lb/hr.

Humidity/Dry Gas (Air). The humidity of the air entering the incinerator will have a not insignificant effect on the heat balance to be calculated further in this chapter. In this case, assume a humidity of 0.01 pound of moisture per pound of dry air. A chart of humidity values is presented in Chapter 4.

Humidity is the flow of moisture into the system with the air supply. Knowing the air flow and the fractional humidity the humidity can be calculated as 62,645 lb/hr air X 0.01 lb H_2O/lb air = 626 lb/hr moisture.

Total H_2O is that moisture exiting the system. It is equal to the sum of three moisture components, moisture in the feed plus moisture of combustion plus humidity, or 1800 lb/hr + 2142 lb/hr + 626 lb/hr = 4568 lb/hr total H_2O.

Total Dry Gas exiting the system is equal to the sum of the dry gas generated by the combustion of the volatiles, with stoichiometric air, plus the flow of excess air into the system. Thus, 31,500 lb/hr dry gas + 34,803 lb/hr excess air = 66,303 lb/hr total dry gas.

HEAT BALANCE

Heat, as mass, is conserved within a system. The heat exiting a system is equal to the amount of heat entering that system. Table 9-2 presents a quantitative means of establishing a heat balance for an incinerator. The heat balance must be preceded by a mass flow balance, as was discussed previously. The result of a heat balance is a determination of the incinerator outlet temperature, outlet gas flow, supplemental fuel requirement, and total air requirement. The total heat into the incinerator was calculated previously in the mass flow computations. By deter-

Table 9-2. Heat Balance.

	Example w/Fuel Oil	Example w/Gas
Cooling Air Wasted, lb/hr	9,000	
°F	450	
BTU/lb	94	
MBH	0.85	
Ash, lb/hr	400	
BTU/lb	130	
MBH	0.05	
Radiation, %	1.5	
MBH	0.63	
Humidity, lb/hr	626	
Correction (@970 BTU/lb), MBH	−0.61	
Losses, Total, MBH	0.92	
Input, MBH	42.00	
Outlet, MBH	41.08	
Dry Gas, lb/hr	66,303	
H_2O, lb/hr	4,568	
Temperature, °F	1,899	
Desired Temp., °F	2,000	
MBH	43.26	
Net, MBH	2.16	
Fuel Oil, Air Fraction	1.20	1.10
Net BTU/gal	57,578	406
gph	37.51	5,320
Air, lb/gal	125.06	0.791
lb/hr	4,691	4,208
Dry Gas, lb/gal	125.54	0.748
lb/hr	4,709	3,979
H_2O, lb/gal	8.75	0.103
lb/hr	328	548
DG w/FO, lb/hr	71,012	70,282
H_2O w/FO, lb/hr	4,896	5,116
Air w/FO, lb/hr	67,336	66,853
Outlet, MBH	46.32	46.40
Reference t, °F	60	60

mining how much of this total heat produced is present in the exhaust gas, the exhaust gas temperature can be calculated. If the calculated exhaust gas temperature is equal to the desired exhaust gas outlet temperature the process is autogenous and supplemental fuel is not required. If the desired temperature is lower than the actual outlet temperature additional air must be added (or additional water, if this is possible) to lower the outlet temperature to that desired. This condition is also autogenous since supplemental fuel is not added to the system.

For the case where the actual outlet temperature is less than the desired outlet temperature supplemental fuel must be added. The products of combustion must include the products of combustion of the supplemental fuel.

Flue gas properties are assumed to be identical to those of air.

As with Table 9-1, Table 9-2 will be described on a step-by-step basis.

Cooling Air Wasted. Assume the incinerator shell is cooled by a flow of 2000 SCFM of air. A standard cubic foot of air weighs 0.075 lb/cubic foot, therefore the mass air flow is 2000 ft^3/min $\times$ 60 min/hr $\times$ 0.075 lb/ft^3 = 9000 lb/hr. It is further assumed that this flow is wasted, i.e., discharged to the atmosphere.

The temperature of the air at the discharge point is assumed to be 450°F.

From Table 4-1 (Chapter 4) the enthalpy of air at 450°F, the cooling air discharge temperature, is 94 BTU/lb.

The total heat loss due to the wasted cooling air is the quantity of cooling air discharged multiplied by its enthalpy, 9000 lb/hr $\times$ 94 BTU/lb = 0.85 MBH (0.85 million BTU per hour).

Ash generated is 400 lb/hr, from the mass flow table.

The heating value of ash can be calculated from the methods developed in Chapter 8. In lieu of an analysis of the heating value of the ash component, a figure of 130 BTU/lb is reasonable. The heat loss through ash discharge is therefore 400 lb/hr ash $\times$ 130 BTU/lb ash = 0.05 MBH.

Radiation. The heat lost by radiation from the incinerator shell can be approximated as a percentage of the total heat of combustion. Table 9-3 lists typical

Table 9-3. Furnace Radiation Loss Estimates.

Furnace Rate, MBH	Radiation Loss, % of furnace rate
< 10	3
15	2.75
20	2.50
25	2
30	1.75
> 35	1.50

values of radiation loss. For this case, with a heat release of 42 MBH a radiation loss of 1.5% is used.

The total loss by radiation is equal to 42 MBH released × 1.5% or 0.63 MBH.

Humidity is water vapor within the air. The humidity component of the air, 626 lb/hr, is taken from the mass flow sheet.

When considering the heat absorbed by the moisture or water within the incinerator exhaust stream humidity has released its heat of vaporization because it is in the vapor phase. The other moisture components, moisture in the feed and moisture of combustion, enters the reaction as a liquid and the heat of vaporization is released by the reaction. To simplify these moisture calculations the heat of vaporization of humidity moisture at 60°F, 960 BTU/lb, is added to the total heat capacity of the flue gas. Therefore the correction factor is 626 lb/hr humidity × 960 BTU/lb = 0.61 MBH.

Total Losses of an incinerator are the sum of the heat discharged as cooling air plus the heat lost in the ash discharge plus the radiation loss. To this is added the correction for humidity. In this case the total loss is 0.85 MBH + 0.05 MBH + 0.63 MBH − 0.61 MBH = 0.92 MBH.

Input to the system is the heat generated from the combustible, or volatile, portion of the feed. From the mass flow table this figure is 42.00 MBH.

Outlet heat content is that amount of heat exiting in the flue gas. The heat left in the flue gas is the heat generated by the feed (input) less the total heat loss, 42.00 MBH − 0.92 MBH = 41.08 MBH.

Dry Gas is 66,303 lb/hr, from the mass flow sheet.

H_2O is 4568 lb/hr, from the mass flow sheet.

Temperature is the temperature of the exhaust gas where the heat content of the dry gas flow plus the heat content of the moisture flow exiting the incinerator equals the outlet MBH. From the methods of Chapter 4:

With 66,303 lb/hr dry gas and 4,568 lb/hr moisture at 1,800°F and at 1,900°F the dry gas enthalpy is 453.24 BTU/lb and 481.57 BTU/lb, respectively, and the moisture enthalpy is 1,948.02 BTU/lb and 2,007.17 BTU/lb respectively. Therefore the total calculated enthalpy is

$$
\begin{array}{ll}
1800°F & 38.95 \text{ MBH} \\
X & 41.08 \text{ MBH (Outlet)} \\
1900°F & 41.10 \text{ MBH}
\end{array}
$$

By interpolation:

$$X = 1800 + (1900 - 1800)\frac{41.08 - 38.95}{41.10 - 38.95} = 1899°F$$

Therefore the exhaust gas temperature is 1899°F.

Desired Temperature is the temperature to which it is desired to bring the products of combustion. For this example a temperature of 2000°F will be used.

At this temperature the heat content of the wet gas stream is that of the dry gas and that of moisture at 2000°F or 66,303 lb/hr dry gas × 510.07 BTU/lb + 4,568 lb/hr moisture × 2,067.42 BTU/lb = 43.26 MBH.

Net MBH is the amount of heat that must be added to the flue gas to raise its heat content to the desired level. (In this case the desired heat level is evaluated at 2000°F). The net MBH, therefore, is the desired less the outlet MBH, 43.26 MBH − 41.06 MBH = 2.16 MBH.

Note that this analysis will continue based on the use of No. 2 fuel oil as supplemental fuel. Table 9-4 lists fuel oil combustion parameters. Table 9-5 likewise is a listing of combustion parameters for natural gas. The incinerator analysis of Tables 9-2 and 9-6 includes listings for the use of natural gas as supplemental fuel. The values for natural gas are calculated in similar manner to the calculations for fuel oil.

Fuel Oil, Air Fraction is the total air fraction required for combustion of supplementary fuel. The total air normally required for efficient combustion of gas fuel is from 1.05–1.15 and for light fuel oil is 1.10–1.30. In this case assume No. 2 fuel oil is used for supplemental heat, with a total air supply of 1.20.

Net BTU/GAL. When fuel is combusted the products of combustion must be heated to the desired flue gas temperature. The amount of heat required to heat these combustion products must be subtracted from the total heat of combustion to obtain the effective heating value of the fuel. As the temperature to which the fuel products must be raised increases, the net heat available from the fuel decreases. From Table 9-4, bringing the products of combustion of fuel oil, with 1.2 total air, to 2000°F, a net heating value of 57,578 BTU/gal is available.

The gallons of fuel oil required to provide the heat required to bring the exhaust gas temperature from its actual temperature, 1899°F, to the desired temperature, 2000°F, is equal to the net MBH required divided by the net BTU/gallon available, 2.16 MBH net ÷ 57578 BTU/gal = 37.51 gph.

Air required for combustion of fuel oil, with 1.2 total air, from Table 9-4, is 125.06 lb/gal of fuel oil.

The fuel combustion air flow is the unit flow multiplied by the fuel quantity, 125.06 lb air/gal × 37.51 gal fuel oil = 4691 lb/hr air.

Dry Gas - from Table 9-4 the dry gas produced from combustion of fuel oil with 1.2 total air is 125.54 lb/gal of fuel oil.

The dry gas flow rate is 125.54 lb/dry gas/gal × 37.51 gal/hr fuel oil = 4709 lb/hr.

H_2O produced from combustion of fuel oil with 1.2 total air and 0.013 humidity is 8.75 lb/gal fuel oil from Table 9-4.

Table 9-4.   No. 2 Fuel Oil, 139,703 BTU/gal, 7.6 lb/gal.

Total Air:	1.1	1.2	1.3
lb air/gal	114.640	125.062	135.483
lb dry gas/gal	115.115	125.537	135.958
lb H_2O/gal	8.615	8.751	8.886
Temp., °F	Heat Available, BTU/gal		
200	126,210	125,707	125,206
300	123,016	122,255	121,495
400	119,802	118,780	117,760
500	116,562	115,277	113,995
600	113,283	111,732	110,184
700	109,965	108,146	106,328
800	106,029	103,885	101,742
900	103,197	100,829	98,463
1,000	99,747	97,099	99,747
1,100	96,255	93,325	90,397
1,200	92,721	89,505	86,291
1,300	89,147	85,643	82,140
1,400	85,535	81,738	77,943
1,500	81,887	77,796	73,707
1,600	78,205	73,817	69,431
1,700	74,487	69,799	65,115
1,800	70,746	65,757	60,771
1,832	69,538	64,452	59,369
1,900	66,971	61,679	56,389
2,000	63,175	57,578	51,984
2,100	59,341	53,445	59,349
2,192	55,813	49,628	44,385
2,200	55,507	49,294	43,084
2,300	41,637	45,114	38,594
2,400	47,750	40,916	34,085
2,500	43,852	36,706	29,562
2,600	39,914	32,453	24,995
2,700	35,938	28,162	20,388

The moisture flow rate from combustion of fuel oil is 8.75 lbs H_2O/gal fuel oil × 37.51 gal fuel oil/hr = 328 lb/hr.

DG W/FO is the total quantity of dry gas exiting the system. It is equal to the dry gas produced from combustion of the waste plus the dry gas produced from fuel combustion 66303 lb/hr + 4709 lb/hr = 71012 lb/hr dry gas.

H_2O W/FO is the total quantity of moisture exiting the system, that calculated in the mass flow sheet plus the contribution from combustion of supplementary fuel, 4568 lb/hr + 328 lb/hr = 4896 lb/hr.

Air W/FO is the total amount of air entering the incinerator, calculated from

Table 9-5. Natural Gas, 1000 BTU/cubic foot,
0.050 lb/cubic foot.

Total Air:	1.05	1.10	1.15
lb air/CF	0.755	0.791	0.827
lb dry gas/CF	0.712	0.748	0.784
lb H$_2$O/CF	0.103	0.103	0.104
Temp., °F	Heat Available, BTU/gal		
200	861	860	857
300	839	837	834
400	817	814	810
500	794	790	785
600	772	767	761
700	749	743	736
800	722	715	707
900	702	694	685
1000	678	670	660
1100	654	644	633
1200	629	619	607
1300	605	593	580
1400	579	567	553
1500	554	541	526
1600	529	514	498
1700	503	487	470
1800	477	460	442
1832	468	451	433
1900	450	433	414
2000	424	406	385
2100	397	378	356
2192	372	352	329
2200	370	350	327
2300	343	322	298
2400	316	294	269
2500	289	265	239
2600	261	237	210
2700	233	208	179

the mass flow sheet, plus that needed for supplemental fuel combustion, 62645 lb/hr + 4691 lb/hr = 67336 lb/hr.

Outlet MBH, the total heat value of the flue gas exiting the incinerator, is the sum of the heat content of the gas prior to adding supplemental fuel plus the heat addition of the supplemental fuel. The supplemental fuel adds 37.51 gal/hr × 139,703 BTU/gal = 5.24 MBH to the flue gas. Therefore, the flue gas outlet contains 41.08 MBH + 5.24 MBH = 46.32 MBH.

As a check on this figure, the outlet temperature will be calculated using the

Table 9-6. Flue Gas Discharge.

	Example w/Fuel Oil	Example w/Gas
Inlet, °F	2,000	2,000
Dry Gas, lb/hr	71,012	70,282
Heat MBH	46.32	46.60
BTU/lb dry gas	652	660
Adiabatic t, °F	176	176
H_2O Saturation, lb/lb dry gas	0.4411	0.5511
lb/hr	39,135	38,732
H_2O Inlet, lb/hr	4,896	5116
Quench H_2O, lb/hr	34,239	33,616
gpm	68	67
Outlet Temp., °F	120	120
Raw H_2O Temp., °F	60	60
Sump Temp., °F	147	147
Temp. Diff., °F	87	87
Outlet, BTU/lb dry gas	111.65	111.65
MBH	7.93	7.85
Req'd Cooling, MBH	38.39	38.55
H_2O, lb/hr	441,264	443,103
gpm	883	886
Outlet, ft³/lb dry gas	16.515	16.515
ft³/min	19,546	19,345
Fan Press., in. WC	30	30
Outlet, ACFM	20,987	20,771
Outlet, H_2O/lb dry gas	0.08128	0.08128
H_2O, lb/hr	5,772	5,713
Recirc. (Ideal), gpm	68	67
Recirc. (Actual), gpm	548	538
Cooling H_2O, gpm	883	886

flue gas flow (71012 #/hr dry gas and 4896 #/hr moisture):

1900°F	44.02 MBH
X	46.32 MBH
2000°F	46.34 MBH

$$X = 1900 + (2000 - 1900) \frac{46.32 - 44.02}{46.34 - 44.02} = 1999°F$$

This calculation of flue gas outlet temperature, 1999°F, is a good approximation of the desired gas outlet temperature, 2000°F.

Reference t is the datum temperature for enthalpy. It is the temperature at which feed, supplemental fuel, and air enter the system, 60°F for this example.

FLUE GAS DISCHARGE

To meet the rigorous air pollution codes in effect today wet gas scrubbing equipment is often necessary. Table 9-6, the flue gas discharge table, provides a method of calculating gas flow volumes exiting a wet scrubbing system as well as scrubber flow quantities. This table can also be used when calculating volumetric flow from a dry flue gas system.

Table 9-6 entries are as follows:

Inlet. insert the incinerator outlet temperature, 2000°F, from the heat balance table. This temperature is the inlet temperature of the flue gas processing system.

Dry Gas is the total flow of dry gas exiting the incinerator. From the heat balance table, this figure is 71012 lb/hr.

Heat is the total heat exiting the incinerator in the flue gas, in MBH. From the heat balance table this figure is 46.32 MBH.

The heat is calculated in terms of the dry gas component of the flue gas, as BTU/lb dry gas. From the entries for total heat and dry gas flow the heat is 46.32 MBH ÷ 71012 lb/hr = 652 BTU/lb DG.

Adiabatic t. When one pound of water evaporates it absorbs approximately 1000 BTU, without a change in temperature. This heat absorption is called the *heat of vaporization*, or *latent heat*. Latent heat is opposed to sensible heat, which is the heat required for a change in temperature without a change in phase. For evaporated water (steam) the sensible heat is approximately 0.5 BTU per pound of steam for every rise of one degree Fahrenheit.

The adiabatic temperature of the flue gas is the quench temperature. Quenching of a gas is defined as use of the latent heat of water (or other liquid) to decrease the gas temperature. This process does not involve the addition or removal of heat, only the use of the heat of vaporization of the quench liquid, i.e., water. The term adiabatic defines a process where heat is neither added nor removed from a system.

Considering the properties of dry flue gas equal to the properties of dry air, listed in Table 4-6 (Chapter 4), note that a maximum amount of moisture can be held in dry air at a particular temperature. This table lists saturation moisture quantities, volume and enthalpy at saturation, as a function of temperature. The temperature at which the enthalpy of the saturated dry flue gas (dry air) is equal to the enthalpy calculated above, BTU/lb dry gas, is the adiabatic temperature of the system. There has been no transfer of heat from the system, only the conversion of latent heat in the quench water to sensible heat in the dry flue gas.

In this example the adiabatic temperature is found in Table 4-5 as that temperature where the dry flue gas (saturated mixture) will have an enthalpy of approximately 652 BTU/lb, 176°F.

H_2O *Saturation.* The quenched flue gas, at the adiabatic temperature (176°F in this example) will contain an amount of moisture equal to the maximum amount of moisture that it can hold, saturation. From Table 4-5 the saturation moisture, in lb H_2O per lb. dry gas (air) is read opposite the adiabatic temperature. In this case, for 176°F adiabatic temperature the saturation moisture is 0.5511 lb H_2O/lb dry gas.

The moisture flow is the saturation moisture multiplied by the dry gas flow, 0.5511 lb H_2O/lb dry gas × 71012 lb dry gas/hr = 39135 lb H_2O/hr.

H_2O Inlet. The moisture component of the flue gas exiting the incinerator is inserted here from the heat balance table: 4896 lb/hr.

Quench H_2O is the moisture required for quenching the incoming flue gas to its adiabatic temperature. This is equal to the saturated moisture content of the flue gas less the moisture initially carried into the system with the flue gas. This figure is equal to H_2O saturation (39,135 lb/hr) less H_2O inlet (4,896 lb/hr), which in this example is equal to 34,239 lb/hr.

The conversion factor from lb/hr of water to gpm (8.34 lb/gal × 60 min/hr) is 500. The quench water required is equal to 34,239 lb/hr ÷ 500 = 68 gpm.

Outlet Temp., the outlet temperature, is that temperature entering the low (negative) pressure side of the ID fan, or, with no ID fan, the temperature within the stack. This temperature is normally selected in the range of 120-160°F. The lower this temperature, the smaller the size of the outlet plume and the lower the volumetric flow of flue gas. For this example 120°F was chosen as the outlet temperature. As can be seen below, a lower outlet temperature would require additional amounts of cooling water.

Raw H_2O Temp. This entry is the temperature of the water available for cooling the flue gas from the adiabatic temperature to the outlet temperature. In this example a raw water temperature of 60°F was chosen.

Sump Temp. Normally a quantity of water in excess of that calculated for quenching and for cooling the flue gas is provided for particulate removal. The excess water is generally collected in a sump where a quiescent period is allowed to permit larger particles within the spent water to settle to the sump floor, eventually to be drained. The temperature of the water in the sump must be ascertained in order to determine the effective cooling rate of the water flow. The sump temperature is a practical impossibility to forecast accurately, but an empirical relationship has been established. The sump temperature is assumed equal to the adiabatic temperature divided by 1.2. In this example the sump temperature is 176 ÷ 1.2 = 147°F.

Temp. Diff. The temperature differential of note is the difference in temperature between the raw water entering the cooling tower (or scrubber) and the temperature exiting the tower, the sump temperature. Sump temp. less raw H_2O temp. is in this example 147°F - 60°F = 87°F, the temperature differential.

Outlet. The gas exiting the scrubber system is designed to be at the outlet temperature, 120°F in this example, saturated with moisture. The outlet enthalpy is inserted from Table 9-4 for the outlet temperature chosen: 111.65 BTU/lb DG.

The total heat in the outlet flue gas is its enthalpy multiplied by its flow i.e., 111.65 BTU/lb dry gas × 71012 lb dry gas/hr = 7.93 MBH.

Req'd Cooling. As noted previously the flue gas is initially quenched, without heat addition or removal, to its adiabatic temperature. To reduce the adiabatic temperature of the flue gas to the desired outlet temperature a supply of cooling water is required. This cooling water must remove the heat content at adiabatic conditions relative to the heat content at outlet conditions. The required cooling is therefore the heat inlet less the outlet MBH. For this example 46.32 MBH heat inlet less 7.93 MBH outlet equals 38.39 MBH required cooling, i.e., with removal of 38.39 MBH from the saturated flue gas stream the flue gas temperature will fall from 176°F to 120°F.

H_2O. The moisture flow referred to is that required to achieve the desired cooling effect. With $Q = WC\Delta t$ or $W = Q/C\Delta t$ and W the lb/hr cooling water, Q the cooling load in BTU/hr, C the specific heat of water (1 BTU/lb-°F), and Δt the temperature difference of the cooling water across the flue gas stream (sump water temperature less raw water temperature) the required cooling water flow can be calculated. For this example:

$$W = 38.39 \text{ MBH} \div 1 \text{ BTU/lb-°F} \div 87°F$$

$$= 441{,}264 \text{ lb/hr.}$$

The flow in gallons per minute is that in lb/hr divided by 500,441,264 ÷ 500 = 883 gpm.

Outlet. The outlet volumetric flow is obtained with use of Table 4-4. The specific volume, ft³ mixture per lb dry gas (air), is found in this table for the outlet temperature. For this example, with an outlet of 120°F the specific volume is 16.515 ft³/lb dry gas.

The volumetric flow is equal to the specific volume multiplied by the dry gas flow divided by 60 min/hr, i.e., 16.515 ft³/lb dry gas × 71,012 lb/hr ÷ 60 min/hr = 19,546 ft³/min.

Fan Press. Induced draft fans used to clean the incinerator of gases may require a relatively high pressure. The actual differential pressure across the fan, inserted here, will be used to modify the volumetric flow value. For this example the fan pressure is 30 inches WC.

Outlet ACFM. The volumetric flow immediately prior to entering the ID fan will experience an expansion because of the fan suction. The volumetric flow correction is as follows: multiply the CFM determined above by the ratio $(407 + p) \div 407$, where p is the pressure across the fan. The figure 407 is atmospheric pressure (14.7 psia) expressed in inches of water column (14.7 psia ÷ 62.4 lb H_2O/ft³ × 1728 in³/ft³ = 407 inches WC). In this example the corrected, or actual, volumetric flow entering the ID fan is $[(407 + 30) \div 407] \times 19{,}546$ ft³/min = 20,987 ft³/min actual flow (ACFM).

Outlet. From Table 4-4, insert the saturation humidity, lb H_2O/lb dry gas (dry air) corresponding to the outlet temperature. For this example, with an outlet temperature of $120°F$, the humidity is 0.08128 lb H_2O/lb dry gas.

The total moisture exiting the stack is the saturation humidity multiplied by the dry gas flow. For this case, 0.08128 lb H_2O/lb dry gas $\times$ 71012 lb dry gas/ hr = 5772 lb H_2O/hr.

Recirc. (Ideal). The recirculation flow is that amount of flow required for quenching, as calculated above (68.5 gpm for this example). The temperature of this water flow is not critical. This flow is used adiabatically, where the latent heat (not a temperature change) in the water flow reduces the temperature of the flue gas. It is termed a recirculation flow because spent scrubber water from the scrubber sump can be recirculated to the Venturi for use at $140°F$ or greater, instead of a cooler flow of water.

Recirc. (Actual). In practice the ideal flow is inadequate to fully clean the gas stream of particulate. Ideal quenching requires intimate contact of each molecule of water with gas, instantaneous evaporation, and instantaneous heat transfer between the moisture and the gas, none of which occurs. To compensate for actual versus ideal conditions an empirical factor is used. In this case this factor is 8. Therefore, for actual recirculation flow use the ideal flow multiplied by eight, 68.5 gpm $\times$ 8 = 548 gpm.

Cooling H_2O. Insert the flow of cooling water calculated above, in this case, 883 gpm.

Chapter 10
On-Site Solid Waste Disposal

The disposal of solid waste is the most visible of problems associated with waste in this country. The per capita generation of solid waste is approximately 5 lb per day in the United States, over half a million tons per day from domestic sources alone. The many varieties of incinerators in general use for the on-site disposal of solid waste are presented in this chapter. Solid waste incinerators designed for central station disposal are discussed in a later chapter.

SOLID WASTE INCINERATION

Solid waste incinerators are usually categorized according to the nature of the material which they are designed to burn, i.e., refuse or industrial waste. However, more than one waste type can often be burned in a given unit.

Incinerators for destruction of solid waste are the most difficult class of incinerators to design and operate, primarily because of the nature of the waste material. Solid waste can vary widely in composition and physical characteristics, making the effects of feed rates and parameters of combustion very difficult to predict. Solid waste incinerators most often burn wastes over a range of low and high heat values, i.e., from wet garbage with an as-received heat value as low as 2,500 Btu/lb, to plastic wastes, over 19,000 Btu/lb. Materials handling, firing, and residue removal equipment are more critical, cumbersome, expensive, and difficult to control with these than with other types of incinerators.

TYPES OF SOLID WASTE INCINERATORS

There are eight main types of solid waste incinerators:

1. Open burning.
2. Single-chamber incinerators.
3. Teepee burners.
4. Open pit incinerators.
5. Multiple-chamber incinerators.

6. Controlled air incinerators.
7. Central station disposal.
8. Rotary kiln incinerators.

Controlled air incineration, central station disposal and rotary kilns will be discussed in subsequent chapters.

OPEN BURNING

Open burning is the oldest technique for incineration of wastes. Basically it consists of placing or piling waste materials on the ground and burning them without the aid of specialty combustion equipment.

This type of system is found in most parts of the United States. It results in excessive smoking and high particulate emission, and presents a fire hazard.

Open burning has been utilized to dispose of high-energy explosives such as dynamite or TNT. For proper incineration, the waste is placed on an asbestos pad which is in turn placed over gravel, in a cleared location, remote from populated areas.

SINGLE CHAMBER INCINERATORS

Single chamber incinerators will, in general, not meet the air pollution emission standards that have been developed over the past ten to fifteen years. A typical single chamber incinerator is shown in Fig. 10-1. Solid waste is placed on the grate and fired. These incinerators have also been manufactured in top-loading (flue loading) configuration for apartment house waste disposal, as shown in Fig. 10-2, firing waste in 55 gallon drums or wire baskets or in a concrete or refractory lined structure with a cast iron grate, etc. This equipment may or may not have a firing system to ignite the waste. As with open burning, smoking and excessive air pollution emissions can occur.

Attempts have been made to control emissions to reasonable levels by the addition of an afterburner, as shown in Fig. 10-2. Note that both an afterburner and a draft control damper have been added to control the combustion process. Most of the air emissions in single chamber incinerators result from incomplete burning of the waste. Normally a temperature of $1400°F$ is required, at a retention time of 0.5 seconds, and the afterburner is used to obtain these combustion parameters in the exiting off-gas. Tables 10-1 and 10-2 show stack emissions and other parameters of operation and design from tests performed on flue-fed incinerators similar to that pictured in Fig. 10-2 Table 10-1 represents conditions without an afterburner in operation while Table 10-2 reflects conditions with an afterburner firing Note the dramatic change in particulate emissions with and without provision of an afterburner.

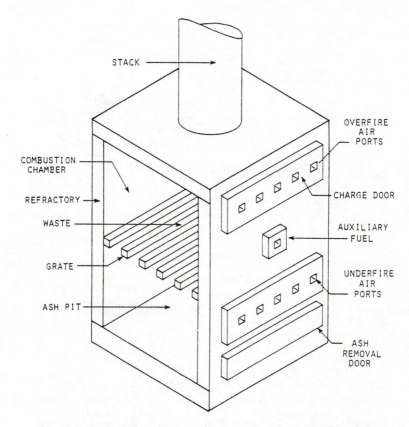

Fig. 10-1. Single chamber incinerator. *Source:* Reference 10-5, Page 2–8.

A *jug incinerator* is another type of single chamber unit. A typical jug incinerator is shown in Fig. 10-3. This is a specialty incinerator used for the destruction of cotton waste and other waste agricultural products. It is a brick lined vertical cylindrical/conical structure. Waste is fed through the top section of the incinerator and falls to its floor, which may or may not be provided with grates. Waste is pneumatically conveyed to the incinerator charging system and the transfer air is the only combustion air supplied to the incinerator. Afterburners are provided in the stack to control air emissions, although many such incinerators discharge from their conical top, without provision of a stack or afterburning equipment.

OPEN PIT INCINERATORS

Open pit incinerators have been developed for controlled incineration of explosive wastes, wastes which would create an explosion hazard or high heat release

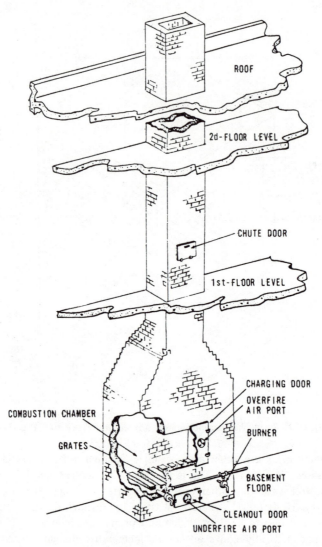

Fig. 10-2. Flue-fed incinerator modified by a roof afterburner and a draft control damper. *Source:* Reference 10-8.

in a conventional, enclosed incinerator. They are constructed as shown in Fig. 10-4 with an open top and a number of closely spaced nozzles blowing air from the open top down into the incinerator chamber. Air is blown at high velocity, creating a rolling action, i.e., a high degree of turbulence. Burning rates within the incinerator provide temperature in excess of 2000°F with low smoke and relatively low particulate emissions discharges.

Table 10-1.   Particulate Emissions From Typical Flue-Fed Incinerators

Test designation		Particulate matter,		Average stack volume, scfm	Grate area, ft^2	Stack height, ft
	lb/ton	gr/scf at 12% CO_2	gr/scf			
1	76	2.27	0.61	458	1.5	25
2	52	1.40	0.13	1,190	16	35
3	48	1.60	0.21	326	9	68
4	37	1.40	----	213	8	32
5	37	1.18	0.20	820	12	80
6	34	1.06	0.21	930	12	80
7	25	0.94	0.18	500	6	54
8	23	0.99	0.10	1,120	12	56
9	23	0.75	0.2	860	12	80
10	19	0.75	0.26	530	4	25
11	17	0.60	0.09	441	20	56
12	7	0.27	0.08	817	8	46

Source: Reference 8

Incinerators of this type may be built either above or below ground. They are constructed with refractory walls and floor or as earthen trenches. The width of an open pit incinerator is normally on the order of 8 feet, with a depth of approximately 10 feet. The length varies from 8 to 16 feet.

Overfire air nozzles are 2–3 inches in diameter, located above one edge of the pit. They fire down at an angle of 25–35 degrees from the horizontal. The incinerator is normally charged from a top loading ramp on the edge opposite the air nozzles. Some units have a mesh placed on their top to contain larger particles of fly ash. Residue cleanout doors are often provided on above-ground incinerators.

For a waste with a heating value of 5000 Btu/lb note the following typical parameters of design for open pit incinerators:

Table 10-2.   Particulate Emissions From a Typical Flue-Fed Incinerator Modified With a Draft Control Damper and a Roof Afterburner

Test designation	Burning rate, lb/hr	Particulate matter			Afterburner efficiency, %	Average oxygen content, %	Average stack volume, scfm	Average outlet temperature, °F
		lb/ton	gr/scf at 12% CO_2	gr/scf				
A	100	5.9	0.20	0.004	80	12.1	760	1,130
B	80	5.2	0.18	0.035	82	11.6	690	1,240
C	68	5.6	0.20	0.034	80	12.7	710	1,130
D	49	1.2	0.15	0.027	85	9.5	590	1,560

Source: Reference 8

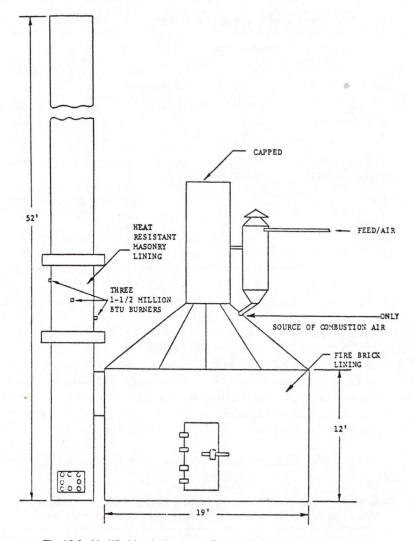

Fig. 10-3. Modified jug incinerator. *Source:* Reference 10-5, Page 4-45.

- Heat release of 3.4 million Btu/hr per foot of length.
- Provision of 100–300% excess air.
- Overfire air of 850 SCFM per foot of pit length at 11 inches WC.

Particulate emissions are normally below 0.25 grains per DSCF, corrected to 12% CO_2, which is unacceptable with regard to current air pollution control standards. (Most current statutes limit air pollution emissions from burning

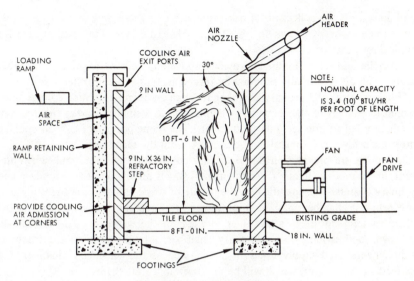

Fig. 10-4. Open-pit incinerator.

refuse to 0.08 grains per DSCF corrected to 12% CO_2.) Other than combustion control by control of overfire air there is no mechanism practicable for control of exhaust emissions. This incinerator, while effective in destruction of some waste, can normally not be used without relaxation of local air pollution emissions requirements.

MULTIPLE-CHAMBER INCINERATORS

In an attempt to provide complete burnout of combustion products and decrease the airborne particulate loading in the exiting flue gas, multiple chamber incinerators have been developed. A first, or primary, chamber is used for combustion of solid waste. The secondary chamber provides the residence time, and supplementary fuel, for combustion of the unburned gaseous products and airborne combustible solids (soot) discharged from the primary chamber.

There are two basic types of multiple chamber incinerators: the retort and the in-line systems.

Retort Incinerator

This unit is a compact cubic-type incinerator with multiple internal baffles. The baffles are positioned to guide the combustion gases through 90 degree turns in both lateral (horizontal) and vertical directions. At each turn ash drops out of the flue gas flow. The primary chamber has elevated grates for discharge of waste and an ash pit for collection of ash residual. A cutaway view of a typical retort type incinerators is shown in Fig. 10-5. Fig. 10-6 gives dimensional data for typical retort units.

Overfire and underfire air is provided above and below the primary chamber grate. This air is normally supplied by forced air fans at a controlled rate.

Flue gas exits the primary chamber through an opening, termed a *flame port*, which discharges to the secondary chamber or to a smaller *mixing chamber* immediately before the secondary chamber. The flame port is actually an opening atop the bridge wall, separating the primary from the secondary chamber.

Air ports are provided in the secondary combustion chamber and, when present, in the mixing chamber. Supplemental fuel is provided in the secondary and primary chambers. Depending on the nature of waste charged, the fuel supply in the primary chamber may be unnecessary after startup, i.e., after bringing the chamber temperature to a level high enough for the waste to ignite and sustain its own combustion. The secondary chamber normally requires a continuous supplemental fuel supply unless it is operated as the secondary chamber of a starved air incinerator, which will be discussed in a later chapter.

As the flue gas enters and exits the secondary combustion chamber larger airborne particles settle out of the gas stream. Temperatures in the secondary chamber are high enough (in the range of $1400°F$ for refuse, pathological, and other carbonaceous waste) to destroy unburned airborne particles. This equipment therefore, has relatively low particulate emissions and in many cases can meet the emission standard of 0.08 grains/DSCF, corrected to 12% CO_2, without additional air pollution control equipment.

In-Line Incinerator

This is a larger unit than the retort incinerator. Flow of combustion gases is straight through the incinerator, axially, with abrupt changes in the direction of flow only in the vertical direction as shown in Fig. 10-7, an in-line incinerator using natural gas as supplemental fuel. Waste is charged on the grate which can be stationary or moving. A moving grate lends itself to continuous burning whereas stationary grates, as with the retort incinerator, are used for batch or semi-continuous operation. In-line incinerators are often provided with automatic ash removal equipment or ash discharge conveyors which also contribute to continuous operation of the incinerator.

As with the retort type, changes in the flow path and flow restrictions in an in-line incinerator provide settling out of larger airborne particles and increase turbulence for more effective burning.

Supplemental fuel burners in the primary chamber ignite the waste whereas secondary chamber supplementary fuel burners provide heat to maintain complete combustion of the burnable components of the exhaust gas.

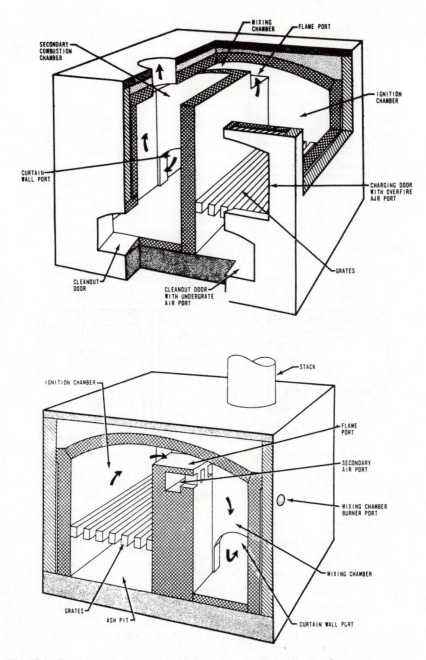

Fig. 10-5. Cutaway of a retort multiple-chamber incinerator. *Source:* Reference 10-7, Page 414.

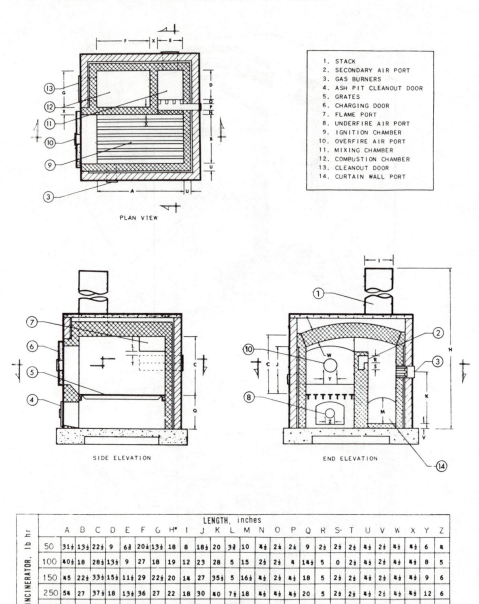

Fig. 10-6. Design standards for multiple-chamber, retort incinerators. *Source:* Reference 10-7, Page 424.

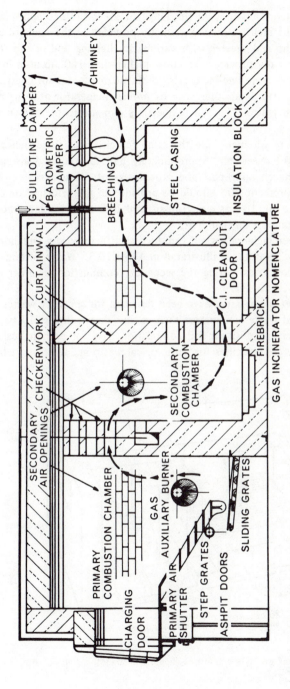

Fig. 10-7. Gas incinerator. *Source:* Incinerator Committee, Industrial & Commercial Gas Section, American Gas Association, New York, NY

The retort incinerator is used in the range of 20 lb/hour to approximately 750 lb/hour. In-line incinerators are normally provided in the range of 500 lb/hour to 2000 lb/hr and greater with automatic charging and/or ash removal equipment not normally provided for units smaller than 1000 lb/hour in capacity. Fig. 10-7 shows an in-line incinerator with moving grates for charging and ash disposal. Fig. 10-8 is another type of in-line incinerator utilizing manual charging, i.e., fixed grates. Typical in-line unit dimensions are shown in Fig. 10-9.

Combustion air requirements are the same for either of these incinerators: approximately 300% excess air. Approximately half the required air enters as leakage through the charging port and other areas of the incinerator. Of the remaining air requirement 70% should be provided in the primary combustion chamber as overfire air, 10% as underfire air, and 20% in the mixing chamber or in the secondary combustion chamber.

Multiple chamber incinerators will produce significantly less emissions than single chamber incinerators, as illustrated in Table 10-3. Water curtains across the path of the flue gases exiting the secondary combustion chamber will decrease emissions even further.

Multiple-chamber incinerators have been designed for specialty wastes. Typical are pathological waste incinerators such as that shown in Fig. 10-10. Table 10-4 lists chemical composition and combustion data for pathological waste. Design factors and gas velocities for pathological waste incinerators are listed in

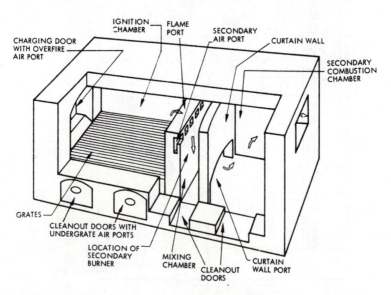

Fig. 10-8. Cutaway of an In-line multiple-chamber incinerator. *Source:* Reference 10-6, Page 172.

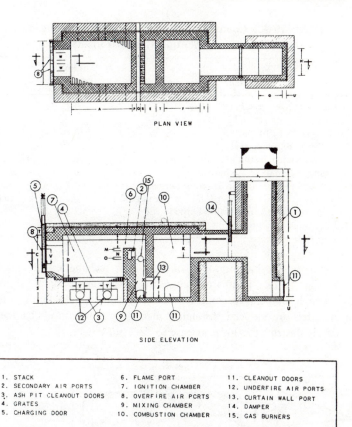

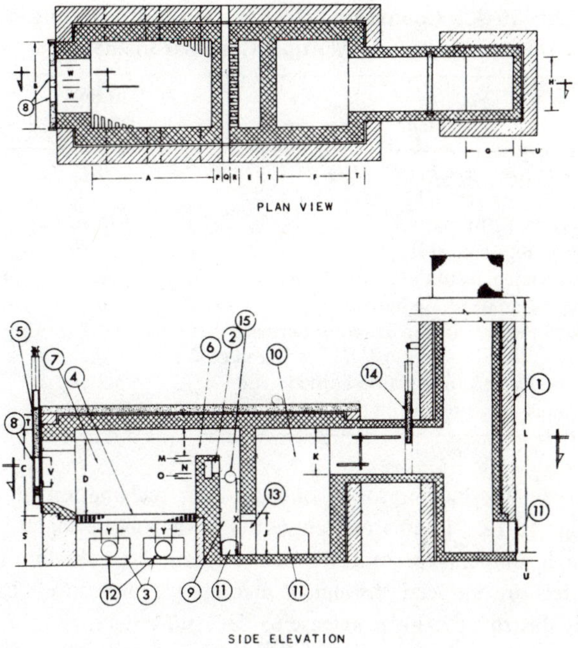
PLAN VIEW
SIDE ELEVATION

1. STACK
2. SECONDARY AIR PORTS
3. ASH PIT CLEANOUT DOORS
4. GRATES
5. CHARGING DOOR
6. FLAME PORT
7. IGNITION CHAMBER
8. OVERFIRE AIR PORTS
9. MIXING CHAMBER
10. COMBUSTION CHAMBER
11. CLEANOUT DOORS
12. UNDERFIRE AIR PORTS
13. CURTAIN WALL PORT
14. DAMPER
15. GAS BURNERS

SIZE OF INCINERATOR, lb/hr	LENGTH, inches																								
	A	B	C	D	E	F	G	H	I	J	K	L*	M	N	O	P	Q	R	S	T	U	V	W	X	Y
750	85½	49½	51½	45	15¾	54	27	27	9½	24	18	32	4½	5	7½	9	2½	2½	30	9	4½	5	11	51	7
1000	94½	54	54	47½	18	63	31½	31½	11	29	22½	35	4½	5	10	9	2½	2½	30	9	4½	7	12	52	8
1500	99	76½	65	55	18	72	36	36	12½	32	27	38	4½	5	7½	9	4½	4½	30	9	4½	8	14	61½	9
2000	108	90	69½	57½	22½	79½	40½	40½	15	36	31½	40	4½	5	10	9	4½	4½	30	9	4½	9	15	63½	10

*Dimension "L" given in feet.

Fig. 10-9. Design standards for multiple-chamber, in-line incinerators. *Source:* Reference 10-7, Page 425.

Tables 10-5 and 10-6, respectively. Air emissions from pathological incinerators based on two test runs with and two runs without afterburner firing are listed in Tables 10-7 and 10-8 respectively. Note the significant decrease in emissions with the afterburner in operation.

A crematory retort is shown in Fig. 10-11. Its operating parameters for typical crematory waste is listed in Table 10-9.

**Table 10-3.   Comparison Between Amounts of Emissions
From Single- and Multiple-Chamber Incinerators**

Item	Multiple chamber	Single chamber
Particulate matter, gr/scf at 12% CO_2	0.11	0.9
Volatile matter, gr/scf at 12% CO_2	0.07	0.5
Total, gr/scf at 12% CO_2	0.19	1.4
Total, lb/ton refuse burned	3.50	23.8
Carbon monoxide, lb/ton of refuse burned	2.90	197 to 991
Ammonia, lb/ton of refuse burned	0	0.9 to 4
Organic acid (acetic), lb/ton of refuse burned	0.22	< 3
Aldehydes (formaldehyde), lb/ton of refuse burned	0.22	5 to 64
Nitrogen oxides, lb/ton of refuse burned	2.50	1
Hydrocarbons (hexane), lb/ton of refuse burned	< 1	—

Another specialty incinerator is one used for reclamation of metal drums, shown in Fig. 10-12. Drums contaminated with volatile materials are passed through a high temperature zone where the volatiles are driven off. These airborne materials are induced through a secondary combustion chamber, where they are fully destroyed prior to release to the atmosphere.

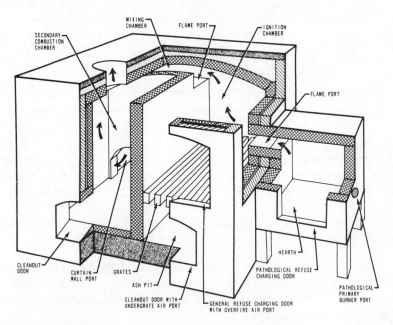

Fig. 10-10. Multiple-chamber incinerator with a pathological-waste retort. *Source:* Figure 7, Page 466.

Table 10-4. Chemical Composition of Pathological Waste and Combustion Data

Ultimate analysis
(whole dead animal)

Constituent	As charged % by weight	Ash-free combustible % by weight
Carbon	14.7	50.80
Hydrogen	2.7	9.35
Oxygen	11.5	39.85
Water	62.1	–
Nitrogen	Trace	–
Mineral (ash)	9	–

Dry combustible empirical formula - $C_5H_{10}O_3$

Combustion data
(based on 1 lb of dry ash-free combustible)

Constituent	Quantity lb	Volume scf
Theoretical air	7.028	92.40
40% sat at 60°F	7.069	93
Flue gas with theoretical air 40% saturated CO_2	1.858	16.06
N_2	5.402	73.24
H_2O formed	0.763	15.99
H_2O air	0.031	0.63
Products of combustion total	8.054	105.92
Gross heat of combustion	8,820 Btu per lb	

Source: Reference 7, Page 420

Table 10-5. Design Factors for Pathological Ignition Chamber (Incinerator Cavity, 25 lb/hr to 250 lb/hr)

Item	Recommended value	Allowable deviation, %
Hearth loading	10 lb/hr ft^2	±10
Hearth length-to-width ratio	2	±20
Primary burner design	$\dfrac{10 \text{ cf natural gas}}{\text{lb waste burned}}$	±10

Source: Reference 7, Page 462

Table 10-6. Gas Velocities and Draft (Pathological Incinerators With Hot Gas Passage Below a Solid Hearth)

Item	Recommended values	Allowable deviation, %
Gas velocities,		
Flame port at 1,600°F, fps	20	±20
Mixing chamber at 1,600°F, fps	20	±20
Port at bottom of mixing chamber at 1,550°F, fps	20	±20
Chamber below hearth at 1,500°F, fps	10	±100
Port at bottom of combustion chamber at 1,500°F, fps	20	±20
Combustion chamber at 1,400°F, fps	5	±100
Stack at 1,400°F, fps	20	±25
Draft,		
Combustion chamber, in, WC	0.25[a]	$\begin{cases} -0 \\ +25 \end{cases}$
Ignition chamber, in, WC	0.05 to 0.10	±0

[a]Draft can be 0.20 in. WC for incinerators with a cold hearth.
Source: Reference 7, Page 462

Table 10-7. Emissions From Two Pathological-Waste Incinerators With Secondary Burners

Test No.	A	B
	Mixing chamber burner Operating	Mixing chamber burner Operating
Rate of destruction to powdery ash, lb/hr	19.2	99
Type of waste	Placental tissue in newspaper at 40°F	Dogs freshly killed
Combustion contaminants,		
gr/scf[a] at 12% CO_2	0.200	0.300
gr/scf	0.014	0.936
lb/hr	0.030	0.360
lb/ton charged	3.120	7.260
Organic acids,		
gr/scf	0.006	0.013
lb/hr	0.010	0.050
lb/ton charged	1.040	1.010
Aldehydes,		
gr/scf	N.A.[b]	0.006
lb/hr	N.A.[b]	0.020
lb/ton charged	N.A.[b]	0.400
Nitrogen oxides,		
ppm	42.70	131
lb/hr	0.08	0.099
lb/ton charged	8.84	2
Hydrocarbons	Nil	Nil

[a]CO_2 from burning of waste used only to convert to basis of 12% CO_2.
[b]Not available.
Source: Reference 7, Page 467

Table 10-8. Emissions From Two Pathological-Waste Incinerators Without Secondary Burners (Source Tests of Two Pathological-Waste Incinerators)

Test No.	A	B
	Mixing chamber burner not operating	Mixing chamber burner not operating
Rate of destruction to powdery ash, lb/hr	26.4	107
Type of waste	Placental tissue in newspaper at 40°F	Dogs freshly killed
Combustion contaminants,		
gr/scf[a] at 12% CO_2	0.500	0.300
gr/scf	0.017	0.128
lb/hr	0.030	0.430
lb/ton charged	2.270	8.040
Organic acids,		
gr/scf	0.010	0.034
lb/hr	0.020	0.110
lb/ton charged	1.514	2.050
Aldehydes,		
gr/scf	0.007	0.010
lb/hr	0.013	0.033
lb/ton charged	0.985	0.617
Nitrogen oxides,		
ppm	14.700	95
lb/hr	0.016	0.082
lb/ton charged	1.210	1.550
Hydrocarbons	Nil	Nil

[a]CO_2 from burning of waste only used to convert to basis of 12% CO_2.
Source: Reference 7, Page 461

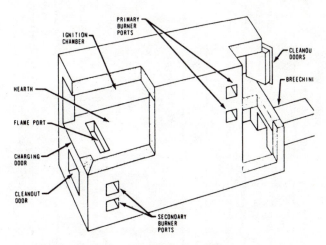

Fig. 10-11. Crematory retort. *Source:* Reference 10-7, Page 467.

Table 10-9. Operating Procedures For Crematory

Phase[a]	Duration, 1-1/2 hr operation, min	Burner settings	Casket	Body		
				Moisture	Tissue	Bone Calcin
Charging[a]	–	Secondary zone on				
Ignition	15	All on	20% burns	–	–	–
Full combustion	30	All on	80% burns	20% evap	10% burns	–
Final combustion	45	All on	–	80% evap	90% burns	50%
Calcining	1 to 12 hr	All off (or small primary on)	–	–	–	50%

OR

Phase	Duration, 2-1/2 hr operation, min	Burner settings	Casket	Body		
				Moisture	Tissue	Bone Calcin
Ignition	15	All on	20% burns	20% evap	–	–
Full combustion	30	Primary off	60% burns	20% evap	–	–
Final combustion	15	All on	20% burns	60% evap	20% burns	50%
	90	All on	–	–	80% burns	–
Calcining	1 to 12 hr	All off (small primary may be on)				

[a] Charge: Casket 75 lb wood
Body 180 lb
    Moisture–108 lb
    Tissue–50 lb
    Bone–22 lb

Source: Reference 7, Page 464

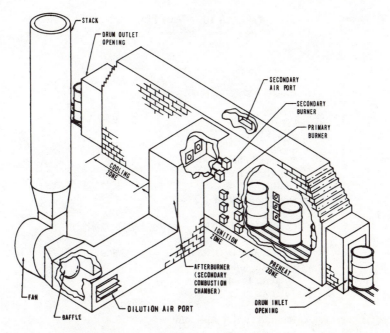

Fig. 10-12. Diagram of a continuous-type drum reclamation furnace with an afterburner. *Source:* Reference 10-7, Page 485.

SOLID WASTE INCINERATION CALCULATIONS

A sugar waste contaminated by caustic is to be incinerated in a solid waste incinerator. For 1000 lb total feed per hour, containing 10% caustic, fuel requirements for heating to 1800°F with 175% excess air will be calculated. The cooling water required to exhaust to 90°F will also be determined. This example will be calculated by the methods of Chapter 9. Tables 10-10, 10-11, and 10-12, are the mass balance, heat balance, and flue gas discharge tables, respectively.

Heat of Combustion of Glucose

From Chapter 8, assuming the sugar is in the form of glucose, $C_6H_{12}O_6$,

$$\begin{array}{ccc} -203.8 & -94.1 & -57.8 \\ C_6H_{12}O_6 + 6O_2 & \longrightarrow 6CO_2 + 6H_2O \\ 180.18 \end{array}$$

$$\Delta H_f \ H_2O(g): \quad -57.8 \times 6 = -346.8$$

$$\Delta H_f \ CO_2: \quad -94.1 \times 6 = -564.6$$

$$\Delta H_f \ C_6H_{12}O_6: \quad -(-203.8) \times 1 = \underline{+203.8}$$

$$\Sigma \Delta H_f = -707.6 \ \text{kcal/mole}$$

Table 10-10. Mass Flow.

	Example (Solid)
Wet Feed, lb/hr	1,000
Moisture, %	2.4
lb/hr	24
Dry Feed, lb/hr	976
Ash, %	7.8
lb/hr	76
Volatile, lb/hr	900
Vol. Htg. Value, BTU/lb	7,077
MBH	6.37
Dry Gas, lb/10 kB	7.89
lb/hr	5,026
Comb. H_2O, lb/10 kB	0.57
lb/hr	363
DG + Comb. H_2O, lb/hr	5,389
100% Air, lb/hr	4,489
Total Air Fract.	2.75
Total Air, lb/hr	12,345
Excess Air, lb/hr	7,856
Humid/Dry Gas (Air), lb/lb	0.01
Humidity, lb/hr	123
Total H_2O, lb/hr	510
Total Dry Gas, lb/hr	12,882

Relating $\Sigma \Delta H_f$ to the weight of glucose present:

$$\frac{1}{180.18 \text{ gram/mole}} \times (-707.6) \, (\text{kcal/mole}) \times 1802 = -7077 \text{ BTU/lb}$$

The negative heating value indicates that this is an exothermic reaction, i.e., heat is released.

Caustic Thermodynamics

Caustic is $Ca(OH)_2$, calcium hydroxide. From Chapter 8:

$$
\begin{array}{ccc}
-239.7 & -151.8 & -57.8 \text{ (steam)} \\
Ca(OH)_2 \longrightarrow & CaO \quad + & H_2O \\
74.28 & 56.08 & 18.02
\end{array}
$$

$$\Delta H_f \; H_2O(g): \qquad\qquad -57.8 \times 1 = -\;57.8$$

$$\Delta H_f \; CaO: \qquad\qquad -151.8 \times 1 = -151.8$$

$$\Delta H_f \; Ca(OH)_2: \qquad -(-239.7 \times 1) = \underline{+239.7}$$

$$\Sigma \Delta H_f = +\;30.1 \text{ kcal/mole}$$

Table 10-11.   Heat Balance.

	Example (Solid)
Cooling Air Wasted, lb/hr	–
°F	–
BTU/lb	–
MBH	–
Ash, lb/hr	76
BTU/lb	656
MBH	0.05
Radiation, %	3
MBH	0.19
Humidity, lb/hr	123
Correction (@970 BTU/lb), MBH	−0.12
Losses, Total, MBH	.12
Input, MBH	6.37
Outlet, MBH	6.25
Dry Gas, lb/hr	12,882
H_2O, lb/hr	510
Temperature, °F	1,651
Desired Temp., °F	1,800
MBH	6.83
Net, MBH	0.58
Fuel Oil, Air Fraction	1.1
Net BTU/gal	70,746
gph	8.20
Air, lb/gal	114.64
lb/hr	940
Dry Gas, lb/gal	115.12
lb/hr	944
H_2O, lb/gal	8.62
lb/hr	71
DG w/FO, lb/hr	13,826
H_2O w/FO, lb/hr	581
Air w/FO, lb/hr	13,285
Outlet, MBH	7.40
Reference t, °F	60

Relating $\Sigma \Delta H_f$ to the weight of caustic present:

$$\frac{1}{74.28 \text{ gram/mole}} \times (30.1)\,(\text{kcal/mole}) \times 1802 = 730 \text{ BTU/lb}$$

This value is positive, indicating that 730 BTU is required to be added to complete the destruction of each pound of caustic The contribution of latent heat of moisture must be subtracted from the total heating value, 730 BTU/lb, because the heat required to raise the moisture present to the furnace temperature is included with the heat balance calculations for the incinerator.

Table 10-12.   Flue Gas Discharge.

	Example (Solid)
Inlet, °F	1,800
Dry Gas, lb/hr	13,826
Heat, MBH	7.40
BTU/lb dry gas	535
Adiabatic t, °F	171
H_2O Saturation, lb/lb dry gas	0.4493
lb/hr	6,212
H_2O Inlet, lb/hr	581
Quench H_2O, lb/hr	5,631
gpm	11.3
Outlet Temp., °F	90
Raw H_2O Temp., °F	70
Sump Temp., °F	143
Temp. Diff., °F	73
Outlet, BTU/lb dry gas	48.212
MBH	0.67
Req'd. Cooling, MBH	6.73
H_2O, lb/hr	92,192
gpm	184
Outlet ft^3/lb dry gas	14.547
ft^3/min	3,352
Fan Press, in. WC	20
Outlet, ACFM	3,516
Outlet, H_2O/lb dry gas	0.03115
H_2O, lb/hr	431
Recirc. (Ideal), gpm	11.3
Recirc. (Actual), gpm	91
Cooling H_2O, gpm	184

Therefore the net heating value of caustic is calculated as follows:

Latent heat of moisture: 970 BTU/lb

$$970 \ (BTU/lb \ H_2O) \times \frac{18.02 \ lb \ H_2O}{74.28 \ lb \ caustic} = 235 \ BTU/lb$$

Net heat required to destruct caustic:

730 BTU/lb gross − 235 BTU/lb H_2O = 495 BTU/lb

For incinerator calculations a moisture and an ash component will be determined, as follows:

Total caustic: 1000 lb hr × 10% = 100 lb hr caustic

The ash generated is CaO, with a molecular weight of 56.08, compared to a molecular weight of 74.28 for caustic. Therefore the quantity of ash produced is:

$$\frac{56.08 \text{ lb ash}}{74.28 \text{ lb caustic}} \times 100 \text{ lb hr caustic} = 75.50 \text{ lb hr ash}$$

The ash absorbs heat as follows:

$$495 \text{ BTU/lb caustic} \times \frac{74.28 \text{ lb caustic}}{56.08 \text{ lb ash}} = 656 \text{ BTU/lb ash}$$

The quantity of moisture within the caustic is proportional to atomic weight:

$$\frac{18.02 \text{ lb } H_2O}{74.28 \text{ lb caustic}} \times 100 \text{ lb caustic} = 24 \text{ lb } H_2O$$

The values developed for wet feed, moisture, ash content and heating value are inserted in Table 10-10. Total moisture and dry gas flow, plus air requirements for waste combustion, are calculated by the methods of Chapter 9.

The ash heating value, 656 BTU/lb, is inserted in the heat balance sheet, Table 10-11. This table plus the flue gas analysis, Table 10-12, have been completed per Chapter 9 calculations.

Chapter 11
Central Disposal Systems

In this chapter burning systems that have been developed for centralized disposal of waste are presented. These facilities include mass burning systems for municipal solid waste disposal and large scale industrial waste disposal systems. Controlled-air incineration systems, which are growing in usage in central disposal facilities, are covered in a later chapter.

THE NEED FOR CENTRAL DISPOSAL

Central disposal of waste is a consideration in the disposal of municipal solid waste. In recent years industrial firms with many plants have looked towards incineration of their wastes in a central plant location as an efficient and economical means of disposal.

In Europe, after the Second World War, the disposal of municipal solid waste at a central location, by incineration, was given impetus by the following factors:

1. Population concentrations and increases required the use of more land for housing and for farming. The use of land for burying refuse was becoming impractical.

2. Technology developed to the point where it became economical to generate energy, i.e., steam and or electric power, from incineration. Economics of scale dictated that the larger the facility the more efficient would be its energy generation potential.

3. In general all utilities, including refuse disposal and electric/steam power generating industries, were state owned. The interests of the electric utility and the refuse disposal authority were therefore common. This conflicts with conditions in the United States, where refuse collection is a public or governmental function and electric power generation is generally a private sector function. Cooperation between these agencies in Europe has promoted the development of energy-producing incinerators. The power utility readily purchases energy from an incineration facility, providing revenues for the incinerator operation.

4. The higher cost of fossil fuel, particularly fuel oil, has helped promote energy generation, hence, central disposal facilities.

Only in the past decade, when the United States came to the realization that the cost and availability of energy was unreliable and out of its control, has a serious attempt been made to generate energy from waste in central collection and incineration facilities.

Table 11-1 lists information on selected mass buring central facilities in the United States. Cost data are included for reference only. The tipping fee (the cost of dumping refuse at the facility) may reflect conditions other than cost.

Table 11-1. Typical Central Mass Burning Facilities, 1981, Municipal Solid Waste.

Location	Startup	Design Rate, tons/day	Initial Cost, $	Process Equipment	Comments
Chicago, IL	1971	1600	25,000,000	Martin	This plant has been in continuous operation burning an average of 1200 TPD since 1971. It has been designed to generate steam. However, a reliable steam customer has only been found in the past few years. A steam pipeline had to be built for this customer and steam is scheduled to be delivered the end of 1981. Early problems included freezing of the air cooled steam condensers during winter months. Ferrous metals are recovered from the incinerator ash. There is no tipping fee; disposal costs are paid by the City of Chicago.
Harrisburg, PA	1973	720	8,300,000	Martin	An average 600 TPD has been processed since the plant began operations in 1973. Since 1978 the plant has been selling steam to Pennsylvania Power & Light for $3.60 per 1000 lb. They are studying the feasibility of recovery of metals from ash. The tipping fee varies from $10.80 to $12.80 per ton.

Table 11-1. *(Continued)*

Location	Startup	Design Rate, tons/day	Initial Cost, $	Process Equipment	Comments
Cataraugus, MA	1976	1500	53,000,000	Von Roll (Wheelabrator-Frye)	Approximately 1150 TPD have been processed since startup. Earlier air pollution and operational problems have been corrected. Steam is produced and sold to the nearby General Electric plant, producing 50% of their steam demand. GE pays for steam on the basis of the equivalent cost fuel for generating its own steam. Ferrous metals are recovered from ash. The current tipping fee is $15.47 per ton.
Akron, OH	1979	1000	56,000,000	Detroit Stoker	This plant averages 600–700 TPD throughput. Refuse feeding conveyors have had operational problems and several solutions are currently being considered. Steam customers include adjacent industry (year-round) and winter heating of local businesses in downtown Akron. The tipping fee is $6.00 per ton.
Nashville, TN	1974	720	25,000,000	Detroit Stoker	A number of operating problems delayed completion of this project and led to the expenditure of additional millions of dollars, particularly in the area of air pollution control. At this time the plant is operating at an average rate of 400 TPD. There is no materials recovery and steam is generated for winter heating in downtown

Table 11-1. (*Continued*)

Location	Startup	Design Rate, tons/day	Initial Cost, $	Process Equipment	Comments
Nashville, TN	1974	720	25,000,000	Detroit Stoker	Nashville. A fixed annual tipping fee of $1,300,000 for the City of Nashville works out to approximately $9.30 per ton.
Quebec City, PQ	1974	1000	25,000,000	Von Roll	This plant has been processing an average of 600 TPD since startup. Until recently the plant experienced severe air emissions problems. However, new equipment has been installed and all the air pollution emissions codes are satisfied. Steam is sold to a local paper company at $3.00 per 1000 lbs. Ferrous metals are recovered from incinerator ash. The current tipping fee is $20.00 per ton.

For instance, a tipping fee may have been established to attract customers from an existing landfill. Likewise, the steam charge may have been established at a lower rate than economics would dictate to encourage sales. A low steam cost would promote steam sales, and such promotion would be necessary, because in the United States incineration is considered an unconventional, untried source of steam.

MUNICIPAL SOLID WASTE

Central station incineration is usually applied to municipal solid waste destruction. The average characteristics of refuse and other wastes are listed in Chapter 8. The actual variation in average waste composition from one country to another is listed in Table 11-2. The "ash" column represents the residual from coal or wood buring for domestic heat in the winter months. For instance, 43% of the composition of refuse in the United States was ash due to household coal burn-

Table 11-2. A Summary of International Refuse Composition.

	Ash	Paper	Organic matter	Metals	Glass	Miscellaneous
United States (1939)	43.0	21.9	17.0	6.8	5.5	5.8
United States (1970)	0	44.0	26.5	8.6	8.8	12.1
Canada	5	70	10	5	5	5
United Kingdom	40–40	25–30	10–15	5–8	5–8	5–10
France	24.3	29.6	24	4.2	3.9	14
West Germany	30	18.7	21.2	5.1	9.8	15.2
Sweden	0	55	12	6	15	12
Spain	22	21	45	3	4	5
Switzerland	20	40–50	15–25	5	5	–
Netherlands	9.1	45.2	14	4.8	4.9	22
Norway (summer)	0	56.6	34.7	3.2	2.1	8.4
Norway (winter)	12.4	24.2	55.7	2.6	5.1	0
Israel	1.9	23.9	71.3	1.1	0.9	1.9
Belgium	48	20.5	23	2.5	3	3
Czechoslovakia (summer)	6	14	39	2	11	28
Czechoslovakia (winter)	65	7	22	1	3	2
Finland	–	65	10	5	5	15
Poland	10–21	2.7–6.2	35.3–43.8	0.8–0.9	0.8–2.4	–
Japan (1963)	19.3	24.8	36.9	2.8	3.3	12.9

Source: Ref. 11-12.

ing in 1939, whereas 30 years later this component, i.e., ash from coal burning, was absent from the refuse.

When burning refuse, the generation of dry gas and moisture from combustion can be estimated as follows:

Dry gas   7.5 lb/10,000 BTU fired

Moisture   0.51 lb/10,000 BTU fired

GRATE SYSTEM

The grate system is one of the most crucial systems within the mass buring incinerator. The grate must transport refuse through the furnace and, at the same time, promote combustion by adequate agitation and good mixing with combustion air. Abrupt tumbling caused by the dropping of burning solid waste from one tier to another will promote combustion. This action, however, may contribute to excessive carryover of particulate matter in the exiting flue gas. A gentle agitation will decrease particulate emissions.

Combustion is largely achieved by injection of combustion air below the grates, i.e., underfire air. Underfire air is also necessary to cool the grates. It is normally

Table 11-3.   Ash Fusion Temperatures.

	Reducing Atmosphere, °F	Oxidizing Atmosphere, °F
Refuse		
Initial deformation	1880–2060	2030–2100
Softening	2190–2370	2260–2410
Fluid	2400–2560	2480–2700
Coal		
Initial deformation	1940–2010	2020–2270
Softening	1980–2200	2120–2450
Fluid	2250–2600	2390–2610

Source: Ref. 11-14.

provided at a rate of approximately 40–60% of the total air entering the furnace. Too low a flow of underfire air will inhibit the burning process and will result in high grate temperatures.

Note the ash fusion temperatures listed in Table 11-3. These temperatures limit the operating temperatures of the grate areas. With insufficient air a reducing atmosphere will result and the ash deformation temperature can be as low as 180°F. If the refuse reaches this temperature slagging will begin, further reducing the air supply by clogging the grates, and forming large, unwieldy clinkers. The ash properties of coal are listed for comparison.

Overfire air is injected above the grates. Its main purpose is to provide sufficient air to completely combust the flue gas and flue gas particulate rising from the grates. Numerous injection points are located on the furnace walls above the grates to provide a turbulent overfire air supply along the furnace length.

Ash and other particles dropping through the grates are termed *siftings*, and must be effectively removed from the system. Siftings can readily clog grate mechanisms, generate fires, and create housekeeping problems if not attended to. Siftings, due to their small particle size, have been found to be more dense than incinerator ash, approximately 1780 lb per cubic yard versus 1040 lb per cubic yard for typical incinerator ash.

GRATE DESIGN

There are a number of different types of grate designs that are used in central waste burning facilities. Each grate system manufacturer provides a unique grate feature attempting to obtain a competitive edge in the marketplace.

The grate system manufacturer should be contacted for design and sizing information for a particular grate design. The following listing describes typical grate systems, both generic grate types and grates specific to certain manufacturers:

Circular Grate

This type is no longer in common usage. As shown in Fig. 11-1, a rotary cone slowly moves a series of rabble arms across a circular grate. The circular grate furnace is charged at short intervals with the fresh charge falling directly on top of a pile of burning refuse. The refuse forms a cone-shaped pile that gradually burns down with combustion largely on the surface of the pile. Combustion air is provided through the grate and through the center cone by forced draft fans. Refuse is moved slowly down to the peripheral dumping grates by the rotating rabble arms. The refuse will burn out to ash on these dumping grates, which must then be cleaned. They can be cleaned manually, through clean-out doors on the periphery of the furnace, or they can be provided with power cylinders. These power cylinders will lower and raise the individual dumping grate sections to expedite ash removal. Fig. 11-2 illustrates a vertical circular furnace utilizing a circular grate. This furnace is automated with automatic ash (residue) collection in a residue hopper, for eventual removal from the system. Waste is combusted in the primary chamber, above the circular grate. The products of com-

A. Rotating Cone
B. Extended Stoking
   Arm (Rabble Arm)
C. Stationary Circular
   Grate
D. Peripheral Dumping
   Grate

Fig. 11-1. Circular grates.

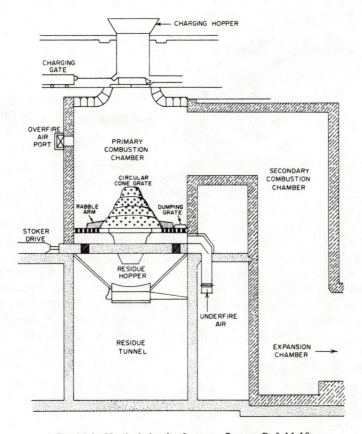

Fig. 11-2. Vertical circular furnace. *Source:* Ref. 11-10.

bustion are fully incinerated in the secondary combustion chamber, which may contain a supplemental fuel burner. This incinerator system discharges relatively high levels of pollutants and as such is not favored for new installations.

Traveling Grate

This type is no longer in common usage. As shown in Fig. 11-3, it is normally not a single grate but a series of grates which are placed in a manner that separates the drying and burning functions of the incinerator. The angled grate receives refuse on a continuous basis from a charging hopper. The refuse is normally dried on this grate and does not begin to burn until it drops onto the horizontal, or burning, grate. The speed of the burning grate is chosen so that by the time the burning refuse reaches the end of the grate it is combusted to ash. The ash falls from this grate to an ash hopper. The grates themselves are con-

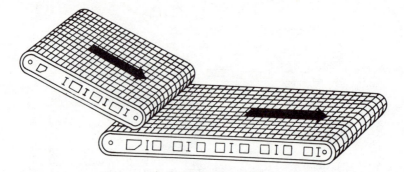

Fig. 11-3. Traveling grates.

tinuous metal-belt conveyors or interlocking linkages. Refuse is not agitated on the grate, but it does experience some turbulence when falling from one grate to another. Fig. 11-4 shows a typical system utilizing a traveling grate, a feeder stoker unit. Refuse is dropped by a grapple bucket into a hopper feeding a charging chute. An inclined stoker feeder moves the waste to a second, horizontal stoker, while the waste is being dried by the surrounding hot flue gases. Waste burns to ash on the horizontal stoker and the ash residue falls into either of two pits, each serviced by a horizontal flight conveyor. Note that sifting hoppers are

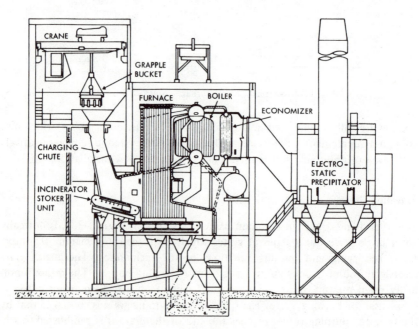

Fig. 11-4. Traveling grate system. *Source:* Ref. 11-16.

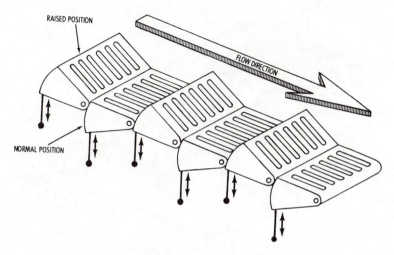

Fig. 11-5. Rocking grates.

provided beneath each of the two grates. Siftings are conveyed to the main ash conveyors where they exit with the bottom ash. The furnace is lined with water walls, tubes integral with the boiler which capture heat for the generation of steam. Waste is charged to this system as received. It is neither shredded nor otherwise processed.

Rocking Grate

As shown in Fig. 11-5, these grate sections are placed across the width of the furnace. Alternate rows are mechanically pivoted or rocked to produce an upward and forward motion, advancing and agitating the waste. The stroke of the grate sections is 5–6 inches. This grate will handle refuse on a continuous basis.

Reciprocating Grate

As shown in Fig. 11-6, this grate consists of sections stacked above each other similar to overlapping roof shingles. Alternate grate sections slide back and forth while adjacent sections remain fixed. Drying and burning is accomplished on single, short but wide grates. The moving grates are basically bars, stoking bars, which move the waste along and help agitate it.

Rotary Kiln

As shown in Fig. 11-7, two traveling grates are initially used for drying the incoming refuse and for initial ignition. The kiln is at the heart of this system. By

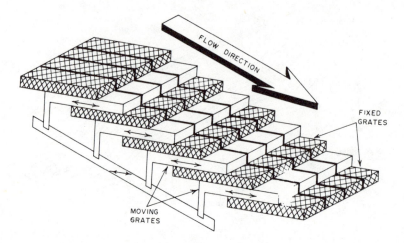

Fig. 11-6. Reciprocating grates.

varying the kiln rotational speed, burnout of the refuse is accurately controlled. The refuse burns out in the kiln and ash is discharged from the end of the kiln to residue conveyors. Some of the flue gases are diverted for drying the incoming refuse. Flue gas can be passed through a waste heat boiler for energy recovery.

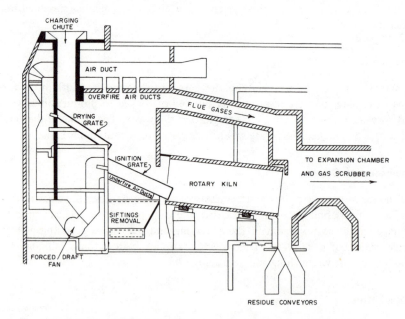

Fig. 11-7. Municipal rotary kiln incineration facility. *Source:* Ref. 11-8.

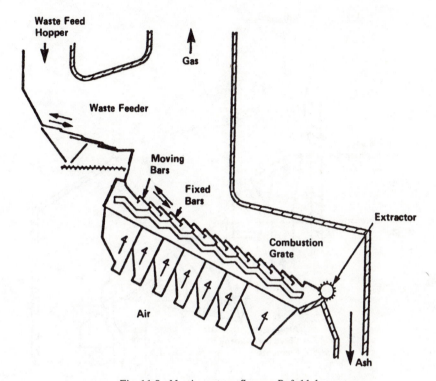

Fig. 11-8. Martin system. *Source:* Ref. 11-1.

Martin System[1]

As shown in Fig. 11-8, this system utilizes reverse reciprocating grates. As the grates move forward, and then reverse, there is continuous agitation of the waste. The bars making up the grates are hollow, allowing air to circulate within them and keep them relatively cool.

Von Roll System[2]

As shown in Fig. 11-9, a series of reciprocating grates are used to move refuse through the furnace. The first grate section dries the refuse, the second is a burning grate, and burnout to ash takes place on the third grate.

---

[1]Proprietary system, marketed in the United States by Ogden Corporation.
[2]Proprietary system, marketed in the United States by Wheelabrator Frye.

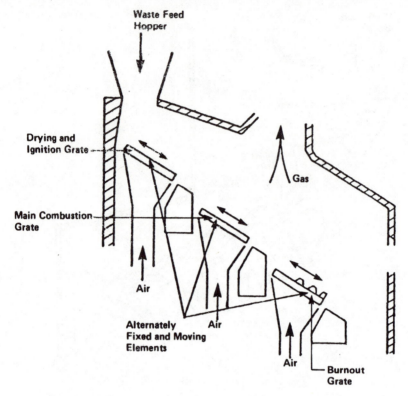

Fig. 11-9. Von Roll system. *Source:* Ref. 11-1.

VKW System[3]

As shown in Fig. 11-10, is a variation of the traveling grate concept. A series of drums are utilized as grates. The drums rotate slowly, agitating the waste and moving it along to subsequent drums. Air passes through openings in these drums, or roller grates, as underfire air. Both speed of rotation on the roller grates and quantity of underfire air per roller grate are variable.

Alberti System[4]

As illustrated in Fig. 11-11, this grate system has a single-section grate constructed of fixed and moving elements arranged, as shown, in a series of steps. Feed is

---

[3]Proprietary system of VKW, Düsseldorf, West Germany, marketed in the United States by Browning-Ferris Industries, Houston, TX.
[4]Proprietary system.

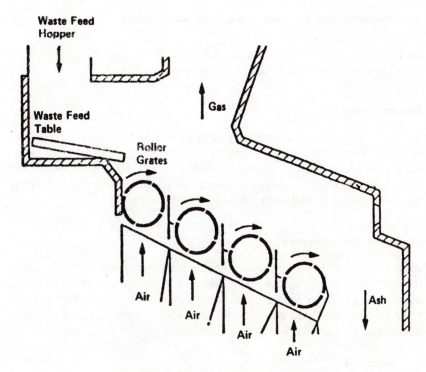

Fig. 11-10. VKW system. *Source:* Ref. 11-1.

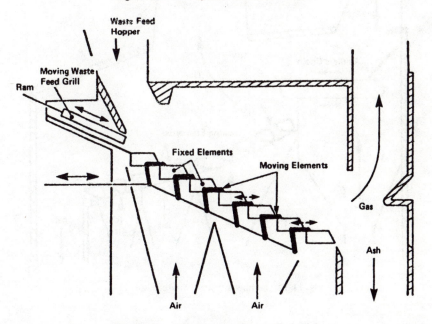

Fig. 11-11. Alberti system. *Source:* Ref. 11-1.

rammed from the feed hopper to the grates. The fixed grate contains the refuse while the moving elements agitate the waste, driving it down to the next grate.

Esslingen System[5]

As shown in Fig. 11-12, a traveling grate is used to feed a rocking grate system. It is normally provided with a single-grate section composed of semicircular rocking elements. Each movement of these elements promotes transport and agitation of the waste. Underfire air passes through the rocking elements, keeping them cool and providing for combustion of the waste.

The Heenan Nichol System[6]

As shown in Fig. 11-13, this system utilizes grates composed of three or more sections which are arranged in steps. Each pair of elements moves in a rocking manner so that at any moment half of the elements are moving. All odd numbered elements are linked to each other, as are all the even numbered elements. The rocking action moves and agitates the waste.

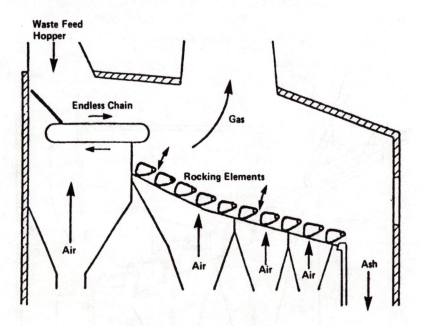

Fig. 11-12. Esslingen system. *Source:* Ref. 11-1.

[5,6]Proprietary system.

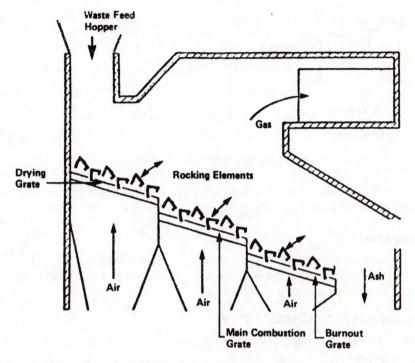

Fig. 11-13. Heenan Nichol system. *Source:* Ref. 11-1.

CEC System[7]

This system, illustrated in Fig. 11-14, utilizes a single-section grate. Two or more grate sections can be arranged in parallel. The grate is constructed of successive sliding, rocking, and fixed elements. The sliding and rocking elements are synchronized so that the sliding elements move over the rocking elements when the rocking elements are retracted. The sliding elements, therefore, promote transport of the waste while the rocking elements provide the required agitation.

Bruun and Sorensen System[8]

As shown in Fig. 11-15, this system utilizes a series of rollers, up to six in each of its three sections. Odd-numbered rollers turn clockwise while even-numbered rollers turn counter-clockwise. Underfire air passes through the rotary grates, or

---

[7] Proprietary system of Carbonisation Enterprise et Céramique.
[8] Proprietary system.

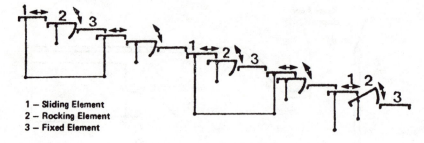

1 – Sliding Element
2 – Rocking Element
3 – Fixed Element

Fig. 11-14. CEC system. *Source:* Ref. 11-1.

drums. The action of the drums provides good agitation and the slope of the grate to the horizontal promotes the transport of waste along the grate.

Volund System[9]

This system is shown in Fig. 11-16. It utilizes a rotary kiln for controlled burning of waste. Reciprocating grates are used for waste drying and initial combustion. Burnout takes place within the kiln.

ASSOCIATED DISPOSAL SYSTEMS

There are refuse burning systems in use which cannot strictly be classified as grate systems.

Suspension Burning

An incinerator coupled to a refuse processing system is pictured in Fig. 11-17. Refuse is shredded and air classified into light and heavy fractions. The light fraction is blown into the boiler through a pneumatic charging system. Fig. 11-18 shows the air distribution within the furnace, and around the waste feed. Waste that is not burned in suspension will drop onto the shredder stoker, a variation of the traveling grate, and will burn out. Ash not airborne, produced by suspension burning, will also drop onto the spreader stoker. The stoker moves slowly, discharging its ash load to an ash hopper for ultimate disposal. Heavier components of the refuse that are not incinerated are composed mainly of metals and glass. These materials can, in certain instances, be marketed.

Fibreclaim System[10]

This system is a variation of pulping processes used in the paper industry. Municipal waste is mixed with water in a hydropulper, a mechanical device resembling

---

[9]Proprietary system, marketed in the United States by Waste Management, Inc., Chicago, IL. A similar system is manufactured by International Incinerators, Columbus, GA.
[10]Proprietary system, marketed in the United States by Black Clawson, Inc., Everett, WA.

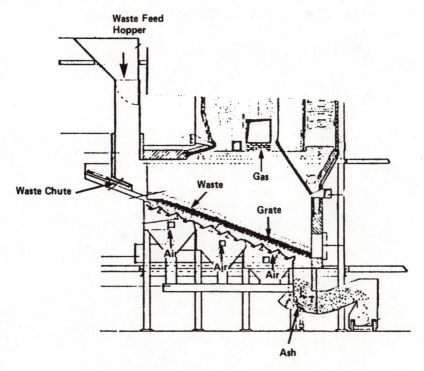

Fig. 11-15. Bruun and Sorensen system. *Source:* Ref. 11-1.

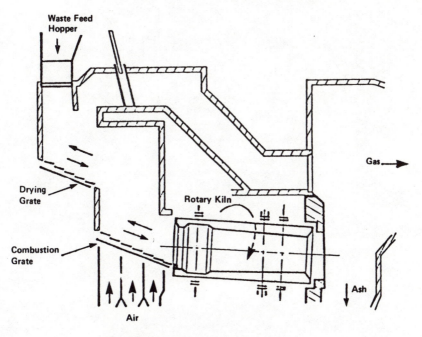

Fig. 11-16. Volund system. *Source:* Ref. 11-1.

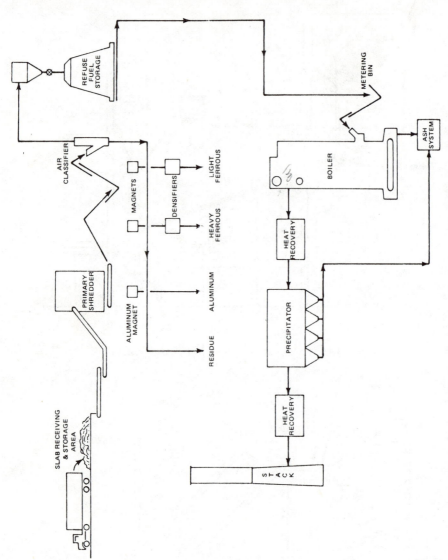

Fig. 11-17. *Source:* AKRON Recycle Energy System, Teledyne National, Akron, OH.

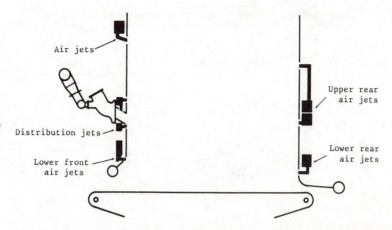

Fig. 11-18. Air and feed distribution. *Source:* Babcock & Wilcox, North Canton, OH.

a giant food blender. Pulp material is generated, and a residue remains after the pulp is removed. Ferrous metals are extracted from the residue and the balance of the residue returns to the process. Glass and other fines are removed from the pulp in a cyclone and the pulp goes through a drying process. The dried material is blown into a furnace where it burns in suspension generating steam. This system is installed in the Hempstead Resource Recovery Facility, which has been the focus of concern for the generation of dioxins from the destruction of waste.

INCINERATOR CORROSION PROBLEMS

Severe corrosion has been found in three major areas of the mass burning incinerator system.

Scrubber Corrosion

The acidic components of the flue gas present corrosion problems in wet scrubbing equipment, which will be discussed in a later chapter.

Corrosion of Grates

This problem was alluded to earlier in this chapter. With insufficient air flow, a reducing atmosphere can occur and ash can soften, or fluidize. Fluid ash can be exceedingly corrosive, readily attacking cast iron or steel.

Table 11-4 illustrates the corrosion rate as a function of temperature for two steel alloys in widespread use in grate construction. Temperature alone greatly increases the corrosive rate, the amount of material lost per month of service.

Table 11-4.   Gas-Phase Corrosion at Elevated Temperatures.

Alloy	Temperature	Corrosion Rate, mils per month
A106	800	0.9
A106	1000	8
A106	1200	36
T11	800	0.8
T11	1000	6
T11	1200	29

Source: Ref. 11-11.

At $1200°F$ over $\frac{3}{8}$ inch of material is "wasted" or lost from the steel structures, for instance.

Fireside Corrosion

There are two modes of corrosion that affect boiler tubes. Low-temperature or dewpoint corrosion is metal wastage caused by sulfuric or hydrochloric acid condensation. Chlorides and sulfides within the refuse (chlorides are present in plastics) will partially convert to free chlorine, hydrogen chloride, sulfur dioxide, and sulfur trioxide in the exhaust gas stream. These gases will condense at temperatures below $300°F$, and their condensate or liquid phase will be hydrochloric and sulfuric acid, both of which will attack steel. It is important, therefore, that the temperature of the boiler tubes, constructed of steel, be kept above $300°F$. This is a function of the temperature of the steam or hot water generated. The boiler tubes will be at a temperature close to that of the circulating fluid, and this $300°F$ rule therefore limits the minimum temperature of the steam or hot water generated.

High-temperature corrosion is a more complex problem. Table 11-5 lists steam pressure, temperature and external tube temperatures for an assortment of incinerator systems burning municipal solid waste. At temperatures exceeding $700°F$ a complex reaction takes place between the sulfide and chlorine/chloride bearing flue gas and the steel boiler tube, as illustrated in Fig. 11-19. Chlorine reacts with the iron in the tube wall to produce ferrous chloride which, upon contact with oxygen in the flue gas, converts to iron oxide. The iron oxide (rust) will leave the surface of the steel, causing wastage of the steel surface. Other components of the refuse which become airborne, such as alkali salts, will promote this corrosion. For incinerators operating above $700°F$, metal temperatures (most of the incinerator tubes noted in Table 11-5 operate at temperatures in excess of $700°F$) special refractory-lined water walls must be utilized to protect the tubes from this metal wastage.

Table 11-5. Nominal Operating Conditions of Water-Wall Incinerators.

Location	Steam Pressure, psig	Steam Temp, F	Metal Temp, F (Approx.)
Milan, Italy	500	840	890
Mannheim, Germany	1800	980	1030
Frankfurt, Germany	960	930	980
Munster, Germany	1100	980	1030
Moulineaux, France	930	770	820
Essen Karnap, Germany	--	930	980
Stuttgart, Germany	1100	980	1030
Munich, Germany	2650	1000	1050
Rotterdam, Netherlands	400	680	730
Edmonton, England	625	850	900
Coventry, England	275	415	465
Amsterdam, Netherlands	600	770	820
Montreal, Canada	225	395	445
Chicago (N. W.), Illinois	265	410	460
Oceanside, New York	460	465	515
Norfolk, Virginia	175	375	425
Braintree, Massachusetts	265	410	460
Harrisburg, Pennsylvania	275	460	510
Hamilton, Ontario	250	400	450

Source: Ref. 11-11.

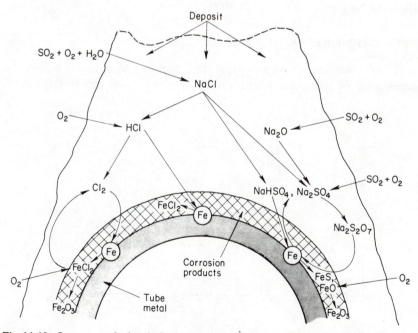

Fig. 11-19. Sequence of chemical reactions explaining corrosion on incinerator boiler tube. *Source:* Ref. 11-11.

Table 11-6. Performance of Alloys in Fireside Areas*.

Alloy	Resistance to Wastage		
	300-600 F	600-1200 F	Moist Deposit
Incoloy 825	Good	Fair	Good
Type 446	Good	Fair	Pits
Type 310	Good	Fair	SCC**
Type 316L	Good	Fair	SCC
Type 304	Good	Fair	SCC
Type 321	Good	Fair	SCC
Inconel 600	Good	Poor	Pits
Inconel 601	Good	Poor	Pits
Type 416	Fair	Fair	Pits
A106-Grade B (Carbon Steel)	Fair	Poor	Fair
A213-Grade T11 (Carbon Steel)	Fair	Poor	Fair

* Arranged in approximate decreasing order.
** Stress-corrosion cracking.

Source: Ref. 11-11.

Table 11-6 indicates the relative performance of various alloys in incinerator fireside areas. Although stainless steels appear to have favorable corrosion resistance, the danger of stainless steel stress corrosion cracking prohibits its use in pressure vessels such as boilers and high-temperature hot water heaters.

Figure 11-20 further illustrates the rate of corrosion of carbon steel by chloride attack as a function of metal temperature.

INCINERATOR REFRACTORY SELECTION

Table 11-7 illustrates the types of problems that can be expected within the various areas of a large incinerator system associated with refractory selection.

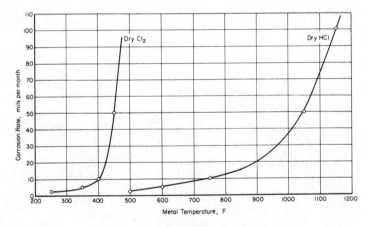

Fig. 11-20. Corrosion of carbon steel in chlorine and hydrogen chloride. *Source:* Ref. 11-11.

Table 11-7. Suggested Refractory Selection for Incinerators.

Incinerator part	Temperature (°F) range	Abrasion	Slagging	Mechanical shock	Spalling	Fly ash adherence	Recommended refractory
Charging gate	70–2600	Severe, very important	Slight	Severe	Severe	None	Superduty
Furnace walls, grate to 48 in. above	70–2600	Severe	Severe, very important	Severe	Severe	None	Silicon carbide or superduty
Furnace walls, upper portion	70–2600	Slight	Severe	Moderate	Severe	None	Superduty
Stoking doors	70–2600	Severe, very important	Severe	Severe	Severe	None	Superduty
Furnace ceiling	700–2600	Slight	Moderate	Slight	Severe	Moderate	Superduty
Flue to combustion chamber	1200–2600	Slight	Severe, very important	None	Moderate	Moderate	Silicon carbide or superduty
Combustion chamber walls	1200–2600	Slight	Moderate	None	Moderate	Moderate	Superduty
Combustion chamber ceiling	1200–2600	Slight	Moderate	None	Moderate	Moderate	Superduty
Breeching walls	1200–3000	Slight	Slight	None	Moderate	Moderate	Superduty
Breeching ceiling	1200–3000	Slight	Slight	None	Moderate	Moderate	Superduty
Subsidence chamber walls	1200–1600	Slight	Slight	None	Slight	Moderate	Medium duty
Subsidence chamber ceiling	1200–1600	Slight	Slight	None	Slight	Moderate	Medium duty
Stack	500–1000	Slight	None	None	Slight	Slight	Medium duty

Source: Ref. 11-13.

The nature of the hot gas stream within the incinerator can create significant detrimental effects on grates, walls, ceilings and other areas within the furnace enclosure. Some of the more common concerns in refractory selection are discussed here.

Abrasion

This is the effect of impact of moving solids within the gas stream or of heavy pieces of materials charged into the furnace upon refractory surfaces. Fly ash also causes abrasive effects. Abrasion is the wearing away of refractory, or any surface, under direct contact with another material with relative motion to its surface.

Slagging

When a portion of charged material, usually ash, metals or glass within an incinerator, reaches a high enough temperature, deformation of that material will occur. The material will physically change to a more amorphous state and may begin to flow, as a heavy liquid. When the temperature of the material is then reduced below that required for deformation, the material will solidify into a hard slag. This process can take place when high temperatures are experienced on a grate. When the grate section moves through a lower-temperature zone a slag may form. Molten ash may become airborne, then attach to a refractory or a metal surface within the furnace which is cooler than the air stream. Slag will then form on this surface. This slag can be acidic (as a result of silicon, aluminum, or titanium oxides released from the burning waste) or it can be basic (due to the generation of oxides of iron, calcium, magnesium, potassium, sodium, or chromium), and the selection of refractory must be compatible with these materials to help insure long refractory life. (For acidic slag, fireclay or high-alumina and/or silica firebrick would be used. Chrome, magnesite, or forsterite brick would be used for basic slags.)

Mechanical Shock

The impact of falling refuse can cause mechanical shock, as can constant vibration caused by grate bushings or supports and vibration set up by turbulent flow adjacent to air inlet ports.

Spalling

The flaking away of refractory surface, or spalling, is most commonly caused by thermal stresses or mechanical action. Uneven temperature gradients can

Table 11-8. Typical Steam Generation.

Solid Waste Type	MSW	MSW	MSW	RDF
Steam temperature, °F	620	500	465	400
Steam pressure, psig	400	225	260	250
Steam production, tons/ton refuse	3.6	1.4-3.0	1.5-4.3	4.2

Note:
   MSW: Municipal solid waste.
   RDF: Refuse derived fuel, i.e., shredded MSW Less metals.
Source: Ref. 11-17.

cause local thermal stresses in brick which will degrade the refractory surface, causing a spalling condition. The more common type of mechanical spalling is caused by rapid drying of wet brickwork. The steamed water does not have an opportunity to escape the brick surface through the natural porosity of the refractory, but expands rapidly, causing cracking and spalling of the brick.

Fly Ash Adherence

As noted above, fly ash can have fluid properties within the hot gas stream, and adhere to refractory or other cooler surfaces within the furnace chamber. Fly ash accumulation can result in corrosive attack on these surfaces. Heavy accumulations will interfere with the normal surface cooling effects within the furnace resulting in a decrease of furnace refractory life.

As noted above, Table 11-7 lists various areas within an incinerator and describes the severity of the above problems at each of these locations. Normal temperature ranges are noted as well as recommended refractory types.

Table 11-9. Steam Production Related to MSW Quality.

	As-Received Heating Value, BTU/lb				
	6500	6000	5000	4000	3000
Refuse:					
% moisture	15	18	25	32	39
% noncombustible	14	16	20	24	28
% combustible	71	66	55	44	33
Steam generated, tons/ton refuse	4.3	3.9	3.2	2.3	1.5

Source: Ref. 11-18.

HEAT RECOVERY OR WATHING OF HEAT

Steam can be generated by utilization of incinerator waste heat, as shown in Table 11-8, which lists steam production rates. Variations in waste heating value will produce variations in steam generation as shown in Table 11-9. But these quantities must be weighed against the overall implications of a boiler installation for waste burning. Note the following comparison between an energy recovery system and an incinerator without provision for heat recovery:

With Heat Recovery	Without Heat Recovery
Reduced gas temperatures and volumes due to absorbtion of heat by heat recovery system	Hotter gas temperatures
Moderate excess air	High excess air required to control furnace temperatures
Moderate size combustion chamber	Large refractory-lined combustion chamber to handle high gas flow
Smaller air and induced draft fans required for smaller gas volume.	Higher gas volumes due to higher temperatures, requiring larger air and gas flow equipment
Steam facilities including integral water wall, boiler drums and boiler auxiliary equipment required	No steam facilities are required
Operations involve boiler system monitoring, adjustments for steam demand, etc.	Relatively simple operating procedures
Steam tube corrosion is possible as well as corrosion within exhaust gas train	Corrosion possible in the exhaust gas train
Licensed boiler operators are required to operate incinerator	Conventional operators are satisfactory
Considerable steam credits possible, including in-plant energy savings in addition to salvage	Only credits are possible salvage of the equipment after its useful life

Chapter 12
Pyrolysis and Controlled Air Incineration

Pyrolysis is a thermal process related to oxidation, or burning. In this chapter the mechanisms of pyrolysis are discussed and equipment utilizing pyrolysis and starved air reactions are presented.

GENERAL DESCRIPTION

Pyrolysis is the destructive distillation of a solid, normally carbonaceous, material, in the presence of heat and in the absence of stoichiometric oxygen. It is an endothermic reaction, i.e., heat must be applied for the reaction to occur.

Ideally a pyrolytic reaction will occur as follows, using cellulose:

$$C_6H_{10}O_5 \xrightarrow{\text{heat}} CH_4 + 2CO + 3H_2O + 3C$$

A gas is produced containing methane, CH_4, carbon monoxide, CO, and moisture. The carbon monoxide and methane components are combustible, providing heating value to the off gas. The carbon residual, a char, also has heating value. This is an idealized reaction. No oxygen is added and the original material is pure cellulose, $C_6H_{10}O_5$. In general the initial material is not pure and contains additional components both organic and inorganic. The off gas is a mixture of many simple and complex organic compounds. The char is often a liquid which contains minerals, ash and other inorganics as well as residual carbon or tars.

Pyrolysis as an industrial process has been in use for years and only since the late 1960s has had significant use in destruction of waste materials. The pyrolysis process produces charcoal from wood chips, coke and coke gas from coal, fuel gas and pitch from heavy hydrocarbon still bottoms, etc.

PYROLYSIS PRODUCTS

As noted previously, an ideal, simple, pyrolytic reaction produces an off-gas with combustible content and a char. In an actual case, however, the feed materials are complex and the products generated from the pyrolytic process are likewise complex and varied. Depending on the temperature of the reaction, the pressure in the reaction chamber, and the amount of air (or oxygen) allowed to enter the chamber, the reaction products will vary in composition.

The product yields from the pyrolysis of typical solid waste charges are listed in Table 12-1. For this particular unit the temperature in the pyrolysis chamber varied from 950 to 1650°F, with a waste retention time of 12–15 minutes. The processed municipal waste had the majority of its glass and metal content removed. Note that the off-gas produced is greatest at the higher temperature whereas tars and other liquors generated are greatest at the lower temperature.

The gas produced in these reactions will have a heating value of approximately 300–400 BTU per standard cubic foot. Ammonia is produced from the nitrogen component of the waste feed and much of this ammonia combines with sulfur in the waste to produce ammonium sulfate.

Table 12-2 lists the composition of the solid residues, and the residue heating value, from the processes of Table 12-1.

Typical off-gas compositions for various reaction temperatures are given in Table 12-3. This gas has sufficient heating value to maintain combustion and it can be used as a low grade fuel. It does contain particulate matter and trace contaminants, such as organic acids, and these components must be removed before the pyrolysis gas can be used remote from the reactor system. In general, pyrolysis gas is utilized solely in the afterburner of the pyrolysis system.

THE PYROLYSIS SYSTEM

An idealized pyrolysis system for disposal of mixed waste is shown in Fig. 12-1. The waste received is sorted for removal of glass, metal and cardboard, all of which has possible resale value. The waste stream enters a shredder (grinder) and the shredded material passes through a magnetic separator where residual ferrous metal is removed, for resale.

The balance of the waste stream will be fed into the reactor from a feed hopper. The hopper discharge and the feeder must be provided with air locks to minimize the infiltration of air (oxygen) which will degrade the pyrolysis reaction.

Shredding is a necessary step, not only to allow metals removal but to provide a uniform size feed of relatively small particles to the reactor.

The converter is heated externally, as shown. Other types of pyrolytic reactors are designed to allow sufficient air infiltration to provide some burning within the reactor, generating enough heat internally to sustain the process.

Gas, exiting the reactor, is collected in a storage tank where organic acids and other organic compounds condense and are eventually discharged. 30–40% of the gas is required to heat the pyrolytic reactor; the balance of the gas stream can be used for other processes. In this generalized scheme a significant portion of the heating value of the off-gas is contained within the condensables. If the

Table 12-1. Yields of Products from Pyrolysis of Municipal and Industrial Waste.

Refuse	Pyrolysis Temp. F	Yields, Weight-Percent of Waste							Yields per Ton of Waste				
		Residue	Gas	Tar	Light Oil in Gas	Free Ammonia	Liquor	Total	Gas Cubic Feet	Tar, gallons	Light Oil in Gas Gallons	Liquor, gallons	Ammonium Sulfate, pounds
Raw municipal waste	930	9.3	26.7	2.2	0.5	0.05	55.8	94.6	11,509	4.8	1.5	133.4	17.9
	1380	11.5	23.7	1.2	0.9	0.03	55.0	92.3	9,628	2.6	2.5	131.6	23.7
	1650	7.7	39.5	0.2	0.0	0.03	47.8	95.2	17,741	0.5	0.0	113.9	25.1
Processed municipal waste	930	21.2	27.7	2.3	1.3	0.05	40.6	93.2	11,545	5.6	3.7	96.7	16.2
	1380	19.5	18.3	1.0	0.9	0.02	51.5	91.2	7,380	2.2	2.6	122.6	28.4
	1650	19.1	40.1	0.6	0.2	0.04	35.3	95.3	18,058	1.4	0.6	97.4	31.5
Industrial-sample A	930	36.1	23.7	1.9	0.5	0.05	31.6	93.9	9,563	4.1	1.4	75.2	12.5
	1380	37.5	22.8	0.7	0.9	0.03	30.6	92.5	9,760	1.5	2.6	73.0	19.5
	1650	38.8	29.4	0.2	0.6	0.04	21.8	90.8	12,318	0.5	1.6	51.1	21.7
Industrial-sample B	930	41.9	21.8	0.8	0.6	0.03	29.5	94.6	9,270	1.7	1.6	70.2	20.4
	1380	31.4	25.5	0.8	0.8	0.03	31.5	90.0	10,952	1.8	2.2	74.9	21.2
	1650	30.9	31.5	0.1	0.5	0.03	29.0	92.0	14,065	0.02	1.4	68.5	22.9

Source: Reference 2

Table 12-2. Chemical Analyses* of Solid Residues from Pyrolysis of Municipal and Industrial Waste.

Refuse	Pyrolysis Temp. F	Proximate, percent				Ultimate, percent					Heating Value, BTU/lb	Million BTU/Ton
		Moisture	Volatile Matter	Fixed Carbon	Ash	Hydrogen	Carbon	Nitrogen	Oxygen	Sulfur		
Raw Municipal Waste	930	2.6	4.4	29.6	66.0	0.4	32.4	0.5	0.5	0.2	5020	10.04
	1380	2.2	7.4	51.4	41.2	0.8	54.9	1.1	1.8	0.2	8020	16.04
	1650	1.0	4.7	31.7	63.6	0.3	36.1	0.5	0.0	0.2	5260	10.52
Processed Municipal Waste	930	1.7	4.8	56.7	38.5	0.6	57.7	0.8	2.1	0.3	8800	17.60
	1380	1.3	13.4	34.6	52.0	0.8	41.9	0.8	4.4	0.1	6080	12.16
	1650	1.2	3.3	53.5	43.2	0.5	53.4	0.7	1.8	0.4	8090	16.18
Industrial-Sample A	930	0.9	2.6	15.2	82.2	0.3	17.0	0.1	0.2	0.2	2520	5.04
	1380	1.2	5.1	17.9	77.0	0.5	19.4	0.2	1.8	0.2	2900	5.08
	1650	0.1	2.5	12.9	84.6	0.3	14.8	0.2	0.0	0.2	2180	4.36
Industrial-Sample B	930	0.3	3.0	9.7	87.3	0.2	11.8	0.1	0.4	0.2	1660	3.32
	1380	1.0	3.6	16.6	79.8	0.3	19.5	0.2	0.0	0.2	2680	5.36
	1650	0.2	6.4	16.2	77.4	0.4	19.3	0.3	2.4	0.2	2810	5.62

*Moisture on as-received basis, all other data on dry basis.
Source: Reference 2

Table 12-3.   Pyrolysis Gas Composition.

	Pyrolytic Temperature, °F			
	900	1200	1500	1700
Gas composition, volume percent				
Carbon monoxide	33.6	30.5	34.1	35.3
Carbon dioxide	44.8	31.8	20.6	18.3
Hydrogen	5.6	16.5	28.6	32.4
Methane	12.5	15.9	13.7	10.5
Ethane	3.0	3.1	0.8	1.1
Ethylene	0.5	2.2	2.2	2.4
Heating Value (HHV), BTU/SCF	312	403	392	385

Source: Ref. 12-7.

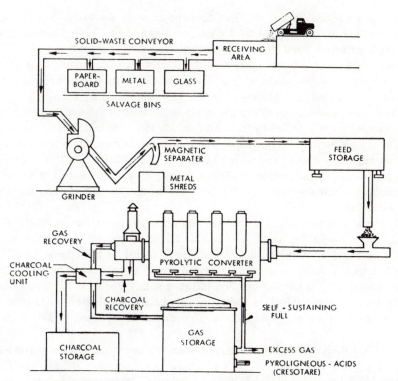

Fig. 12-1. Pyrolytic waste conversion. *Source:* Reference 12-5, Page 240.

gas is heated and the discharged char residue is cooled, the condensables will remain in the gaseous state. As the gas cools and the condensables leave the gas stream the gas heating value will decrease. Therefore, for maximum energy reclamation from the gas it is important that the gas be kept in a heated state as long as possible—at least long enough for the gas to reach the furthest gas burner. Storage should be minimized because the condensables will leave the gas stream relatively readily in any quiescent area.

The residual solid material is termed charcoal. This is, ideally, a desired byproduct of this reaction.

A stack is shown immediately downstream of the converter. Upon startup of the process, when an outside source of heat is required to initiate the reaction (not shown), the initial off-gas is basically composed of steam, carbon dioxide, entrapped air, and trace amounts of carbon monoxide. These components can be vented through the stack until the process stabilizes and pyrolysis gas is produced.

PUROX

The Union Carbide Corporation has developed a pyrolysis reactor for destruction of refuse or other mixed solid waste, as shown in Fig. 12-2. The reactor is a vertical refractory lined chamber with waste feed charged from its top.

Oxygen is injected in the bottom of the reactor, hence the name PUROX©, Pure Oxygen. Oxygen reacts with a portion of the waste to produce the heat required to sustain this process. The hot gases generated by burning rise up through the waste and pyrolyze the waste as it cools. In the upper portion of the reactor the gas is cooled further as it dries the incoming material.

The off gas exiting the furnace is relatively clean, normally at a temperature of approximately 200°F, and with a heating value of 300 BTU/SCF. The use of oxygen instead of air results in a higher heating value (because of the absence of nitrogen which would act as a dilutant) and the absence of nitrogen oxides in the gas stream.

The energy in the off-gas is normally equivalent to approximately 80% of the heat energy within the waste. The residue from this process is quenched in a water bath and is characteristically granular in nature. It is sterile and free of any biologically active material because it has gone through a hot molten state. Also, because of the burning reaction included in this process, it contains a relatively small carbon residual, as shown in Table 12-4. The volume of solid residue is 2–3% of the incoming refuse volume. This reactor has been built in 200 ton/day units.

The TORRAX© (Total Reduction) system for pyrolysis of solid waste has originally been developed by the Carborundum Corporation and is marketed as the ANDCO© and ANDCO-TORRAX© system. The reactor is shown in Fig. 12-3.

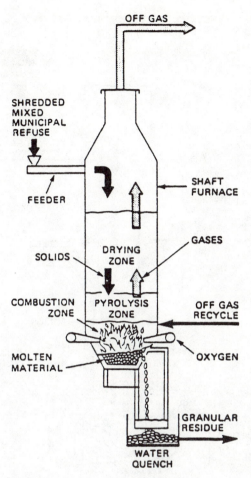

Fig. 12-2. Purox© Reactor. *Source:* Union Carbide Corp., Charleston, WV.

Table 12-4. Purox© Granular Residue.

Component	Weight Percent
FeO	9.0
Fe_2O_3	1.7
MnO	0.7
SiO_2	63.1
$CaCo_3$	1.6
CaO	13.7
Al_2O_3	9.2
TiO_2	0.1

Source: UNION CARBIDE Corp., South Charleston, VA

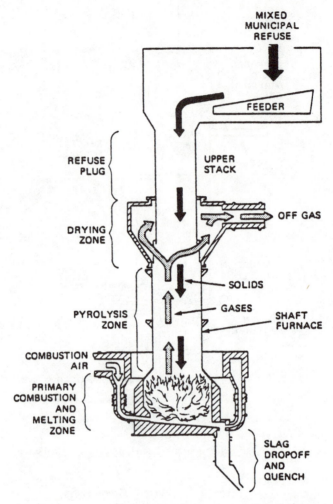

Fig. 12-3. TORRAX© Reactor. *Source:* ANDCO-TORRAX Corp.

Waste is not shredded or otherwise classified. It is charged into the top of the reactor, and by virtue of its weight and compactability forms a plug, preventing flow of gases through it.

Air is injected at the bottom of the reactor to promote sufficient burning of the waste to generate heat to sustain the process. Gases rise through the reactor, heating and pyrolyzing the charge, drying the fresh material, and exiting the reactor beneath the refuse plug.

Gas exits the reactor at temperatures in the range of 800–1000°F and has a heating value of approximately 120–150 BTU/SCF. Normally a gas cannot sustain combustion with a heating value less than 150 BTU/SCF. The low heating value of this gas, therefore, makes it in general uneconomical for export. Because of its high temperature and its low heating value it becomes economical to fire this off-gas on site, adjacent to the reactor. The gas can be used for generation of steam or hot water, although supplemental fuel may be required to sustain combustion. The off-gas can be used to provide combustion air for the reactor, and generate steam, increasing system efficiency.

The solid residual normally amounts to only 3-5% by volume, 15-20% by weight of the incoming feed. The residual is continuously removed (or tapped) from the reactor and is immediately quenched. It is granular in form and typically has a composition as listed in Table 12-5.

OTHER PYROLYSIS SYSTEMS

There are many studies in progress not far distant from those of the traditional alchemists. Instead of yellow gold the goal is black gold—petroleum and petroleum-derived products. Currently processes and systems are exemplary if they just dispose of waste, generating innocuous residual materials without creating nuisance odor or budget overruns. But new processes will undoubtedly follow the perfectly sound theory of oil from waste (one hydrocarbon from another),

Table 12-5. TORRAX© Solid Residue.

Component	Typical % by Weight	Range, %
SiO_2	45.0	32.0–58.0
Al_2O_3	10.0	5.5–11.0
TiO_2	0.8	0.5–1.3
Fe_2O_3	10.0	0.5–22.0
FeO	15.0	11.0–21.0
MgO	2.0	1.8–3.3
$CaCO_3$	1.1	0–1.5
CaO	8.0	4.8–12.1
MnO	0.6	0.2–1.0
Na_2O	6.0	4.0–8.6
K_2O	0.7	0.4–1.1
Cr_2O_3	0.5	0.1–1.7
CuO	0.2	0.1–0.3
ZnO	0.1	0–0.3

Particle density	174.7 lb/cubic foot
Residue density	87.4 lb/cubic foot
Screen size	$4\% > 3\frac{1}{2}$ mesh
	$2\% < 30$ mesh

Source: ANDCO-TORRAX Corp.

and perhaps a workable system will be found to generate the potential of one barrel of oil per ton of waste by pyrolysis. A related process, starved-air, or controlled air combustion has been successfully developed.

CONTROLLED AIR INCINERATION

In the early 1960's a new intrinsically simple type of incinerator started gaining in popularity as public attention focused on air pollution from waste burning. The Modular Combustion Unit (MCU) has become an economical and efficient system for on-site and central destruction of waste. These incinerators are also known as Starved Air and Controlled Air Units. Their operation is based on controlling furnace air injection to promote pyrolytic reactions.

THEORY OF OPERATION

The MCU consists of two major furnace components, as shown in Figure 12-4, a primary chamber and a secondary chamber. Waste is charged into the primary chamber and a carefully controlled flow of air is introduced. Only enough air is provided to allow sufficient burning for heating to pyrolysis to occur. Typically 70% to 80% of the stoichiometric air requirement is introduced into the primary chamber.

The off gas generated by the pyrolysis reaction will contain combustibles and this gas is burned in the secondary chamber, which is sized for sufficient residence time to totally destruct organics in the off gas. As in the primary chamber a carefully controlled quantity of air is introduced into the secondary chamber but in this case excess air, 140% to 200% of the off-gas stoichiometric requirement, is maintained to effect complete combustion. Normally gas cleaning de-

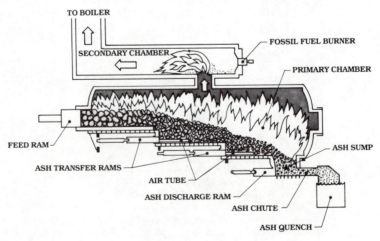

Fig. 12-4. Starved air incinerator. *Source:* Reference 12-10, Page 2-7.

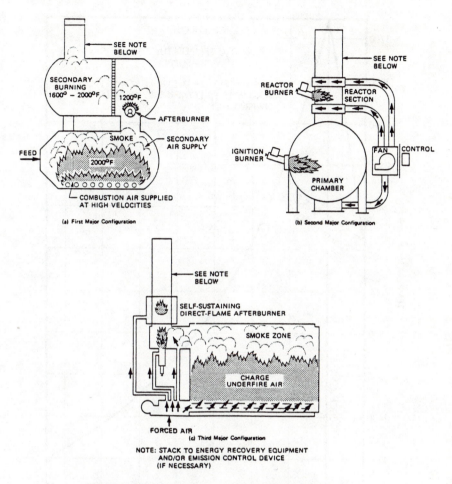

Fig. 12-5. Modular controlled-air incinerator configurations.

vices such as wet scrubbers or electrostatic precipitators are not required. The burnout of the off gas in the secondary chamber is usually sufficient to clean the gas to meet air emissions standards.

Figure 12-5 illustrates the variety of configurations currently marketed for controlled air incineration. They all have a starved-air primary section and a secondary or afterburner chamber.

CONTROL

As can be seen in Figure 12-6, the temperature is directly related to the excess air provided. Temperature therefore is normally utilized to control air flow in both primary and secondary chambers.

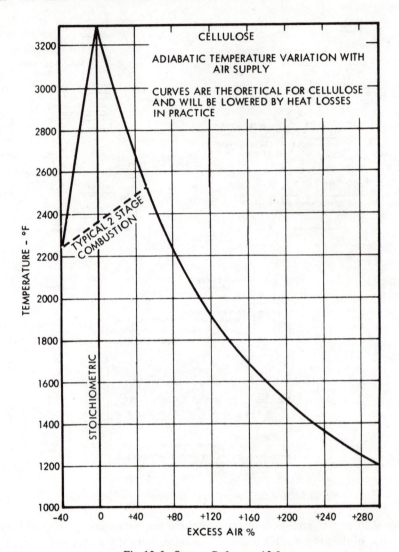

Fig. 12-6. *Source:* Reference 12-5.

Below stoichiometric the temperature of the reaction increases with an increase in air flow. As more air is provided more combustion will occur therefore more heat will be released. This heat release will result in higher temperatures produced.

Control of the primary chamber operation, therefore, where less than complete oxidation is required, is as follows:

- With higher temperatures, decrease air flow
- With lower temperatures, increase air flow

The secondary chamber is designed for complete combustion, greater than stoichiometric air supplied. At stoichiometric conditions all of the combustible material present will combust completely. Additional air will act to quench the off gas, i.e., will lower the resulting exhaust gas temperature. Therefore, control of the secondary chamber operation is as follows:

- With higher temperatures, increase air flow
- With lower temperatures, decrease air flow

MCU's are normally provided with temperature detectors which automatically control fan damper positioning to provide the required chamber air flow.

INCINERABLE WASTES

The MCU's were originally developed for the destruction of trash. They are applicable for other solid waste destruction and their secondary chamber can be used for destruction of gaseous or liquid waste in suspension. Work has been done in the incineration of sewage sludge with minicipal refuse in a MCU.

The nature of the MCU process is such that turbulence of the waste feed is minimal. Materials requiring turbulence for effective combustion such as powdered carbon or pulp wastes are not appropriate candidates for incineration in a MCU.

AIR EMISSIONS

Compared to other incineration methods, the air flow in the primary chamber, firing the waste, is low in quantity and is low in velocity. The low velocity and near absence of turbulence of the waste results in minimal amounts of particulate carried along in the gas stream. Complete burning is accomplished in the secondary chamber and the resulting exhaust gas is clean and practically free of particulate matter, i.e., smoke and soot. The MCU can usually comply with exhaust emissions standards without the use of supplemental gas cleaning equipment such as scrubbers or baghouses.

WASTE CHARGING

Smaller units, under 750 pounds per hour, are normally batch fed. Waste is charged over a period of hours and after a full load has been placed in the chamber the chamber is sealed and the waste is fired.

Figure 12-7 illustrates a typical hopper/ram assembly designed to minimize the quantity of air infiltration into the primary chamber when charging. Figure 12-8 illustrates a double ram charging system which allows a more continuous feed than the single ram. Note that the furnace charging door is not open until

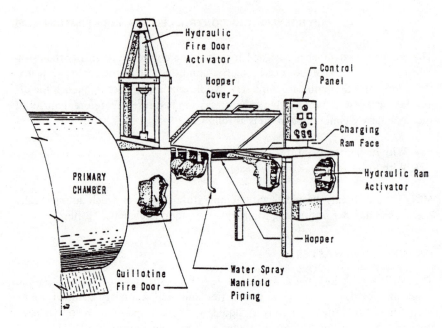

Fig. 12-7. Typical standard hopper/ram assembly.

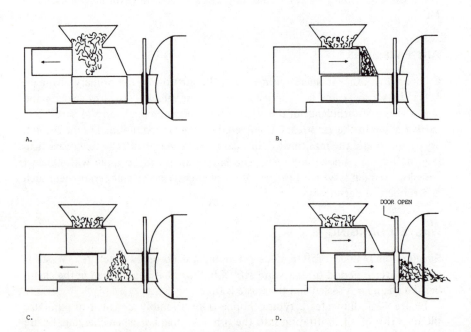

Fig. 12-8. Double-ram type charging system.

the hopper is sealed by the upper ram, preventing air infiltration from the hopper.

Larger units are usually provided with a continuous waste charging system, a screw feeder or a series of moving grates.

ASH DISPOSAL

As with waste charging, MCU's are provided with both manual and automatic discharge systems. With smaller units, after burnout the chamber is opened and ash residue is manually raked out. With continuous operating units such as that shown in Figure 12-4 ash is continually discharged, normally into a wet well, where it is transferred to a container or truck by means of a drag conveyor.

CAPACITY lbs./hr.	A	B	C	CC*	D	E	F	FF*	G	H	J	JJ*	CHUTE CAP. cu. yd.
100-160	8'-6"	4'-6"	5'-0"	9'-0"	6'-4"	3'-6"	9'0"	13'-0"	1'-4"	11'-0"	2'-0"	6'-0"	0.5
220-350	9'-6"	5'-6"	6'-0"	10'0"	7'-0"	4'-6"	11'-0"	15'-0"	1'-7½"	11'-0"	2'-0"	6'-0"	0.5
320-525	9'-6"	6'-0"	6'-6"	10'-6"	7'-0"	5'-6"	12'-6"	16'-6"	1'-9"	11'-0"	2'-6"	6'-6"	1.0
430-700	10'-0"	6'-6"	7'-0"	11'-0"	8'-0"	5'-6"	13'-0"	17'-0"	2'-2½"	11'-0"	2'-6"	6'-6"	1.0
640-1050	11'-6"	7'-6"	8'-0"	12'-0"	8'-0"	6'-0"	15'-0"	19'-0"	2'-6"	13'-0"	2'-6"	6'-6"	2.0
870-1400	12'-0"	8'-0"	8'-6"	12'-6"	9'-0"	6'-6"	16'-0"	20'-0"	2'-9½"	13'-0"	3'-6"	8'-0"	2.0
1300-2100	12'-6"	9'-6"	10'-0"	14'-0"	9'-6"	8'-6"	19'6"	23'-6"	3'-5"	13'-0"	3'-6"	8'-6"	3.0
1950-3200	14'-6"	10'-6"	11'-0"	15'-0"	10'-0"	9'-6"	21'-6"	25'-6"	3'-11"	13'-0"	4'-0"	9'-0"	3.0
2400-3900	16'-0"	11'-0"	11'-6"	15'-6"	12'-0"	9'-6"	22'-0"	26'-0"	4'-5"	14'-0"	5'-6"	9'-6"	4.0
2900-4700	18'-0"	11'-6"	12'-0"	16'-0"	14'-0"	9'-6"	22'-6"	26'-6"	4'-9"	14'-0"	5'-6"	9'-6"	4.0

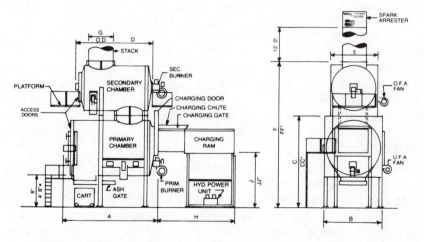

Fig. 12-9. Typical MCU system. *Source:* Morse-Boulger, Inc., Queens, NY.

Table 12-6.   MCU Energy Generation.

Burning Rate Pounds/Hour	Waste Feed Btu/Pound	Energy Generation Mode	Energy Generation Rate	Manufacturer
700	6000	Steam, 100 psig	2550 lbs/hr	Morse Boulger
1000	6000	Steam, 100 psig	3850 lbs/hr	Morse Boulger
1400	6000	Steam, 100 psig	5100 lbs/hr	Morse Boulger
3200	6000	Steam, 100 psig	11600 lbs/hr	Morse Boulger
4700	6000	Steam, 100 psig	17000 lbs/hr	Morse Boulger
1280	6285	Steam, 160 psig	2025 lbs/hr	George L. Simonds
1650	8500	Steam, 150 psig	3500 lbs/hr	George L. Simonds
1050	6240	Steam, 125 psig	1900 lbs/hr	George L. Simonds
650	6500	Steam, 150 psig	2500 lbs/hr	George L. Simonds
1800/15 gph	8500 Trash/ Waste Oil	Steam, 150 psig	8000 lbs/hr	George L. Simonds
1000	6500	Steam, 100 psig Hot Water, 105° Δt	3458 lbs/hr 66 gpm	Smokatrol
1500	6500	Steam, 100 psig Hot Water, 105° Δt	5187 lbs/hr 86 gpm	Smokatrol
2000	6500	Steam, 100 psig Hot Water, 105° Δt	6916 lbs/hr 132 gpm	Smokatrol
2500	6500	Steam, 100 psig Hot Water, 105° Δt	8645 lbs/hr 165 gpm	Smokatrol
1250	4500 7000	Steam, 150 psig	3200 lbs/hr 4950 lbs/hr	Consumat
2100	4500 7000	Steam, 150 psig	5400 lbs/hr 8400 lbs/hr	Consumat
6250	4500 7000	Steam, 150 psig	16100 lbs/hr 25000 lbs/hr	Consumat
8400	4500 7000	Steam, 150 psig	21600 lbs/hr 33600 lbs/hr	Consumat

Source: Selected Manufacturers' Data

ENERGY RECLAMATION

Waste heat utilization is a viable option provided with MCU's. The hot gas exiting the secondary chamber is relatively clean. Boiler or heat exchanger surfaces placed within this gas stream will therefore be subject to minimal particulate matter carryover and attendant problems of erosion and plugging.

TYPICAL SYSTEMS

Dimensional data of a typical MCU (Morse Boulger, Inc.) system is shown in Figure 12-9. Its internal configuration is similar to that of the unit in the upper left of Figure 12-5. Its charging system is similar to that of Figure 12-7.

Table 12-6 lists typical MCU systems which are provided with energy generation systems. These are basically "standard" models which are normally modified to the customers specific waste, heat recovery mode, or other needs.

Chapter 13
Sludge Incineration

Sludges are non-Newtonian liquids with relatively high solids content. The particle sizes of the solids in a sludge is very small as compared to those in what is defined as a slurry, which is basically a liquid containing solids of large particle size. The types of incineration equipment utilized for sludge disposal are presented in this chapter.

SLUDGE INCINERATION EQUIPMENT

Incineration provides ultimate volume reduction when used in combination with other treatment processes, particularly dewatering equipment (vacuum filters, belt filters, centrifuges, or filter presses). Until recently the cost of incineration has been prohibitive at many installations because of other, more economical disposal options: readily available land disposal, sludge lagoons, and ocean disposal. Increasing environmental constraints and decreasing land availability have increased the viability of sludge incineration.

Sludge parameters that have the most influence over incineration are moisture content, percent volatiles and inerts (or non-combustibles), and calorific value. Moisture content is important because of its thermal load on the incinerator. Volatiles and inerts, which affect the net heating value of sludge, can be controlled to some extent by treatment processes such as degritting, mechanical dewatering, and sludge digestion. The majority of combustibles in biological or organic sludges are present as volatiles, in the form of grease or light hydrocarbons, with the balance of the combustibles appearing as fixed carbon. Volatile percentage and moisture content can vary a great deal, so that equipment must normally be designed to handle a wide range of values. The types of incinerators commonly in use burning sludge wastes include:

- Multiple hearth.
- Fluid bed.
- Electric furnace.
- Cyclone furnace.

- Rotary kiln.
- Watergrate furnace.

THE MULTIPLE HEARTH INCINERATOR

The multiple hearth incinerator (also known as the Herreshoff furnace) is the most prevalent incinerator for the disposal of sewage sludge in this country. It was developed specifically for sludge burning. It has been adapted to carbon regeneration and recalcining and has numerous industrial applications.

Sludge cake is introduced at the top of the furnace (see Fig. 13-1). The furnace interior is composed of a series of circular refractory hearths, one above the other. The hearths are self-supporting off the refractory lined cylindrical

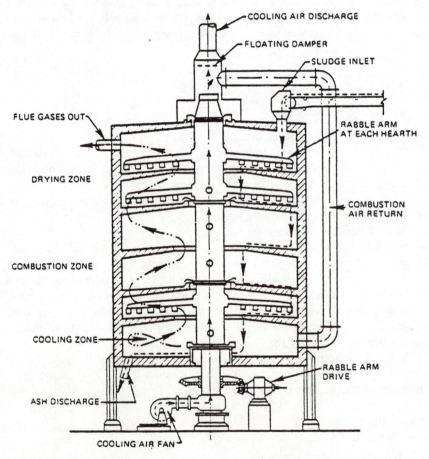

Fig. 13-1. Cross section of a typical multiple hearth incinerator. *Source:* Ref. 13-3.

wall of the furnace. They are numbered No. 1 as the top hearth, No. 2 as the next to top, etc. There are from five to nine hearths in a typical furnace.

A vertical center shaft is positioned in the center of the furnace. Rabble arms are attached to the center shaft above each hearth. The center shaft, along with the rabble arms, rotates relatively slowly, at approximately one rpm. Teeth on the rabble arms are positioned so as to move sludge through the furnace.

Every other hearth has a large annular opening between the hearth and the center shaft. These are called *in hearths*. The teeth on the rabble arms of these hearths will *rabble* sludge to the center of the hearth, where the sludge will fall off the edge of the refractory landing on the hearth below, an *out hearth*.

There are a series of openings on the outside or periphery of the out hearth. The inside of an out hearth is fairly close to the center shaft. A *lute ring cover* is a collar located above the hearth, attached to and moving with the center shaft, preventing sludge from dropping adjacent to the center shaft. Teeth on the rabble arms on the out hearths move sludge to the drop holes on the outside of the hearth, where the sludge drops to the hearth below, an in hearth. This process repeats until sludge, or ash, reaches the bottom hearth, the floor of the furnace, where it discharges from the furnace.

Teeth on each hearth agitate the sludge, exposing new surfaces of the sludge to the gas flow within the furnace. As sludge falls from one hearth to another, it again has new surfaces exposed to the hot gas. The upper hearths of the furnace comprise a drying zone where the filter cake gives up moisture while cooling the hot flue gases. Flue gas exits the top hearth of the furnace at 800-1000°F. The center hearths are the burning zone where temperatures can reach 1700-1800°F. Burnout of sludge to ash is accomplished in the lower hearths of the furnace.

The center shaft and rabble arms are hollow. Air is passed through the center shaft, which is constructed to distribute the air to each of the rabble arms, and discharge it through the top of the shaft. This air is utilized for shaft cooling, and after passing through the center shaft it reaches temperatures of 200-450°F. Often this heated air is recycled into the furnace as preheated combustion air.

Excess air of 100-125% must be provided to insure adequate burnout of sludge. Some 20% of the ash content of the sludge is airborne and gas cleaning equipment must be provided for its capture. Occasional odor problems will exist (see subsequent chapters for discussion of air/odor emissions), which may require installation of afterburning equipment.

FLUID BED INCINERATION

The fluid bed furnace was developed for catalyst recovery in oil refining by Standard Oil early in this century. The first fluid bed used for incineration of sewage sludge was installed in 1962. Its use is gaining in popularity for sludges and liquids in the United States.

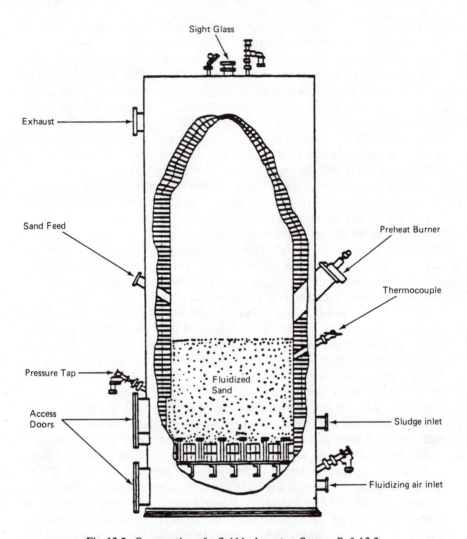

Fig. 13-2. Cross section of a fluid bed reactor. *Source:* Ref. 13-3.

As illustrated in Fig. 13-2, the fluid bed furnace is a cylindrical refractory lined shell with structure on its bottom surface to support a sand bed.

Air is introduced at the fluidizing air inlet at pressures on the order of 3.5–5 psig. The air passes through openings (tuyeres) in the bed plate supporting the sand and creates fluidization of the sand bed.

Air can be introduced cold or, as is usually the case, preheated by the exiting flue gas (see Fig. 13-3). The sand bed is maintained at approximately 1500°F.

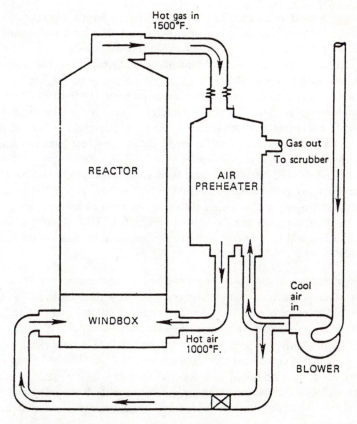

Fig. 13-3. Fluidized bed system with air preheater. *Source:* Ref. 13-3.

It expands 30–60% in volume when fluidized as compared to the unfluidized condition.

Sludge cake or other waste is normally introduced within the fluid bed. The fluidizing air flow must be carefully controlled to prevent sludge from floating on top of the bed. Fluidization provides maximum contact of air with sludge surface for optimum burning. The drying process is practically instantaneous. Moisture flashes into steam upon entering the hot bed. One design of the fluid bed furnace injects waste material from the top center of the furnace. With relatively wet waste, much of the water will evaporate within the freeboard and the waste will fall onto and disappear within the fluid bed at a much lower moisture content.

The furnace itself is an extremely simple piece of equipment with no moving parts. The large amount of sand within the furnace is an effective heat sink. The furnace can be shut down with minimal heat loss. It is a relatively tight system and the sand will retain heat to allow startup after a weekend shutdown with need for only one or two hours of heating. The sand bed should be at least 1200°F before most types of organic sludges are introduced.

Ash and some sand become airborne and exit the furnace in the flue gas. The gas cleaning system has to be sized for this relatively large particulate loading.

Because of the intimate mixing of air and sludge in the fluid sand bed, excess air requirements are low, 20–40%. The large volume within the furnace above the sand bed is normally maintained at 1200–1500°F. Residence time of the flue gases at these temperatures is sufficient to obtain complete burn out and elimination of odors for most organic wastes.

Sludge is force-fed into the furnace with either positive displacement pumps or screw/plunger type feeders. Sludge feeding has had its problems because of the tendency of the sludge within the feeder to dry and harden during periods when the furnace is maintained hot without sludge feed. (Hot standby conditions.)

A fluid bed will have an agglomorative property depending on the nature of the waste burned:

Agglomorative. A condition where the bed volume increases with time. Ash and/or other products of combustion are retained within the bed. The agglomorative bed must be tapped on a regular basis to discharge bed material to prevent the buildup of excessive volume of bed material.

Non-agglomorative. A bed which loses material with time. Sand, ash, and other products of combustion are airborne and exit the fluid bed furnace in the flue gas. Sand must be added to a non-agglomorative bed to maintain bed level. Sewage sludge, for instance, is non-agglomorative when incinerated and requires the addition of approximately 5% of bed sand volume every 300 hours of operation.

Fuel is used for startup, reheat, and, depending on the properties of the sludge, for incineration. It can be injected within the bed or sprayed on top of the bed. The different manufacturers each utilize their own method of introduction of supplementary fuel.

The fluid bed furnace normally has only one major item of air moving equipment, the forced draft fan (or fluidizing air blower). The fan is sized to blow the flue gas through the gas scrubbing systems. This necessitates that the reactor be pressurized and that it be tight to prevent spurious leakage of hot gas.

THE ELECTRIC FURNACE

The electric or radiant heat (or infrared) furnace is basically a conveyor belt system passing through a long rectangular refractory-lined chamber, as shown in Fig. 13-4.

Combustion air is introduced at the discharge end of the belt, shown as the

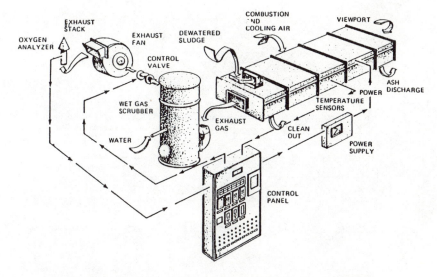

Fig. 13-4. Radiant heat furnace. *Source:* SHIRCO, Inc., Dallas, TX.

viewport. It is often heated with an external preheater recuperating exhaust heat. However, the air will pick up heat from the hot burned sludge as sludge and air travel in countercurrent to each other.

Supplemental heat is provided by electric infrared heating elements within the furnace. Cooling air prevents local hot spots in the immediate vicinity of the heaters and is used as secondary combustion air within the furnace.

The conveyor belt is continuous woven wire mesh made of steel alloy chosen to withstand the 1300–1500°F encountered within the furnace. The refractory is not brick but ceramic felt, as shown in Fig. 13-5. It does not have a high capacity for holding heat and can therefore be started up from cold condition relatively quickly.

Sludge fed onto the belt is immediately leveled to a depth of approximately one inch. There is no other sludge handling mechanism. The belt speed and travel is sized to provide burnout of the sludge without agitation. This feature results in a very low level of particulate emissions.

Usually a low-energy gas scrubber, such as a cyclonic scrubber, is all that is required to clean the flue gas. Excess air requirements are 20–30%.

Supplemental fuel, in this case electric, is required for startup and, with wet sludges, to maintain combustion. Unfortunately, the power needed for startup results in a large connected load. In areas of the country where there are high demand charges for electric power this system is economically impractical. It is competitive where demand, charges are low and where sludge burning is autogenous, requiring no supplemental power.

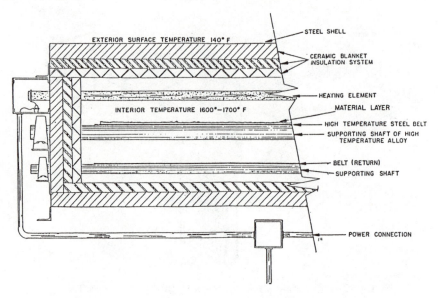

Fig. 13-5. Radiant heat furnace, cross section. *Source:* SHIRCO, Inc., Dallas, TX.

CYCLONIC FURNACE

The cyclonic furnace is a single hearth unit where the hearth moves and the rabble teeth are stationary (see Fig. 13-6). Sludge is rabbled towards the center of the hearth where, as ash, it is discharged.

Air is introduced at tangential burner ports on the shell of the furnace. The furnace is a refractory-lined cylindrical shell with a domed top. The air, heated with the immediate introduction of supplemental fuel, creates a violent swirling pattern which provides good mixing of air and sludge feed. The air, later flue gas, swirls up vertically in cyclonic flow through the discharge flue in the center of the domed roof.

Sludge is fed into the furnace with a screw feeder and is deposited on the periphery of the rotating hearth. A progressive cavity pump is normally used to feed sludge.

Temperatures within the furnace are 1500–1600°F. These furnaces are relatively small and can normally be placed in operation, at operating temperature, within an hour's time.

A variation of the cyclonic furnace is shown in Fig. 13-7. This is a horizontal cyclone furnace. Ash is discharged with the flue gas. Sludge is pumped into the furnace tangentially from the furnace wall. Air, as above, is introduced at tangential burner ports creating a cyclonic effect.

There is no hearth, only the furnace shell and refractory. The sludge deten-

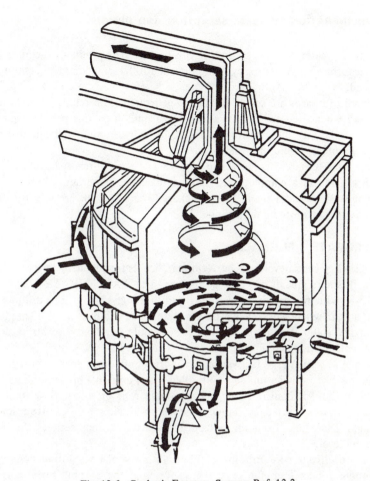

Fig. 13-6. Cyclonic Furnace. *Source:* Ref. 13-3.

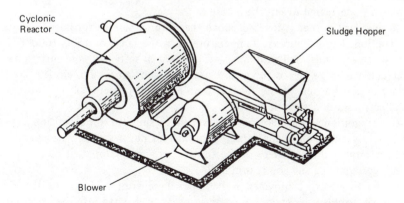

Fig. 13-7. Skid-mounted cyclonic reactor system. *Source:* Ref. 13-3.

tion time is greater than ten seconds in this furnace. The products of combustion exit the furnace in vortex or cyclonic flow at 1500°F and complete combustion is assured.

Cyclonic furnaces are suited for sludge generated from relatively small influent flows when used for sewage sludge, two million gallons per day and less. They are relatively inexpensive, mechanically simple units, well suited for on-site sludge and/or liquid disposal.

Horizontal furnaces, as shown in Fig. 13-7, can be purchased skid mounted for installation as a complete independent package, requiring only utility and feed connections and a stack. Many packaged units like this are sold for onsite sludge and other waste liquid disposal.

WATERGRATE© FURNACE

A furnace has been developed by Neptune/Nichols for the incineration of waste scums, floatable grease, oil, and skimmings. The WATERGRATE©, shown in Fig. 13-8, with typical characteristics listed in Table 13-1, utilizes a liquid surface as a grate where floatable wastes are maintained and burned. The base liquid is normally water. However, where a waste would not float on water a denser liquid, such as brine, would be used.

The WATERGRATE© is a vertical cylindrical refractory lined steel structure. Its lower structure is an unlined tank filled with liquid to a predetermined level. The liquid level is controlled by an adjustable weir or an overflow standpipe. Floatable wastes are fed by screw conveyor or by pumps to a bottom inlet to the tank. They float from the bottom inlet to the burning surface, the top of the liquid level.

A motor-driven rake rotating at approximately one revolution per minute continuously breaks up the crust which forms at the burning surface. It also rotates to expose new surfaces for burning. Normally the rake rotation can be increased or decreased to suit the nature of the waste.

An auxiliary burner is located above the burning surface for initial heat-up and ignition. Normal furnace temperature is in the range of 1500–1600°F. In general, greases, oils, and skimmings will collect in sufficient concentration to burn autogenously, i.e., without the need for supplemental fuel. The liquid surface at the bottom of the furnace is an excellent reflector of heat promoting effective waste destruction.

Air for ignition and primary combustion air are supplied to the ignition chamber (the lower portion of the furnace) through air nozzles which, arranged tangentially around the periphery of the furnace, provide cyclonic air motion promoting good mixing and good burning.

A refractory baffle is provided in the upper chamber of the WATERGRATE© to control turbulence. Secondary combustion air is injected into the upper portion of the furnace just beneath the baffle.

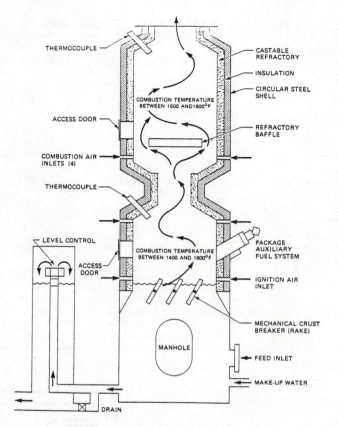

Fig. 13-8. The Nichols WATERGRATE©. *Source:* Nichols Engineering & Research Corp., Belle Mead, NJ.

An induced draft fan is included in the WATERGRATE© package to provide driving pressure for the system. A quencher and wet scrubber is normally used for cleaning the exhaust gas.

ROTARY KILN

The rotary kiln system is the most nearly universal of waste disposal systems. It can be used for a wide variety of solid and sludge waste disposal, and for the incineration of liquid and gaseous waste.

The heart of the system, the rotary kiln, is a horizontal cylinder, lined with refractory, which turns about its horizontal axis. Its angle to the horizontal, or rake, is normally less than 2–3%. The kiln, as shown in Fig. 13-9, rotates, continually exposing the waste material surfaces to the heat and oxygen in the gas

Table 13-1. WATERGRATE© Characteristics.

Capacity, grease, lb/hr moisture-free	100	400	900	1600	2500
Rake drive, HP	$\frac{1}{4}$	$\frac{1}{4}$	$\frac{1}{4}$	$\frac{1}{4}$	$\frac{3}{4}$
Burners					
Number	1	1	2	2	2
Size, BTU/hr, each	800,000	800,000	800,000	1,000,000	1,500,000
Combustion air fan					
capacity, SCFM	900	2800	6200	10,700	18,000
Horsepower	5	15	25	40	75
Induced draft fan					
capacity, ACFM	1,600	5,700	12,700	21,900	36,000
Horsepower	$7\frac{1}{2}$	20	40	75	125
Water requirements					
Quencher, gpm	7	23	51	88	144
Scrubber, gpm	4	12	26	45	72
Make-Up, gpm	10	25	50	100	100
Dimensions					
Plan area	20' × 20'	20' × 20'	25' × 25'	35' × 30'	45' × 40'
Height	25'	30'	40'	45'	55'

Source: Nichols Engineering & Research Corp., Belle Meade, NJ.

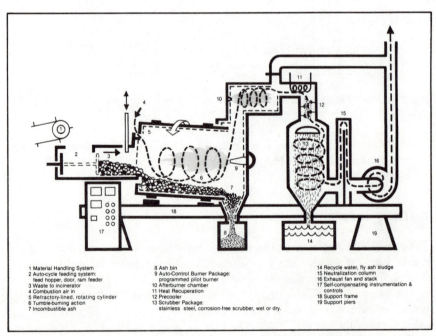

1 Material Handling System
2 Auto-cycle feeding system: feed hopper, door, ram feeder
3 Waste to incinerator
4 Combustion air in
5 Refractory-lined, rotating cylinder
6 Tumble-burning action
7 Incombustible ash

8 Ash bin
9 Auto-Control Burner Package: programmed pilot burner
10 Afterburner chamber
11 Heat Recuperation
12 Precooler
13 Scrubber Package: stainless steel, corrosion-free scrubber, wet or dry.

14 Recycle water, fly ash sludge
15 Neutralization column
16 Exhaust fan and stack
17 Self-compensating instrumentation & controls
18 Support frame
19 Support piers

Fig. 13-9. Rotary kiln. *Source:* C-E Raymond, Combustion Engineering, Inc., Chicago, IL.

flow. The speed of rotation is variable, normally running in the range of 0.25–1.5 rpm. The peripheral speed of the kiln (outer surface), normally is within the range of 1–5 feet per minute.

The retention time of material within the kiln is a function of the kiln speed, rake and physical parameters of the unit.

An approximation of residence time can be calculated from the following equation:

$$t = \frac{3.82L}{RDs}$$

where

L = kiln length, feet.
D = kiln inside diameter, feet.
R = rotation, revolutions per minute.
s = rake, inches per foot of length.
t = mean residence time, minutes.

Normally, the ratio of length to diameter for kilns used in waste disposal range from 2:1 to 10:1.

Example

Calculating residence time for a kiln rotating at 0.75 rpm (R) with a 1% slope (s = 0.12 inch per foot of length), with a 4 foot inside diameter (D) and 12 foot length (L):

$$t = \frac{(3.82)(12 \text{ ft})}{(0.75 \text{ rpm})(4 \text{ ft})(0.12 \text{ inch/ft})} = 127 \text{ minutes}$$

By inspection note that a doubling of rotation (R) would halve the retention time, and halving the rake (s) would double the retention time.

The rotary kiln is supported, as pictured, on two points (or up to four points in larger units), by two trunnions. The rake is adjusted by raising or lowering one trunnion support with respect to the other and normally, once a rake is set it is not altered. The change in retention time is, therefore, controlled by varying the rotational speed of the kiln.

In the system pictured in Fig. 13-9 material is ram fed into the kiln. As the kiln rotates the waste burns to ash, which is shown discharged to an ash bin. Burners for startup, and to provide a supplementary source of heat, can be mounted on either end of the kiln although in this illustration it is mounted near the kiln exit.

Liquids which will burn in suspension, as well as gases, can be injected into the kiln from the front or rear face, although these nozzles are normally placed at the kiln entrance. Liquid and gaseous waste can also be injected into an afterburner for destruction.

Sludges, slurries and solids are dropped on the kiln hearth. For some materials the rake is set at zero or at a negative value to increase retention time. Lips are designed for these kilns in a manner that will allow material flow from the hearth to the ash discharge system with a negative rake. The forward lip would be higher than the rear lip. When the waste/ash reaches the height of the rear lip it will fall over the lip, discharging from the kiln.

Many kilns, particularly of European manufacture, utilize a molten slag to aid in emissions control. Lips are placed appropriately to contain a liquid slag on the kiln hearth. The kiln is heated to a temperature high enough to melt waste residual, and additives are sometimes added to the waste to lower the temperature at which it becomes fluid. A liquid ash will capture much of the particulate generated within the kiln, reducing the particulate emissions.

The kiln rotation and the attendant turbulence in the waste will produce a significant particulate load. The molten slag feature can reduce this loading however, as shown in Fig. 13-9, air pollution control systems are normally provided.

The kiln interior can be smooth, but when increased turbulence is desired baffles, longitudinal flights, circumferential ridges (or dams), or a combination of these features can be provided.

One problem with the rotary kiln is the difficulty of sealing its ends. As rotating equipment with a pair of stationary shrouds on its ends, the stationary/rotating interface is prone to leakage. Fig. 13-10 illustrates kiln seals for two

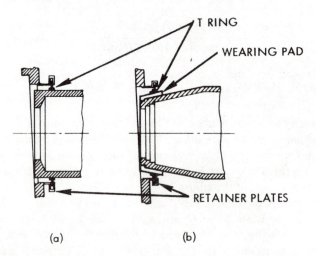

(a)   (b)

Fig. 13-10. Kiln seal arrangements. (a) Single-floating-type-feed-end air seal. (b) Single-floating-type air seal on air-cooled tapered feed end. *Source:* Ref. 13-13.

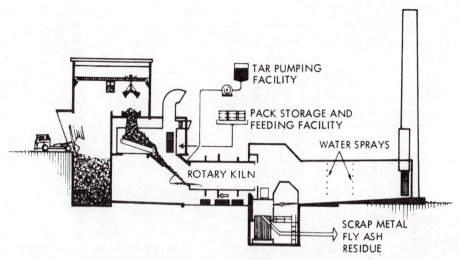

Fig. 13-11. Typical major industrial rotary kiln incineration facility. *Source:* Ref. 13-12.

different types of kilns. The T-ring is made of a flexible, heat-resistant material and normally wears, needing planned routine replacement.

Figure 13-11 illustrates a typical industrial rotary kiln facility. Note that the kiln is used for destruction of solid waste (charged by the open bucket shown), waste packs or drums (dropped into the kiln vertically above the kiln entrance), and tar waste (a sludge pumped to the kiln entrance).

As pictured, sprays inject water into the exiting flue gas in an effort to reduce the particulate emissions level.

A rotary kiln will release 15,000–40,000 BTU per cubic foot. To illustrate sizing of a kiln which is to burn 2000 lbs per hour of a sludge cake with an as-received heating value of 2200 BTU per lb, note the following calculations:

Heat release: $\qquad$ 2000 lb $\times$ 2200 BTU/lb = 4.4 MBH from waste

Heat release (assumed): $\qquad$ 25000 BTU/hr-ft³ in kiln

Required volume: $\qquad$
$$V = \frac{4.4 \times 10^6 \text{ BTU/hr}}{25000 \text{ BTU/hr-ft}^3}$$
$$= 176 \text{ ft}^3$$

For a kiln whose length (L) is approximately 3 times its diameter (D),

$$L = 3D$$
$$V = \frac{\pi D^2}{4} \times L = \frac{\pi D^2}{4} \times 3D = \frac{3\pi D^3}{4}$$
$$D = \left(\frac{4V}{3\pi}\right)^{1/3} = \left(\frac{4 \times 176 \text{ ft}^3}{3\pi}\right)^{1/3} = 4.21 \text{ ft}$$

Take D = 4 ft:

$$L = 3D = 3 \times 4 \text{ ft} = 12 \text{ ft}$$

Therefore the kiln size required is 4 ft inside diameter by 12 feet long. The actual heat release is:

Actual volume: $\dfrac{\pi}{4}(4)^2 \times 12 = 150.796 \text{ ft}^3$

Release: $\dfrac{4.4 \times 10^6 \text{ BTU/hr}}{150.796 \text{ ft}^3} = 29,178 \text{ BTU/hr-ft}^3$

SLUDGE INCINERATOR CALCULATIONS

Table 13-2, the mass flow sheet; Table 13-3, the heat balance sheet; and Table 13-4, the flue gas discharge sheet, represent the incineration of 12000 lb/hr of sludge in an electric furnace. The wet feed has a moisture content of 74%, an ash content of 43% and has the following constituents:

Carbon	64.3%
Hydrogen	8.2%
Sulfur	2.2%
Oxygen	21.0%
Nitrogen	4.3%

From Chapter 8, using Du Long's approximation, the heating value is calculated as follows:

$$Q = 14544(C) + 62028(H_2 - 0.125O_2) + 4050(S)$$

$$= 14544(0.643) + 62068(0.082 - 0.125 \times 0.21) + 4050(0.022)$$

$$= 12,898 \text{ BTU/lb}$$

Products of combustion are calculated as follows:

From the methods of Chapter 4, to calculate combustion parameters:

$$28.02$$

$$C \ + \ O_2 \ + 3.7619\,N_2 \longrightarrow \ CO_2 \ + 3.7619\,N_2$$

$$12.01 \quad 32.00 \qquad\qquad\quad 44.01 \quad 105.408$$

Table 13-2. Mass Flow.

	Example (Elect.)
Wet Feed, lb/hr	12,000
Moisture, %	74
lb/hr	8,880
Dry Feed, lb/hr	3,120
Ash, %	43
lb/hr	1,342
Volatile, lb/hr	1,778
Vol. Htg. Value, BTU/lb	12,898
MBH	22.93
Dry Gas, lb/10 kB	7.494
lb/hr	17,184
Comb. H_2O, lb/10 kB	.568
lb/hr	1302
DG + Comb. H_2O, lb/hr	18,486
100% Air, lb/hr	16,708
Total Air Fract.	1.2
Total Air, lb/hr	20,050
Excess Air, lb/hr	3,342
Humid/Dry Gas (Air), lb/lb	.01
Humidity, lb/hr	201
Total H_2O, lb/hr	10,383
Total Dry Gas, lb/hr	20,526

For .643 lb C,

$$\frac{44.01}{12.01} \times 0.643 = 2.356 \text{ lb } CO_2$$

$$\frac{105.408}{12.01} \times 0.643 = 5.643 \text{ lb } N_2$$

$$
\begin{array}{ccccc}
 & & & & 28.02 \\
S & + & O_2 & + 3.7619\,N_2 \longrightarrow & SO_2 & + 3.7619\,N_2 \\
32.06 & & 32.00 & & 64.06 & 105.408
\end{array}
$$

For .022 lb S,

$$\frac{64.06}{32.06} \times 0.022 = 0.044 \text{ lb } SO_2$$

$$\frac{105.408}{32.06} \times 0.022 = 0.072 \text{ lb } N_2$$

$$2.02 \qquad\qquad\qquad\qquad 28.02$$

$$2H_2 = O_2 + 3.7619\,N_2 \longrightarrow 2H_2O + 3.7619\,N_2$$

$$4.04 \quad 32.00 \qquad\qquad 36.04 \quad 105.408$$

For .082 lb H_2,

$$\frac{36.04}{4.04} \times 0.082 = 0.732 \text{ lb } H_2O$$

$$\frac{105.408}{4.04} \times 0.082 = 2.139 \text{ lb } N_2$$

Table 13-3. Heat Balance.

	Example (Elect.)
Cooling Air Wasted, lb/hr	—
°F	—
BTU/lb	—
MBH	—
Ash, lb/hr	1,342
BTU/lb	130
MBH	0.17
Radiation, %	3
MBH	0.69
Humidity, lb/hr	201
Correction (@970 BTU/lb), MBH	−0.19
Losses, Total, MBH	0.67
Input, MBH	22.93
Outlet, MBH	22.26
Dry Gas, lb/hr	20,526
H_2O, lb/hr	10,383
Temperature, °F	1,166
Desired Temp., °F	1,200
MBH	22.63
Net, MBH	0.67
Fuel Oil, Air fraction	—
net BTU/gal	—
gph	(196)
Air, lb/gal	—
lb/hr	—
Dry Gas, lb/gal	—
lb/hr	—
H_2O, lb/gal	—
lb/hr	—
DG w/FO, lb/hr	20,526
H_2O w/FO, lb/hr	10,383
Air w/FO, lb/hr	20,050
Outlet, MBH	22.63
Reference t, °F	60

Table 13-4.   Flue Gas Discharge.

	Example (Elect.)
Inlet, °F	1,200
Dry Gas, lb/hr	20,526
Heat, MBH	22.63
BTU/lb dry gas	1,103
Adiabatic t, °F	187
H_2O Saturation, lb/lb dry gas	0.9271
lb/hr	19,030
H_2O Inlet, lb/hr	10,383
Quench H_2O, lb/hr	8,647
gpm	17.3
Outlet Temp., °F	120
Raw H_2O Temp., °F	60
Sump Temp., °F	156
Temp. Diff., °F	96
Outlet, BTU/lb dry gas	111.65
MBH	2.29
Req'd Cooling, MBH	20.34
H_2O, lb/hr	211,875
gpm	424
Outlet, ft^3/lb dry gas	16.515
ft^3/min	5,650
Fan Press, in. WC	28
Outlet, ACFM	6,039
Outlet, H_2O/lb dry gas	0.08128
H_2O, lb/hr	1,668
Recirc. (Ideal), gpm	17
Recirc. (Actual), gpm	138
Cooling H_2O, gpm	424

The total dry gas produced per pound of fuel (or sludge volatiles), including the nitrogen in the fuel, is equal to:

$$N_2 = (0.043 + 5.643 + 0.072 + 2.139)$$
$$= 7.897 \text{ lb}$$
$$CO_2 = 2.356 \text{ lb}$$
$$\underline{SO_2 = 0.044 \text{ lb}}$$

$$\text{Total} = 10.297 \, \frac{\text{lb dry gas}}{\text{lb fuel}} \times \frac{\text{lb fuel}}{12{,}898 \text{ BTU}}$$

$$= 7.494 \text{ lb dry gas}/10 \text{ kB}$$

The moisture produced is equal to:

$$H_2O = 0.732 \frac{lb\ H_2O}{lb\ fuel} \times \frac{lb\ fuel}{12,898\ BTU}$$

$$= 0.568\ lb\ H_2O/10\ kB$$

An electric incinerator can require only 20% excess air, which is used in this illustration. There is no cooling air with this type of furnace and the radiation loss was assumed to be equal to 3% of the sludge cake heat input.

Completing Table 13-2 per the methods of Chapter 8, insertions can be made in Table 13-3. For a desired outlet temperature of 1200°F the equivalent of 0.67 MBH of supplemental fuel is required. Converting to electrical energy:

$$0.67 \times 10^6\ (BTU/hr) \times (kW\text{-}hr/3412\ BTU) = 196\ kW$$

Therefore 196 kW of electric energy is required to heat the sludge cake and its products of combustion to 1200°F. Electric energy requires no additional air, unlike fossil fuel; therefore the products of combustion are the same as those prior to the introduction of supplemental heat.

The remainder of Tables 13-3 and 13-4 were completed in accordance with the methods of Chapter 8.

Chapter 14
Liquid Waste Destruction

Increasingly, public attention is focusing on liquid wastes as the most objectionable of waste streams. The fear of buried drums leaking liquid waste or liquid waste seeping into surface waters has resulted in statutes limiting the disposal of many of these waste streams. The existing technology for incineration of liquid waste is presented in this chapter.

LIQUID WASTE PROPERTIES

As discussed in a previous chapter, the line between liquid and non-liquid is not always well defined. A material is considered a liquid, for purposes of incinerator design, if it can be pumped to a burner and atomized, i.e., fired in suspension. In general a material can be pumped if its viscosity is less than 10,000 SSU. Atomization is a function of nozzle type. With the appropriate nozzle design liquids with up to 5000 SSU viscosity can be fired in suspension.

Beside viscosity, other factors are important in liquid incinerator selection and design:

- *Heating value.* Can the liquid sustain combustion or is auxiliary fuel required?
- *Aqueous content.* A liquid composed of over 60% water is considered an aqueous waste.
- *Halogen fraction.* A liquid with a chloride, bromide, or fluoride component requires careful attention to material selection and gas cleaning system design.
- *Metallic salts.* Firing of wastes with metallic salt components often produces salt residue in the furnace and exhaust gas train and may cause severe corrosion of refractory.
- *Sulfonated waste.* Sulfur in a waste will produce acidic corrosion and material selection and exhaust gas cleaning must be carefully controlled.
- *Organic waste.* This contains carbon and hydrogen, and can also contain oxygen, nitrogen, sulfur, and halogens.
- *Cyclic and polycyclic organics.* These are organic compounds character-

ized by the presence of benzene-type rings. They are difficult to effectively incinerate because of the thermal stability of this ring structure.

FURNACE INJECTION

Liquid waste can be the prime fuel source, injected into a cold furnace through standard nozzles. Its heating value would have to be sufficiently high to maintain combustion at desired furnace temperatures.

With aqueous waste or other low-heating-value waste, injection is normally outside the flame envelope. The prime fuel, that fuel firing to establish and maintain furnace temperatures, is allowed to burn completely and release its complete heating value. In-flame injection of a low-BTU waste would tend to cool the flame and thus interfere with efficient fuel burning.

Liquid waste can be injected within a flame front if its heating value is high enough to add and not remove a net heating value to the flame, normally a minimum of 5000 BTU/lb.

In the injection of fuel or waste within a furnace, care must be taken to avoid flame impingement on furnace walls. Flame impingement indicates that excessively high temperatures are reaching refractory surfaces. This will tend to reduce their life. In addition, unburned residue will collect on impingement areas. For every pound of carbon left as an unburnt residual on a furnace wall over 14,000 BTU is lost to furnace heat release. Also, the presence of a residual coating will promote refractory corrosion.

LIQUID INJECTION NOZZLES

Liquid fuels must be vaporized before combustion can occur. The degree of atomization and fuel–air mixing is directly related to burning efficiency. A number of different types of nozzles or burners have been developed for efficient burning of the varying types of fuels/liquid waste streams generated today.

Mechanical Atomizing Nozzles

These are the most common types of burner nozzles in current use. Typical mechanical atomizing nozzles are illustrated in Figs. 14-1 and 14-2. Fuel is pumped into the nozzle at pressures of 75–150 psig through a small fixed-orifice discharge. The fuel is given a strong cyclonic or whirling velocity before it is released through the orifice. Combustion air is provided around the periphery of the conical spray of fuel produced. The combination of combustion air introduced tangentially into the burner and the action of the swirling fuel produce effective atomization. Normal turndown ratios are in the range of 2.5:1 to 3.5:1. By utilizing a return flow line for fuel oil the turndown ratio can be

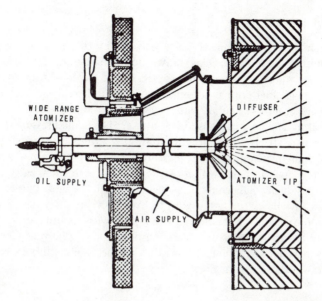

Fig. 14-1. Mechanical atomizing nozzle. *Source:* Combustion Engineering-Superheater, Inc., New York, NY.

increased to as high as 10 : 1. Typical burner capacities are in the range of 10–100 gallons per hour. A major disadvantage of this type of burner/nozzle is its susceptibility to erosion and pluggage from solids components of the fluid stream. Flames tend to be short, bushy, or low velocity, and this results in slower combustion, requiring relatively large combustion chamber volumes. This burner is applicable for fluids with relatively low viscosity, under 100 SSU.

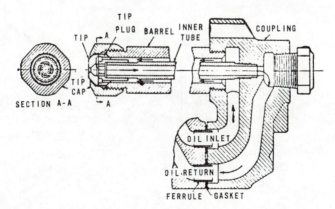

Fig. 14-2. A wide-range mechanical-atomizing assembly with central oil return line. *Source:* Combustion Engineering-Superheater, Inc., New York, NY.

Rotary Cup Burners

As shown in Fig. 14-3, atomization is provided by throwing fuel centrifugally from a rotating cup or plate. Oil is thrown from the lip of the cup in the form of conical sheets which breaks up into droplets by the effect of surface tension. No air is mixed with fuel prior to atomization. Instead, it is introduced through an annular space around the rotary cup. Normally a common motor drives the oil pump, rotating cup, and combustion air blower. The liquid pressure required for this burner is relatively low, since atomization is a function of cup rotation and combustion air injection, not fuel pressure. This low pressure requirement and the relatively large openings within the burner fuel path allow passage of fluids with relatively high solids content, as high as 20% by weight. Burner capacities range from low flows (under 10 gph) to over 250 gph. They have a turndown ratio of approximately 5:1 and can fire liquids with viscosity up to 300 SSU. Rotary cup burners are sensitive to combustion air flow adjustment. Insufficient air flow will result in fuel impingement on furnace walls, while excessive combustion air will cause a flame-out.

External Low-Pressure Air-Atomizing Burners

The major portion of the combustion air requirement is provided at 1-5 psig near the burner tip. Air is injected externally to the fuel nozzle and is directed to the liquid stream to produce high turbulence and effective atomization. The liquid pressure necessary for operation is only enough for positive delivery, normally less than $1\frac{1}{2}$ psig. The quantity of atomizing air required decreases with increased atomization pressure and may range from 400 to 1000 SCF per

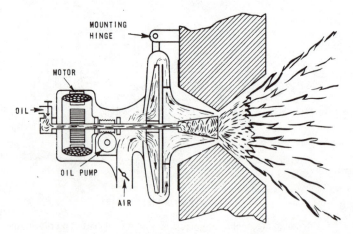

Fig. 14-3. Rotary cup oil burner. *Source:* Preferred Utilities, Inc., Danbury, CT.

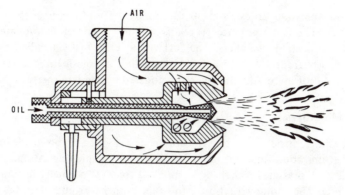

Fig. 14-4. Low-pressure, air-atomizing oil burner. *Source:* Hauck Manufacturing Company, Lebanon, PA.

gallon of fuel. Secondary combustion air is provided around the periphery of the atomized liquid mixture. The flame is relatively short because of the high amount of air provided at the burner (atomization and secondary combustion air). The short flame allows design of smaller combustion chambers. These burners normally operate with liquids in the range of 200-1500 SSU and can handle solids concentrations in the liquid of up to 30%. Fig. 14-4 illustrates a low-pressure air-atomizing burner. A small quantity of the air flow passes around the fuel discharge to aid in optimization of the fuel flow pattern.

External High-Pressure Two-Fluid Burners

The atomizing fluid, air or steam (or nitrogen or other gas), impinges the fuel stream at high velocity to generate small particles that encourage quick vaporization and effective atomization of tars and other heavy liquids. A typical burner is shown in Fig. 14-5. The steam requirement is 2-5 pounds per gallon of fuel, whereas the air requirement is 50-200 SCF per gallon of fuel. The required

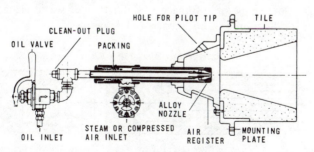

Fig. 14-5. High-pressure, steam- or air-atomizing oil burner. *Source:* North American Manufacturing Company, Cleveland, OH.

atomization pressure varies from 30 to 150 psig. Turndown is in the range of 3 : 1 to 4 : 1. The flame produced is relatively long, requiring appropriately constructed combustion chambers. The fuel viscosity normally handled by these burners ranges from 150 to 5000 SSU for either air or steam atomization. A solids content of up to 70% can be accommodated by these burners.

Internal Mix Nozzles

Air or steam is introduced within the nozzle, as in Fig. 14-6, to provide impingement of atomization fluid on the fuel stream prior to spraying. Atomization air is provided at pressures less than 30 psig and steam is normally introduced at 90–150 psig. The turndown ratio for this type burner is from 3 : 1 to 4 : 1. These nozzles cannot tolerate a significant solids content and can handle only low-viscosity fuels, under 100 SSU. This burner is used for clean, low-viscosity liquids. Its advantage is in its low cost compared to other burners.

Sonic Nozzles

These nozzles utilize a compressed gas such as air or steam to create high frequency sound waves which are directed at the fuel stream. This acoustic energy is transferred to the liquid stream and creates an atomizing force, breaking the stream into minute particles. The fuel nozzle diameter is relatively large, allowing passage of solid particulate streams such as slurries and sludges with high particulate content. Little fuel pressurization is required. The spray pattern is not well defined, with finely atomized, uniformly distributed droplets traveling at low velocities. These nozzles are difficult to adjust, have low turndown, and generate an extremely high noise level during operation. A typical sonic nozzle is shown in Fig. 14-7.

LIQUID DESTRUCTION FURNACES

The type, size, and shape of a furnace is a function of waste characteristics, burner design, air distribution, and furnace wall design.

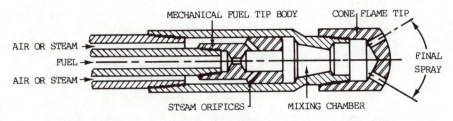

Fig. 14-6. Internal mix nozzle. *Source:* Ref. 14-3.

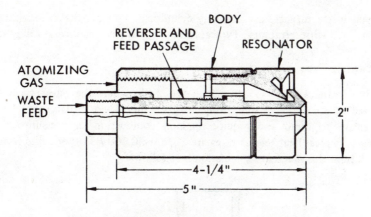

Fig. 14-7. Typical sonic nozzle. *Source:* Ref. 14-11.

As discussed previously, flame impingement on a furnace wall is undesirable creating the potential for refractory corrosion and resulting in lost energy. A furnace must be designed to avoid impingement. Impingement is a function of liquid atomization and vaporization which in turn is dependent on nozzle design, velocity of fluid exiting the burner, air distribution within the furnace, and furnace temperature.

Primary combustion air is that air flow supplied at the fuel burner to combust the prime fuel within the furnace. It is normally distributed through a burner register, an open fan-shaped component normally surrounding the burner nozzle which imparts a circular velocity to the air flow. The register is either fixed or can be adjustable with adjustment arms often located immediately outside the furnace. Secondary air is that air flow necessary for waste combustion and is normally introduced into the furnace downstream of the main flame front. In liquid injection furnaces the secondary air supply is often used not just as combustion air, but to create turbulence within the furnace and to provide a relatively cool flow on the inside refractory surface, keeping the refractory temperature cooler than that of the center of the furnace. The primary and secondary air flows are also introduced in a manner that aids fuel atomization and helps prevent any unburnt materials from impinging on the furnace wall. (Spurious impingement often creates *sparklers*, luminous burnout of volatile particles on the furnace wall).

Liquid destruction furnaces require 5-30% excess air to insure adequate combustion. Another furnace parameter is heat release. Most liquid burners have a heat release rate of 20,000-30,000 BTU per cubic foot per hour. Vortex burners, burners where the primary air flow creates a high velocity cyclonic vortex prior to injection into the furnace chamber, will release 700,000-1,000,000 BTU per cubic foot per hour.

Liquid waste furnaces are normally cylindrical in shape, either horizontal or vertical, lined with refractory. Typical furnace types are described as follows:

Non-Swirling Type

This is a furnace such as the one in Fig. 14-8, where the burner(s) is (are) mounted axially or on the side, firing along a radius. These furnaces are relatively inexpensive to build and require minimal combustion air pressure (smaller blowers). Typical heat release rates are 10,000–30,000 BTU/per cubic foot per hour. A typical furnace size calculation is as follows:

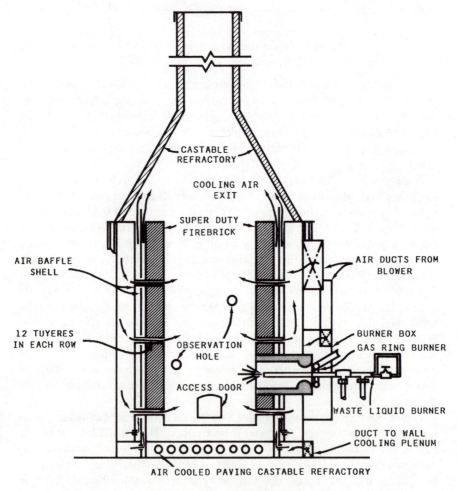

Fig. 14-8. Liquid waste incinerator.

Heat release: $Q = 10,000,000$ BTU/hr
Furnace release rate: $F = 20,000$ BTU/hr-ft^3

Furnace volume: $V = Q/F = \dfrac{10,000,000 \text{ BTU/hr}}{20,000 \text{ BTU/hr-ft}^3}$

                                    $= 500$ ft^3

For a chamber 8 feet long (L), the internal diameter, D, is:

$$V = \frac{\pi}{4} D^2 L$$

$$D = \left(\frac{4V}{\pi L}\right)^{1/2} = \left(\frac{4 \times 500 \text{ ft}^3}{\pi \times 8 \text{ ft}}\right)^{1/2}$$

$$D = 8.92 \text{ ft} = 9 \text{ ft } 11 \text{ in.}$$

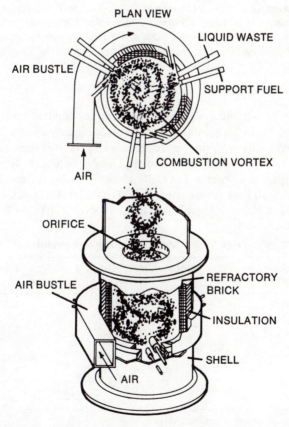

Fig. 14-9. Vortex combustion incinerator. *Source:* Ref. 14-12.

Table 14-1. Axially/Radially Fired Incinerator.

Heat Release, MBH		Length × Width × Height*, feet	Total Weight, lb
Waste	Fuel		
6.3	3.0	9 × 5 × 26	16,000
14.3	7.0	11 × 7 × 26	23,000
24.7	12.0	14 × 8 × 26	32,000

*Overall outside dimensions including stack.
Source: Thermal Research & Engineering Corp., Conshohocken, PA.

This incinerator, or furnace is little more than an empty chamber, refractory lined, with neither baffles nor other changes in the direction of flow. Air jets are often placed in the chamber side walls; these inject compressed air into the furnace, creating increased turbulence. Increasing turbulence within the combustion chamber will increase the burning efficiency, which will be reflected in an increased burning rate, i.e., greater BTU/hr-cubic foot.

Vortex Furnace

Swirl burners or burners firing tangentially into the combustion chamber create a cyclonic or vortex flow within the furnace. Secondary combustion air is also injected tangentially into the furnace to increase the turbulent flow. Fig. 14-9 illustrates a furnace with a series of vortex burners. Secondary combustion air is introduced from an air bustle. The high turbulence within the combustion chamber results in high heat release rates, 40,000–100,000 BTU/hr-cubic foot. Calculating a furnace diameter for a higher rate, say, 60,000 BTU/hr-ft^3, the furnace

Table 14-2. Vertical Vortex Incinerator.

Heat Release, total MBH	Length × Width × Height*, feet	Total Weight, lb
3.0	9 × 6 × 47	7,000
7.0	10 × 7 × 48	9,500
10.0	11 × 7 × 48	13,500
14.0	12 × 8 × 50	17,000
18.0	12 × 8 × 51	19,000
24.0	13 × 9 × 51	23,500
30.0	13 × 10 × 52	29,000
48.0	15 × 12 × 53	41,500

*Overall outside dimensions including stack.
Source: Thermal Research & Engineering Corp., Conshohocken, PA.

diameter will be as follows:

$$V = \frac{D}{F} = \frac{10{,}000{,}000 \text{ BTU/hr}}{60{,}000 \text{ BTU/hr-ft}^3}$$

$$= 167 \text{ ft}^3$$

$$V = \left(\frac{4V}{\pi L}\right)^{1/2} = \left(\frac{4 \times 167 \text{ ft}^3}{\pi \times 8 \text{ ft}}\right)^{1/2}$$

$$D = 5.2 \text{ ft} = 5 \text{ ft } 2 \text{ in.}$$

Comparing axial furnace to vortex furnace size, 8 ft 11 in. versus 5 ft 11 in. diameter, respectively, the vortex furnace is over 40% smaller. The vortex burner design is more complex than the axially fired burner, and high-pressure blowers are required; however, the smaller furnace chamber represents a significantly lower cost.

Tables 14-1 and 14-2 list overall dimensions for typical liquid waste incinerators, axially fired and vortex type.

Chapter 15
Incineration of Gaseous Waste

Many gaseous wastes are characterized by odor or color. Often these characteristics are the result of organic compounds which, properly incinerated, will destruct. In this chapter techniques for destruction of gaseous waste by incineration are presented.

COMBUSTIBLE GAS

Gases having combustible constituents will have a lower explosive limit (LEL) and an upper explosive limit (UEL). The LEL is the least concentration of the gas in air which will sustain gas combustion. At a concentration below the LEL there is insufficient gas present to generate the heat required to sustain combustion.

The UEL is the highest concentration of gas in air at which combustion can be sustained. At a higher concentration of gas to air there will be insufficient air present to sustain combustion.

Table 15-1 lists lower (LEL) and upper (UEL) explosive limits of various liquid and gaseous substances in air. Included in this table is the flash point for some of those materials which are in liquid form. At the flash point, the substances will self-ignite. The flash point, therefore, is a measure of volatility for liquids. The autoignition temperature (see Table 15-2) is that temperature at which a gas will combust. Bringing a gas to its autoignition temperature in a mixture of air between the LEL and UEL of that gas will result in sustained combustion, perhaps rapid combustion (an explosion).

The method of incineration chosen is affected by the LEL, UEL, and ignition temperature. Flares or direct flame incinerators operate best with gases just above the UEL or just below the LEL. Catalytic incinerators are normally used with concentrations no more than 25% of the LEL. The lower the flash point, or autoignition temperature, the less excess air is normally required.

FLARES

Flares are used as a low cost means of disposal of relatively large amounts of gas containing combustible components. They are suited to processes which are not

Table 15-1. Combustibility Characteristics of Pure Gases and Vapors in Air.

Gas or Vapor	Lower Limit, % by volume	Upper Limit, % by volume	Closed Cup Flash Point, °F
Acetaldehyde	4.0	57	−17
Acetone	2.5	12.8	0
Acetylene	2.5	80	−
Allyl alcohol	2.5	−	70
Ammonia	15.5	26.6	−
Amyl acetate	1.0	7.5	77
Amylene	1.6	7.7	−
Benzene (benzol)	1.3	6.8	12
Benzlyl chloride	1.1	−	140
Butene	1.8	8.4	−
Butyl acetate	1.4	15.0	84
Butyl alcohol	1.7	−	−
Butyl cellosolve	−	−	141
Carbon disulfide	1.2	50	−22
Carbon monoxide	12.5	74.2	−
Chlorobenzene	1.3	7.1	90
Cottonseed oil	−	−	486
Cresol, m- or p-	1.1	−	202
Crotonaldehyde	2.1	15.5	55
Cyclohexane	1.3	8.4	1
Cyclohexanone	1.1	−	111
Cyclopropane	2.4	10.5	−
Cymene	0.7	−	117
Dichlorobenzene	2.2	9.2	151
Dichloroethylene (1,2)	9.7	12.8	57
Diethyl selenide	2.5	−	57
Dimethyl formamide	2.2	−	136
Dioxane	2.0	22.2	54
Ethane	3.1	15.5	−
Ether (diethyl)	1.8	36.5	−49
Ethyl acetate	2.2	11.5	28
Ethyl alcohol	3.3	19.0	54
Ethyl bromide	6.7	11.3	−
Ethyl cellosolve	2.6	15.7	104
Ethyl chloride	4.0	14.8	−58
Ethyl ether	1.9	48	−49
Ethyl lactate	1.5	−	115
Ethylene	2.7	28.6	−
Ethylene dichloride	6.2	15.9	56
Ethyl formate	2.7	16.5	− 4
Ethyl nitrite	3.0	50	−31
Ethylene oxide	3.0	80	−
Furfural	2.1	−	140

Table 15-1.   (Continued)

Gas or Vapor	Lower Limit, % by volume	Upper Limit, % by volume	Closed Cup Flash Point, °F
Gasoline (variable)	1.4–1.5	7.4–7.6	−50
Heptane	1.0	6.0	25
Hexane	1.2	6.9	−15
Hydrogen cyanide	5.6	40.0	–
Hydrogen	4.0	74.2	–
Hydrogen sulfide	4.3	45.5	–
Illuminating gas (coal gas)	5.3	33.0	–
Isobutyl alcohol	1.7	–	82
Isopentane	1.3	–	–
Isopropyl acetate	1.8	7.8	43
Isopropyl alcohol	2.0	–	53
Kerosene	0.7	5	100
Linseed oil	–	–	432
Methane	5.0	15.0	–
Methyl acetate	3.1	15.5	14
Methyl alcohol	6.7	36.5	52
Methyl bromide	13.5	14.5	–
Methyl butyl ketone	1.2	8.0	–
Methyl chloride	8.2	18.7	–
Methyl cyclohexane	1.1	–	25
Methyl ether	3.4	18	–
Methyl ethyl ether	2.0	10.1	−35
Methyl ethyl ketone	1.8	9.5	30
Methyl formate	5.0	22.7	− 2
Methyl propyl ketone	1.5	8.2	–
Mineral spirits No. 10	0.8	–	104
Naphthalene	0.9	–	176
Nitrobenzene	1.8	–	190
Nitroethane	4.0	–	87
Nitromethane	7.3	–	95
Nonane	0.83	2.9	88
Octane	0.95	3.2	56
Paraldehyde	1.3	–	–
Paraffin oil	–	–	444
Pentane	1.4	7.8	–
Propane	2.1	10.1	–
Propyl acetate	1.8	8.0	58
Propyl alcohol	2.1	13.5	59
Propylene	2.0	11.1	–
Propylene dichloride	3.4	14.5	60
Propylene oxide	2.0	22.0	–
Pyridine	1.8	12.4	74
Rosin oil	–	–	266

Table 15-1. (*Continued*)

Gas or Vapor	Lower Limit, % by volume	Upper Limit, % by volume	Closed Cup Flash Point, °F
Toluene (toluol)	1.3	7.0	40
Turpentine	0.8	–	95
Vinyl either	1.7	27.0	–
Vinyl chloride	4.0	21.7	–
Water gas (variable)	6.0	70	–
Xylene (xylol)	1.0	6.0	63

Source: Ref. 15-5.

Table 15-2. Autoignition Temperature of Some Common Organic Compounds.

Compound	Temperature, °F	Compound	Temperature, °F
Acetone	1000	Hydrogen	1076
Ammonia	1200	Hydrogen cyanide	1000
Benzene	1075	Hydrogen sulfide	500
Butadiene	840	Kerosene	490
Butyl alcohol	693	Maleic anhydride	890
Carbon disulfide	257	Methane	999
Carbon monoxide	1205	Methyl alcohol	878
Chlorobenzene	1245	Dichloromethane	1185
Cresol	1038	Methyl ethyl ketone	960
Cyclohexane	514	Mineral spirits	475
Dibutyl phthalate	760	Petroleum naphtha	475
Ethyl ether	366	Nitrobenzene	924
Methyl ether	662	Oleic acid	685
Ethane	950	Phenol	1319
Ethyl acetate	907	Phthalic anhydride	1084
Ethyl alcohol	799	Propane	874
Ethyl benzene	870	Propylene	940
Ethyl chloride	965	Styrene	915
Ethylene dichloride	775	Sulfur	450
Ethylene glycol	775	Toluene	1026
Ethylene oxide	804	Turpentine	488
Furfural	739	Vinyl acetate	800
Furfural alcohol	915	Xylene	924
Glycerin	739		

Source: Ref. 15-6.

continuous; continuous gas generation often lends itself towards heat recovery. Incineration by flaring is simply controlled discharge into the atmosphere. Heat recovery, almost by definition, is not possible with a flare.

There are two types of flares, ground level and elevated, or tower flares, in current use. Ground flares can be used where there is sufficient space around the flare to provide for safety of personnel and equipment. The tower flare is used to keep the flame above the level of surrounding equipment and personnel, and to promote dilution of its products of combustion into the air. Radiation from flares should insure that surrounding equipment will not receive more than 3000 BTU/ft^2-hr and that personnel will not receive more than 440 BTU/ft^2-hr continuously, 1500 BTU/ft^2-hr for short term exposure. Temperatures developed in flare systems normally range from 2000 to 2500°F.

As shown in Fig. 15-1, a flare is basically a stack, or open pipe, discharging a combustible gaseous waste directly to the atmosphere with the end of the stack containing a pilot flame (continuously firing), a source of steam and an exit nozzle. Combustion air is provided by the surrounding atmosphere.

Steam is provided for a number of reasons, as follows:

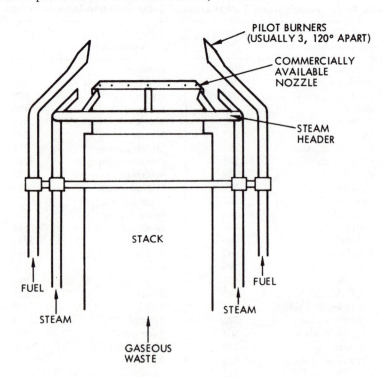

Fig. 15-1. Stack flare equipped with mixing nozzle. *Source:* Ref. 15-5.

- To generate turbulence and momentum, promoting good mixing with the surrounding air.
- As a source of heat to help in cracking complex molecules within the gas stream.
- As a reactant in the combustion process to help oxidize carbon to a gaseous state, simplified as:

$$C + H_2O(steam) \longrightarrow CO + H_2$$

The CO and H_2 produced will readily combust burning clean in the presence of air.

Steam flow will generally be in the range of 0.15–0.50 per lb of hydrocarbon in the gas stream.

Gases with heating values as low as 150 BTU/cubic foot can normally be flared without the addition of supplemental fuel. Below 150 BTU/cubic foot supplementary fuel is normally provided in the form of a gas at the nozzle tip, as shown in Fig. 15-1 as Fuel.

Some gases will burn relatively clean and require little or no steam to promote combustion. Such gases include methane, hydrogen, carbon monoxide, and coke oven gas. Most hydrocarbon gases which are normally flared, however, are heavier, i.e., with two or more carbon atoms in its molecular structure. They invariably require a source of steam in order to burn smokeless (compounds such as olefins, aromatics, and the paraffin group above methane.)

Figure 15-2 shows a tower flare utilizing internal steam injection, known as the Esso flare. A flare system utilizing the Esso flare is shown in Fig. 15-3. Waste gas is piped to a tank with a water level maintained within it. Liquids and condensable vapors contained within the gas are absorbed in the water which discharges, through a seal, for further processing. An automatic pressure sensor monitors the flow of waste gas to automatically control steam flow to the flare. The waste gas discharges from the tank (commonly referred to as a knockout tank) through a flame arrestor into the stack which feeds the flare. The flame arrestor prevents backflashing of flame from the flare to the upstream equipment. A purge gas, such as nitrogen or argon, is usually provided to help clean the liquid systems during periods when the flare is not in use.

Another type of tower flare is the Sinclair flare shown in Fig. 15-4. Steam is provided through a steam ring with numerous openings, $\frac{1}{8}$ inch in diameter, for discharging steam into and around the exiting gas stream. These openings are positioned to provide tangential discharge of steam, promoting high turbulence and air inspiration and mixing. A steel shroud, covered in plastic, reduces the noise and radiant heat discharged to the sides and beneath the flare. Another

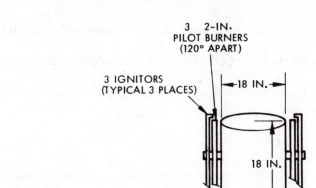

Fig. **15-2.** Esso type flare. *Source:* Ref. 15-5.

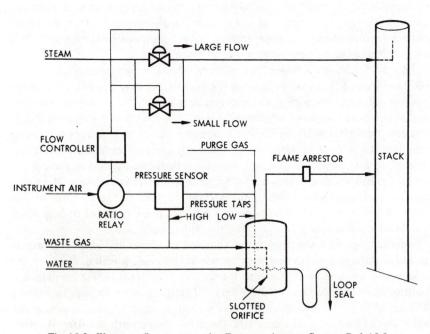

Fig. **15-3.** Waste-gas flare system using Esso type burner. *Source:* Ref. 15-5.

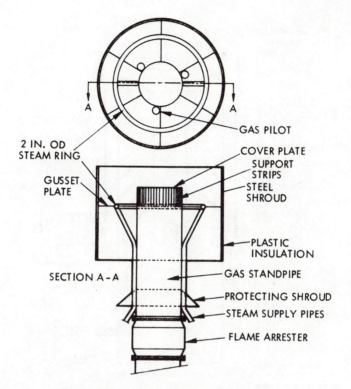

Fig. 15-4. Sinclair type flare. *Source:* Ref. 15-5.

flare, shown in Fig. 15-5 has a steam injector which may or may not be utilized and is provided with insulation for noise reduction.

Figure 15-6 shows a typical ground flare utilizing venturi burner nozzles. Table 15-3 lists capacities of venturi burners as a function of waste gas pressure and burner orifice size. Gas flow properties of natural gas were used to determine the listed figures. Fig. 15-7 shows a ground flare which has two sets of burners, one for low flow and the other for high flow rates of the waste gas. Note the acoustical fence, provided to help attenuate the noise generated by the operating flare.

Water is sometimes used in lieu of steam, as shown in Fig. 15-8, to reduce smoking. It is less costly than steam supply and injection, however it is not as effective as steam. It is used with waste gas at low flow rates and where some smoking can be tolerated.

The innermost stack is required for control of mixing of the water, waste gas and supplemental fuel. The intermediate stack is used to confine the water to aid in mixing with the waste gas. The outer stack confines the flame, directing it upward. Table 15-4 lists water flow requirements for a typical ground flare.

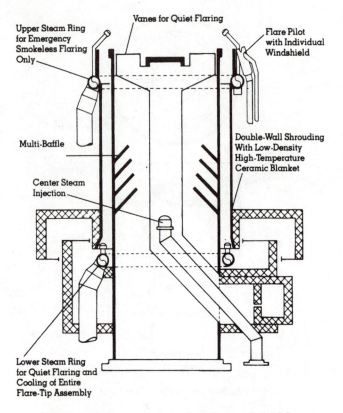

Fig. 15-5. Multi-purpose tower flare. *Source:* Ref. 15-8.

DIRECT FLAME INCINERATION

Direct flame incinerators, also referred to as fume incinerators and gas combustors, are chambers provided with supplemental fuel burners, which provide heat and retention time to destruct gaseous waste materials. Figure 15-9 is a schematic diagram of a direct flame incinerator. A thermocouple in the combustion chamber measures temperature. Appropriate control circuitry alters the rate of supplementary fuel entering the furnace to maintain the desired combustion chamber temperature. These incinerators are applicable for most gaseous waste. Their primary use may be for odor control, toxicity elimination or visible emissions control.

Combustion chamber temperatures are in excess of the autoignition temperature (see Table 15-2) and normally vary, depending on the waste constituents, from 800°F to 1500°F. Table 15-5 lists the efficiency of destruction for gaseous waste composed essentially of hydrocarbon compounds.

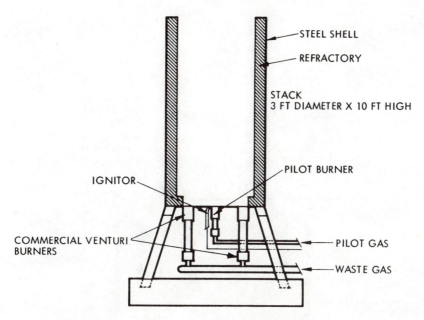

Fig. 15-6. Vertical Venturi type flare. *Source:* Ref. 15-5.

Table 15-3. Venturi Burner Capacities, ft³/hr.

Gas pressure, in. H_2O	$\frac{3}{16}$-in. Orifice	$\frac{7}{16}$-in. Orifice	$\frac{1}{2}$-in. Orifice
2	70		
4	100		
6	123		
8	142		
10	160		
$\frac{1}{2}$ psig	210	1,042	1,360
1 psig	273	1,488	1,900
2 psig	385	2,157	2,640
3 psig		2,654	3,200
4 psig		3,065	3,680
5 psig		3,407	4,080
6 psig		3,742	4,480
7 psig		4,040	4,800
8 psig		4,320	5,160

Basis: 1,000 BTU/ft³ natural gas.
Source: Ref. 15-5.

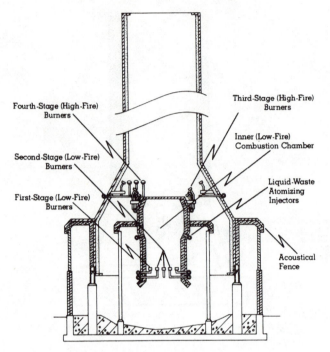

Fig. 15-7. Ground flare. *Source:* Ref. 15-8.

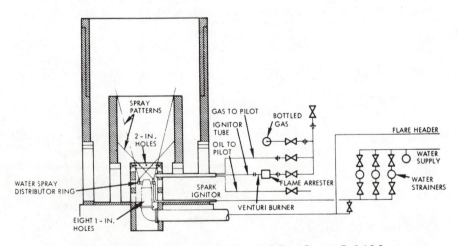

Fig. 15-8. Typical water spray type ground flare. *Source:* Ref. 15-5.

Table 15-4. Water Spray Pressures Required for Smokeless Burning.*

Gas Rate, SCFH	Unsaturates, % by vol	Molecular Weight	Water Pressure, psig	Water Rate, gpm
200,000	0–20	28	30–40	31–35
150,000	30	33	80	45
125,000	40	37	120	51

*The data in this table were obtained with a $1\frac{1}{2}$-inch-diameter spray nozzle in a ground flare with the following dimensions:

	Height, ft	Diameter, ft
Outer stack	30	14
Intermediate stack	12	6
Inner stack	4	12.5

Source: Ref. 15-5.

Retention time is as significant a parameter as temperature. These incinerators are normally designed for combustion chamber sizing to provide 0.25–0.50 seconds retention time, although units have been designed large enough to provide a retention time of 2–3 seconds.

The simplicity of automatic direct flame combustion makes it ideal for combustion control. The configuration of this equipment lends itself to heat recovery. Two modes of heat recovery are outlined in Fig. 15-10. In one case a heat exchanger utilizes the high temperatures in the combustor exhaust to preheat the incoming combustion air. The second case shows a heat exchanger heating a stream for external use. The stream can be gas, water, or water to steam.

Note that the energy requirement of this or any other heat-generating equipment is a function of the temperature to which the products of combustion

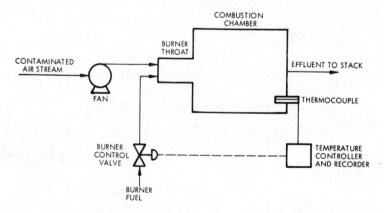

Fig. 15-9. Direct-flame thermal incinerator. *Source:* Ref. 15-5.

Table 15-5.   Direct Flame Combustor Efficiency.

	Hydrocarbon Oxidation	Carbon Monoxide Oxidation	Odor* Destruction
Range of temp., °F	1000–1250	1250–1350	1000–1200
Average temp., °F	1100–1200	1300–1350	1100–1150
Efficiency, %	75–85	75–90	50–99
Range of temp., °F	1000–1300	1300–1450	1100–1300
Average temp., °F	1150–1250	1400–1450	1200–1250
Efficiency, %	85–90	90–99	90–99
Range of temp., °F	1100–1500	–	1200–1500
Average temp., °F	1200–1400	–	1350–1400
Efficiency, %	90–100	–	99+

*For odor generated from hydrocarbon compounds.

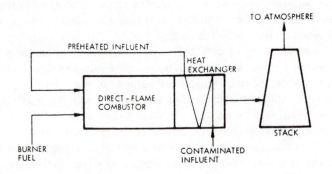

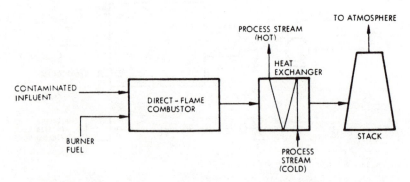

Fig. 15-10. Heat recovery options. *Source:* John Zink Company, Tulsa, OK.

must be raised. Burning at 1400°F in the combustion chamber, without heat recovery, the exiting stream will be at 1400°F. All of the products of combustion must be brought to this temperature. If a heat exchanger were installed to cool the gas outlet temperature, the temperature within the stack, to 1000°F, the products of combustion would only have to be brought to 1000°F although the combustion chamber would still be maintained at 1400°F. A rough calculation of efficiency, based on absolute temperature, is as follows:

With heat exchanger: 1000°F + 460°F = 1460°R outlet

Without heat exchanger: 1400°F + 460°F = 1860°R outlet

Fuel savings with heat exchanger: (1860 - 1460)/1860 = 22%

This figure is a measure of the efficiency of the heat exchanger. It can also be used for cost effectiveness calculations. For instance, if natural gas at $6.00 per million BTU were burned without a heat exchanger, and this incinerator would be in operation for 2000 hours per year burning natural gas at an average rate of 20 000 cubic feet/hr, one years savings with a heat exchanger would be:

20 000 ft³/hr × 2000 hr/year × 1000 BTU/ft³

$$\times \; \$6.00/1\,000\,000 \; hr \times 0.22 \; efficiency = \$52,800 \; per \; year$$

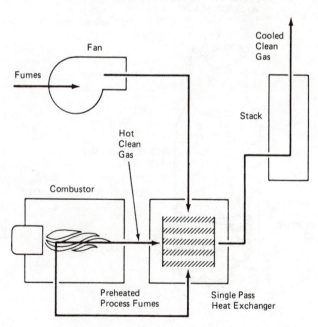

Fig. 15-11. Forced draft direct-flame fume incineration system with a single pass primary heat exchanger. *Source:* Ref. 15-9.

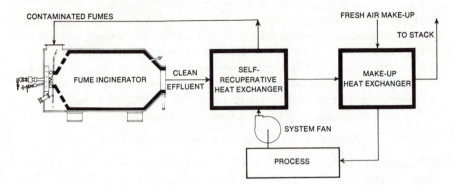

Fig. 15-12. Dual heat exchanger system. *Source:* Peabody International Corporation, Stamford, CT.

An annual savings of over 50 000 dollars is a significant cost savings. Often more than one heat exchanger section is installed, to further increase the savings realized by decreasing the purchased fuel requirement.

Figure 15-11 shows a fume incinerator with a single pass heat exchanger which recovers heat by increasing the temperature of the combustion air entering the

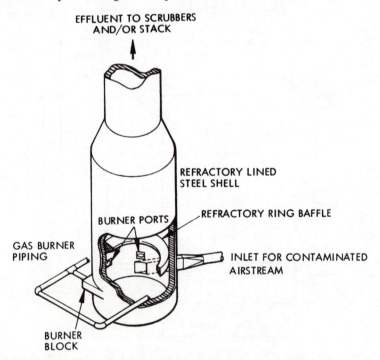

Fig. 15-13. Vertical incinerator. *Source:* Peabody International Corporation, Stamford, CT.

combustor. A dual heat exchange system is shown in Fig. 15-12. One heat exchanger is heating the fresh air to provide heat for process equipment while the second heat exchanger is used to heat the combustion air entering the combustion chamber.

The aforementioned fume incineration equipment were horizontal units where the chamber is horizontal and burners fire along a horizontal axis. Fig. 15-13 shows another type of direct flame combustor, a vertical unit, where waste gas, fuel, and combustion air are introduced at the bottom of the incinerator. The chamber immediately above the burner is designed for the required burnout or retention time for the particular waste being incinerated. Vertical units usually are not suited for heat recovery.

Normally the incinerator is designed for complete destruction of organic components by incineration and particulate matter discharges are almost nonexistent. Where other components are present in the gas, however, such as sulfur or halogens, scrubbers will usually be required.

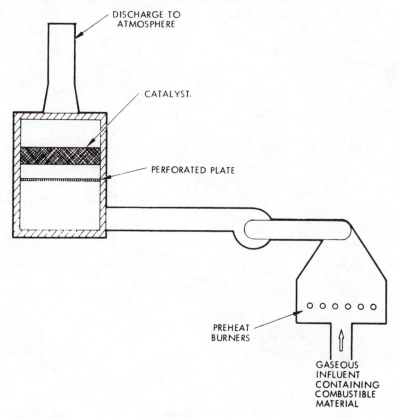

Fig. 15-14. Catalytic incineration without heat recovery. *Source:* Ref. 15-5.

CATALYTIC INCINERATION

Catalytic incinerators normally destruct gaseous waste at low concentrations, less than 25% of the LEL. As shown in Fig. 15-14 heated gas passes through a perforated plate to straighten the flow and then passes through a catalytic material prior to discharge. The catalyst has the property of increasing the rate of oxidation at lower temperatures (i.e., use of a catalyst promotes destruction of gaseous waste at lower temperatures.)

The actual steps in the catalytic reaction are as follows:

- Diffusion of the reactants within the gas stream, plus the gas stream, through the stagnant fluid surrounding the surface of the catalyst.
- Adsorption of the reactants on the catalyst surface.
- Reaction of the reactants to form products (usually oxides).
- Desorption of the products from the catalytic surface.
- Diffusion of the products from the catalyst pores and surface film to the vapor or gaseous phase outside (downstream) of the catalyst.

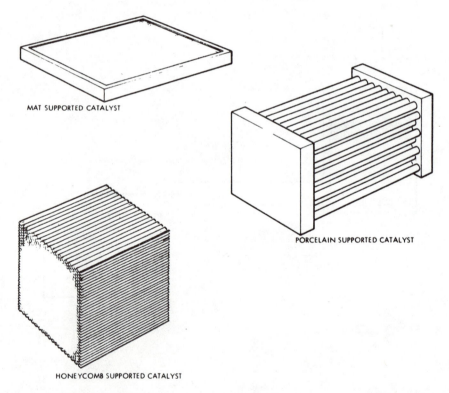

MAT SUPPORTED CATALYST

PORCELAIN SUPPORTED CATALYST

HONEYCOMB SUPPORTED CATALYST

Fig. 15-15. Commercially used catalyst configurations. *Source:* Ref. 15-7.

Catalyst materials normally used are the noble metals, i.e., platinum, palladium, rhodium, etc. Other materials which function as catalyst are copper chromite and the oxides of copper, chromium, manganese, nickel and cobalt.

Figure 15-15 illustrates types of catalyst sections commonly in use. The mat type is similar in appearance to an air filter. It consists of ribbons of nichrome or stainless steel wire to which the catalytic material has been applied. Normally an active metal, such as platinum, is applied to this type of carrier.

The porcelain assembly pictured consists of two end plates which are secured by a center post and a number of tear-shaped rods to which the catalyst material is applied. The porcelain is normally coated with aluminum which, in turn, is coated with the catalyst material. Active oxides, such as copper oxide or active metals are normally utilized in this configuration.

The third configuration is the honeycomb type, constructed of ceramic or other refractory materials. These can utilize most catalyst materials.

Noting their construction, the catalyst bank offers very little resistance to gas flow, often only a fraction of an inch WC.

Catalyst materials are also available in pellet form, stacked in a chamber within a furnace. When stacked, pellets offer relatively large resistance to gas flow.

Catalytic incineration systems must effect intimate mixing of the combustibles within the system. The gas must be brought to the required ignition temperature (see Table 15-6) for the combustible to be burned and good temperature control throughout the catalyst bed is essential. Sufficient oxygen must be present in the gas stream or must be added to it to insure oxidation of the contaminants.

The gas stream must be free of particulate matter to protect the catalyst from fouling. If particulate matter is present, pretreatment of the gas, such as cyclonic

Table 15-6. Comparison of Temperatures Required for 90% Conversion of Combustibles to CO_2 and H_2O.

Combustibles	Ignition Temperature, °F		Difference, °F	Catalyst/ Thermal, %
	Thermal	Catalytic		
Benzene	1,076	575	501	67
Toluene	1,026	575	451	70
Xylene	925	575	163	75
Ethanol	738	575	163	86
Methane	1,170	932	238	85
Carbon monoxide	1,128	500	628	60
Hydrogen	1,065	250	815	47
Propane	898	500	398	71

Table 15-7. Platinum Family Catalyst Contaminants.

Poisons	heavy metals
	phosphates
	arsenic compounds
Suppressants	halogens (elemental and compounds)
	sulfur compounds
Fouling Agents	alumina and silica dusts
	iron oxides
	silicones

Source: Ref. 15-5.

separation or electrostatic precipitators, may be necessary upstream of the catalyst.

Beside particulate fouling, catalysts are sensitive to a number of substances. Table 15-7 lists contaminant components with regard to noble metal catalysts.

The catalytic incinerator chamber is constructed of steel or refractory, depending on the temperatures developed. From 750°F to 1100°F heat resistant steel

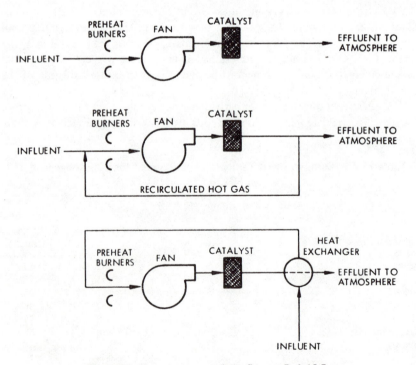

Fig. 15-16. Heat recovery options. Source: Ref. 15-7.

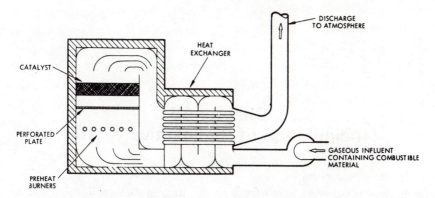

Fig. 15-17. Catalytic incineration with heat recovery. *Source:* Ref. 15-5.

can be used, stainless steel from 1100°F to 1300°F and refractory materials above 1300°F. Steel is normally protected with 4-6 inches of insulation.

As with direct flame incineration, the cost of heat exchange equipment is often more than offset by the savings in supplemental fuel consumption. Fig. 15-16 illustrates two heat recovery options, one utilizing a portion of the exiting hot gas stream directly mixed with the incoming stream to raise its temperature, and another utilizing a non-contact heat exchanger where heat is transferred from the exiting stream to the incoming air/gas flow.

Figure 15-17 is a schematic of a catalytic incinerator with a self-contained heat exchanger. The incoming flow is blown through the outside of a baffled heat exchanger exiting as heated combustion air/gas. After brought to oxidation temperature the gas passes through the catalyst bank and then through the internal section of the heat exchanger. It will give up heat, or temperature, to the incoming flow in proportion to the size of the heat exchanger.

Chapter 16
Incineration of Radioactive Waste

As the quantity of waste from nuclear power plants, medical applications, civilian and military research, etc., increases and public concern heightens, incineration is becoming a reasonable answer to radioactive waste control. Radioactive waste can be substantially reduced in volume and weight by the incineration techniques presented in this chapter.

RADIOACTIVE WASTE

The vast majority of radioactive waste generated is low level radioactive waste (LLW), which is defined as follows:

> Low-level waste comprises that radioactive waste which is not spent fuel or high-level waste and which contains less than 10 nanocuries of transuranics per gram of material.

Transuranic material (TRU) refers to uranium or materials, natural or man-made, with atomic weight equal to or greater than that of uranium.

The International Atomic Energy Agency has defined waste categories, from 1 to 5, based on state (solid, liquid or gas), the concentration of radioactivity, the shielding required, and the complexity of the treatment method necessary. These categories have not found acceptance in the United States.

Generation rates and current accumulation (mid-1980) of LLW is listed in Table 16-1. Of particular interest is that almost 20 years' worth of material, based on current generation, is accumulated and must eventually be disposed of. Ultimate disposal is currently burial, and increasing public concern over burial of radioactive waste and the transportation of them to burial sites have severely limited disposal options. In lieu of disposal, therefore, much of this waste has been accumulated and will be accumulated until viable, reliable methods of ultimate disposal can be found, with minimal impact on public health and safety.

With wastes eventually going to landfill volume reduction becomes of prime

Table 16-1.   Generation of Radioactive Waste.

	Total Accumulation, 1980, 10^6 ft^3	Annual Generation, 1980, 10^6 ft^3
Government	50.8	1.0
Commercial	15.8	>2.0
Fuel cycle		1.2
reactor operation		1.0
fuel fabrication		0.2
other steps		small
Non-fuel cycle		0.78
industrial		0.67
medical		0.64
academic		0.026
Industrial and other		0.11
Research (including pharmaceutical)		
Total	66.6	3.6

Source: Ref. 16-8.

importance. Incineration is an effective means of obtaining volume reduction, as noted in Table 16-2.

Not all LLW is incinerable. They can be roughly categorized into incinerable and non-incinerable, dry and wet waste. Dry waste includes such items as paper, rags, plastics, rubber, wood, glass, and metals. Dry waste can futher be classified as compactible and non-compactible, combustible or incombustible. Most wet wastes are derived from the clean-up of aqueous processes or waste streams prior to recycle or discharge. Such liquid wastes include ion exchange resin slurries, filter sludges, evaporator concentrates, miscellaneous oils, and solvents. In general, liquid LLW will be concentrated by settling, centrifuging, reverse osmosis or other means as a first step in volume reduction.

LLW INCINERATION

As would be expected, radioactive waste incineration requires unique considerations not applicable to other waste streams. Safety is of prime concern. Handling, storage, and feeding of material must be carefully controlled. Tight containers and negative ventilation systems exhausting through high efficiency particulate air filters (HEPA) are necessary prior to feeding and incineration.

Incinerator feeding must be designed to preclude the possibility of the escape of hot gases (which may contain LLW) from the incinerator from back charging into the feeding room when the incinerator charging door is open. This normally requires that the charging system be provided with a series of air locks and

Table 16.2. Development Status of Incinerators Used in the United States to Reduce the Volume of Wastes, 1980.

	Development Status	
Incinerator Type	Nonradioactive wastes[a]	Radioactive wastes
Acid digestion	not applied	pilot plant
Agitated hearth	not applied	under construction[b]
Controlled air	commercial unit	demonstration unit
Cyclone drum	not applied	demonstration unit
Fluidized bed	commercial unit	pilot plant
Microwave/gas plasma	laboratory unit	laboratory unit
Molten glass (joule heating)	commercial unit	test unit
Molten salt	commercial unit	pilot plant
Moving grate	commercial unit	not applied
Multiple heath	commercial unit	not applied
Pyrolysis controlled air)	commercial unit	test unit
Rotary kiln	commercial unit	under construction[b]
Slagging pyrolysis	commercial unit	pilot plant

[a] Includes wastes generated in noncircular applications
[b] Full-scale units.
Source: Ref. 16-8.

multiple feeding chambers to isolate the incinerator gas under all conditions of operation.

Emergency exhaust systems, such as explosion doors used in conventional incineration systems, cannot normally be used when burning LLW. The opening of a door or stack prior to the exhaust gas cleaning train would discharge LLW to the surrounding air. This requirement may preclude incineration of certain wastes which may cause explosions, or the use of supplemental fuel such as natural gas, which is more prone to explosion than other fuels.

Ash and residue from the gas cleaning system will contain LLW more concentrated then when imputted to the system and must be handled with respect to its toxicity. Shielding is necessary when handling these residues and rapidity in their disposal is also necessary to reduce the possibility of their unwanted release.

Liquid feeding must be at a relatively constant rate. Significant variation in flow (or quality) of liquid LLW can lead to "puffing" or pressure surges which will seek an unwanted exit from the incinerator, bypassing the gas cleaning train.

Controls for an incinerator burning radioactive wastes must necessarily be more encompassing, more detailed, more automated and more reliable than those for conventional waste burning incinerators.

Another consideration in radioactive waste incinerator design is reclamation. Many radioactive elements and other sources of radioactivity are extremely expensive. Often their reclamation for reuse is economically feasible, even when the waste has a very low level of radioactivity.

DRUM INCINERATOR

The Mound Laboratory, at Miamisburg, Ohio, has developed a number of types of incineration systems for LLW. One of these, a drum incinerator, is shown in Fig. 16-1. A 55-gallon drum is used as the burning chamber. A basket, loaded with waste in small cartons, is ignited manually and is lowered into the drum. A source of air is injected into the drum tangentially to promote turbulence and provide a cyclonic effect to the combustion process. Temperatures within the drum reach 2000–2200°F. Blowers create sufficient draft to collect all off-gas from the burning material within the drum.

The off-gas passes through a caustic solution deluge chamber where the gas is cooled and neutralized, and most of the particulate matter is removed. The venturi further scrubs the flue gas with caustic, removing the majority of particulate matter remaining in the gas and controlling the tendency for the gas stream to be acidic. A HEPA (high-efficiency particulate air) filter is necessary for essentially complete removal of particulate matter from the gas stream; however, it quickly loses efficiency from moisture within the flowing gas. Therefore it is necessary to dry the flue gas upstream of the HEPA filter. In this case a demister is used to remove entrained moisture particles and a zeolite contactor removes additional particulate, particularly the caustic residual in the gas.

Liquid (spent caustic water solution) from the deluge tank, the Venturi scrubber and the demister are collected in a recycle tank. The spent liquid, or

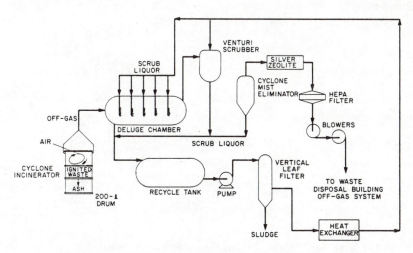

Fig. 16-1. Mound facility cyclone incinerator. *Source:* Ref. 16-8.

liquor, passes through a vertical leaf filter which effectively removes the majority of particulate matter in the form of a sludge. A heat exchanger is provided to heat the caustic and maintain it in solution, preventing its collection on pipes and nozzles within the treatment system. The liquor is then ready for recycle.

The initial volume of LLW is significantly reduced in this process. The LLW is concentrated in a sludge and in the HEPA filter, which of course must be disposed of. But the sludge waste and HEPA filter volume is one tenth the volume of the original waste stream.

CONTROLLED-AIR INCINERATOR

A controlled air incinerator for LLW is shown in schematic in Fig. 16-2. Wastes are hand-sorted to combustible and incombustible fractions. The combustible wastes are shredded, placed in plastic bags and placed in the ram feeding line. Other solid wastes are passed through activity detectors before they are fed to the charging mechanism.

The incinerator operates in the same manner as standard controlled-air or modular combustion units described in Chapter 12. An oxygen deficient atmosphere is provided in the primary chamber and temperatures are controlled in the range of 1450–1850°F. Additional air is provided in the secondary chamber to complete combustion, at temperatures from 1850 to 2200°F. Ash burnout in this system is not as complete as with other incineration systems and the volume reduction, therefore, is not as high as other systems. But the air pollution control system is simpler because incinerator emissions are less than in other systems.

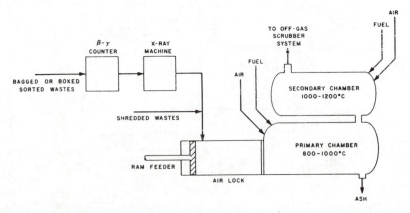

Fig. 16-2. Los Alamos Scientific Laboratory controlled-air incinerator. *Source:* Ref. 13-8.

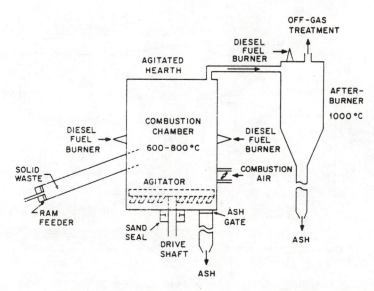

Fig. 16-3. Agitated-hearth incinerator. *Source:* Ref. 16-8.

AGITATED HEARTH

An agitated-hearth incinerator, similar to the mono-hearth incinerator discussed in Chapter 10, has been developed for incineration of LLW. Fig. 16-3 illustrates this type of unit. Solid waste is fed into the incinerator and drops onto its floor. This incinerator can also burn sludge and liquid LLW.

An agitator rotates slowly, turning the waste material surface to increase its contact with combustion air and to help break up cartons and plastic bags. This incinerator is normally designed as a batch unit with 1 hour for burnout plus $\frac{1}{2}$ hour for ash removal.

FLUIDIZED-BED INCINERATOR

This incinerator can handle solids, sludges and liquids. Shown in Fig. 16-4, non-combustibles are separated from the combustible fraction of the waste by an air classifier, after an initial coarse shred. The combustibles (the light fraction) are fine shredded for feeding to the fluid bed. This fraction is charged to the reactor by means of a screw feeder.

The reactor contains sodium carbonate granules which are fluidized by the injection of compressed air and nitrogen. Within the bed the waste is decomposed by partial combustion, and the bed is maintained at a temperature of 1050°F. The air to nitrogen level of the fluidizing gas is adjusted to permit combustion without open-flame burning.

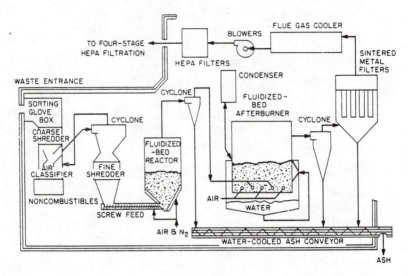

Fig. 16-4. Rocky flats plant fluidized-bed incinerator. *Source:* Ref. 16-8.

Chloride components of the waste stream are neutralized within the sodium carbonate bed to produce sodium chloride, carbon dioxide, and water vapor. Off-gas from the reactor passes through a cyclonic separator where most of the larger particulate exits the gas stream.

In the after-burner chamber combustion air is added to the gas stream as it passes through another fluidized bed. This bed is chromic oxide, an oxidation catalyst which encourages complete combustion of any combustibles within the flue gas. The bed temperature is maintained at approximately 1000°F. A high-pressure water jacket helps control the bed temperature at the desired level.

Flue gas leaving the catalytic after-burner contains fly ash, catalytic dust and small amounts of sodium carbonate and sodium chloride. This particulate matter is essentially removed by passing the flue gas through another cyclonic separator and a bank of sintered metal filters. The flue gas is then cooled to 120°F in a water cooled heat exchanger and is then blown through HEPA filters prior to exiting the system. Ash from the cyclones and the sintered metal filters is collected for ultimate disposal.

ELECTROMELTER SYSTEM

Figure 16-5 shows a proposed molten-glass incinerator for a variety of solid and liquid LLW. It uses technology from the glass-making industry, making use of the Joule effect. Use of the electrical conductivity of molten glass at elevated temperatures to maintain the temperature of the melt is known as Joule heating.

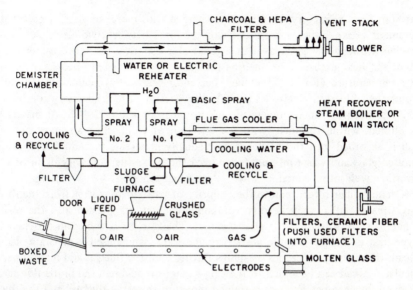

Fig. 16-5. Sketch of the Penberthy molten-glass incinerator (electromelter) system proposed for treating low-level waste. *Source:* Penberthy Electromelt International, Inc., Seattle, WA.

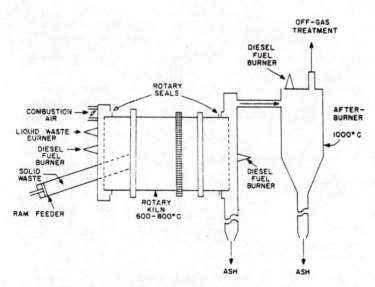

Fig. 16-6. Rotary-kiln incinerator. *Source:* Ref. 16-8.

At start-up a refractory lined vessel is charged with pieces of glass or glass frit (granular glass). Either a sacrificial wire mounted between two electrodes protruding through the glass layer or a removable electrical resistance heater provides the heat necessary for melting the glass. The glass charge heats slowly to the temperature at which the melt becomes electrically conducting. At this point the immersed electrodes provide the electric current necessary for Joule heating, normally at a temperature in the range of 1800–2200°F. As material is burned, within the melt, with the injection of combustion air, most of the ash is absorbed into the molten glass gradually increasing the liquid level. The molten glass and waste mixture is removed through either a gravity drain or an overflow weir into a container where it cools and solidifies.

Ceramic fiber filters remove the majority of particulate matter from the flue gas stream and as they become contaminated they are pushed into the melt where they dissolve into the glass material. The flue gas is cooled in a jacketed type heat exchanger and the heated cooling water can be used to generate plant steam or hot water. A series of sprays further cool the flue gas and remove particulate. Alkali can be added to the spray water for acid removal in the flue gas. A demister chamber removes entrained moisture from the flue gas and the flue gas is heated slightly to reduce its specific humidity in preparation for the HEPA filter sections (they operate best on gas with low-humidity). From the HEPA filters the gas stream is discharged to the atmosphere. Spent water from the flue gas coolers and demister is collected and particulate matter is removed in the

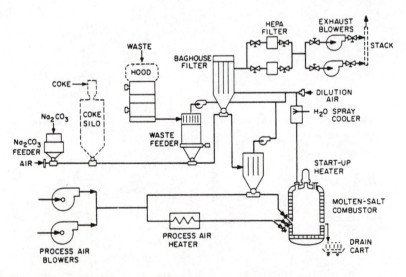

Fig. 16-7. Schematic of the Atomics International molten-salt combustion system. *Source:* Rockwell International, Energy Systems Group, Canoga Park, CA.

form of a sludge. The filter water is reused while the sludge is brought to the furnace where it is disposed of within the melt.

This incinerator is universal in its application to LLW producing a sterile product (glass) that is but a fraction of the initial charge. Even the filters used in this process can be disposed of within the melt.

OTHER SYSTEMS

There is no single design or design criteria for LLW. Each incinerator has been designed for a specific waste application and set of circumstances. As noted in Table 16-2, many types of incinerators have applications for LLW. For instance a rotary-kiln LLW incinerator is shown in Fig. 16-6, a molten salt incinerator for LLW is shown in Fig. 16-7 and a pyrolysis unit is shown in Fig. 16-8.

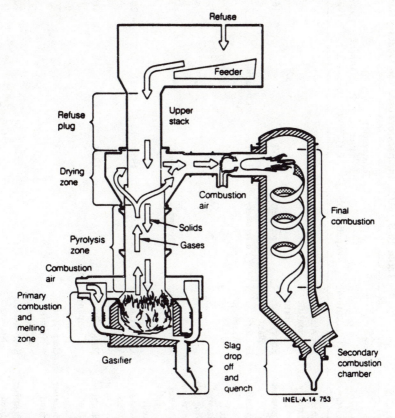

Fig. 16-8. Andco-Torrax slagging-pyrolysis incinerator. *Source:* ANDCO-TORRAX Corp.

Table 16-3. U.S. Incinerators used for Treating Radioactive Wastes.

Type of incinerators	Location	Application	Construction date (yr)	Pretreatment required	Capacity (kg/hr)	Operating temperatures (°C)	Remarks
Single basket; electrically heated	Los Alamos Scientific Laboratory	Pu recovery	1952		<1	800	
Dual chamber	Rocky Flats Plant	Pu recovery	1959		16	1200–1400	Refractory PuO_2 can form
Dual chamber; electrically heated; horizontal moving belt (woven wire)	Hanford, Washington	Pu recovery	1961	Chopping	2	700–800	Maintenance is different
Single basket; electrically heated	Mound Facility	Pu recovery	~1972		23	800	
Single chamber gas fired	Union Carbide Company (Y-12 Plant)		1955		20	870	
Dual chamber; gas fired	National Lead Company (Ohio)		1954		1000	980	
Dual chamber, gas fired	Gulf General Atomics (California)		1963		20	900–1200	
Dual chamber; gas fired	Kerr-McGee Nuclear (Ohio)		1972		80		

Design	Company	Application	Year	Pretreatment	Capacity	Temperature (°C)	Remarks
Single chamber	Babcock & Wilcox Company (Virginia)	U recovery	1972		80	1090	
Vortex, gas fired	General Electric Company (North Carolina)	U waste	1972	Shredding to uniform size	450	1000–1100 (815–980)	Feed is blown tangentially into burning chamber
Dual chamber, excess air burner	Westinghouse Nuclear Fuel Division (South Carolina)	Uranium	1974			650–1200	
Dual chamber; gas fired	Nuclear Fuel Services (Tennessee)	Uranium			270		
Dual chamber; gas fired	Goodyear Atomic (Ohio)	U recovery	1971		68	815–1000	
Dual chamber; gas fired	Oak Ridge Gaseous Diffusion Plant	U recovery	1972			930–1100	
Cyclone air feed	Yankee (Rowe) Nuclear Power Plant	Dry low-level waste	1968		18		
Dual chamber; controlled pyrolysis; vertical retort; rotary grate	Battelle Pacific Northwest Laboratory		Designed 1974	Shredding	15	Retort: 100–1200 Afterburner: 800	Not funded in 1976

Source: Ref. 16-8.

Table 16-4. Some Incinerators used Outside the United States for Treatment of Radioactive Wastes.

Type of Incinerator	Location	Application	Construction date (yr)	Pretreatment required	Capacity (kg/hr)	Operating Temperature (°C)	Remarks
Continuous slagging pyrolysis; movable paddles	Belgium	Solid waste	1975	Sorting and mixing	100	1400–1600	Off-gas passes through sand filters before HEPA
Dual chamber, batch pyrolysis; movable grate	W. Germany (Juelich)	Solid waste	1977	Hand sorting	100	1000	
Vertical retort; afterburning in hot ceramic candle filters; initial gas blast— 300 to 850°C	W. Germany (Karleruhe)	Dry low-level waste	1970	Sorting	60	1000–1100	
	Switzerland (EIR-Wuerenlingen)	Dry low-level waste		Sorting	25–30	1000	
	Japan (Tauruga)	Power plant waste	1977	Sorting	50	1000	
	(Tokai)		1979	Sorting	100	1000	
	Austria (SEAE-Sieheradorf)	Dry low-level waste	1975 (Controlled)	Sorting	60	1000	Hand loaded; no PVC permitted in feed
Vertical retort; dual chamber; batch operation; oil fired	Sweden (Studuvik)		1976	Sorting	200–400		
Vertical two-zone furnace; continuous operation	France (Cadarache)	Solid and liquid waste	1980	Crushing	30 (solid) 10–15 l/hr (liquid)		

Process	Location	Waste type	Year	Preprocessing	Capacity	Temperature (°C)	Off-gas system
Batch pyrolysis (TRECAN)	Canada (Bruce Nuclear Power Develop)	Low-level solid waste and organic liquids	1971	Sorting	2270 kg/batch (solid) ~45 l/hr (liquid)	870–900 (afterburning)	Bag filter in off-gas stream
	(Chalk River)		Under construction		1135 kg/batch		
Dual chamber; controlled air	United Kingdom (Windscale)	Plutonium	1972	Sorting	3–5	900	Wet scrubber system
Continuous dual chamber, electrically heated	France (Marcoule)	Plutonium	1970	Cutting			
Batch dual chamber	France (Marcoule)	Solid waste	Before 1970	Hand sorting	80		Off-gas passes through rotating filter
	(CEM-FAR)	Animal carcasses only			50	900	Bag filter on off-gas stream
Horizontal continuous furnace	France (Gadarache)	Liquids, organo-chlorides, and organo phosphates	1980		50 l/hr		
Horizontal batch furnace; stationary grate; excess air (Wellman)	United Kingdom (Winkley Point)	Low-level power plant waste	1977	Sorting and bagging	70	900	Off-gas system may need modification
Acid digestion	United Kingdom (Maxwell)	Pu recovery	(1981)	Shredding	10		

Source: Ref. 16-8.

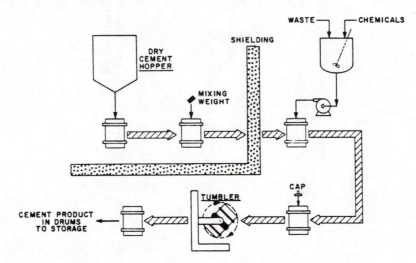

Fig. 16-9. An in-drum method for mixing radioactive waste with cement. *Source:* Ref. 16-3.

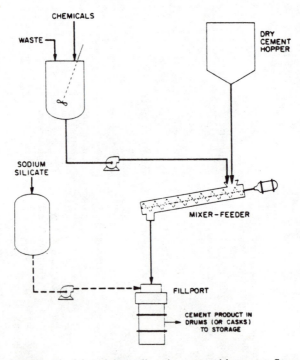

Fig. 16-10. An in-line method for mixing radioactive waste with cement. *Source:* Ref. 16-9.

In all of these incinerators each discharge is carefully controlled to catch any radioactive component before it can enter the air or water environment.

Table 16-3 lists many LLW facilities in use in the United States and Table 16-4 lists foreign LLW facilities.

Equipment for ultimate disposal, i.e., the mixing of LLW with cement, is shown in Fig. 16-9 and Fig. 16-10. Fig. 16-8 illustrates an in-drum mixing method where cement and waste are mixed within a container whereas in Fig. 16-10 the waste and cement are mixed external to the drum. Sodium silicate and other chemicals are often used to aid in solidification of liquids and sludges but will normally not be required with ash or other solid wastes.

Chapter 17
Miscellaneous Waste Destruction Processes

There are a number of waste destruction systems which do not utilize burning processes. Some of these systems are described in this chapter.

MOLTEN SALT INCINERATION

The molten salt process has been developed by Atomics International (division of Rockwell International Corporation) to dispose of a wide variety of wastes: solids, liquids and sludges. It is not a true incineration process because flame is not produced. But it is an oxidation process and a recombinant process where wastes are oxidized and/or chemically altered to innocuous substances.

Soluble alkali salts are used as the bed material. A single salt or a mixture of salts can be used. The more common salts in use are sodium chloride (NaCl), sodium sulphate (Na_2SO_4), sodium phosphate (Na_3PO_4), sodium carbonate (Na_2CO_3), and corresponding calcium salts.

The salt bed is heated to fluidization, its temperature a function of the salt material utilized. Waste is prepared before feeding, if solid waste, by shredding, and is fed directly into the molten salt bed if liquid or sludge. Typical temperatures of the bath are in the range of 1500–1800°F.

A schematic diagram of a molten salt incinerator is shown in Fig. 17-1. Waste components dissolve within the melt producing an off-gas. The off-gas will contain carbon dioxide, moisture (steam), and oxygen and nitrogen from the air supply. It will also contain particulate matter, salt, and other components generated within the melt. After particulate removal the gas is discharged to the atmosphere.

The volume of the melt increases as waste is added to the system and must be tapped periodically. Table 17-1 shows typical combustion products within the melt for particular waste components. Many of these compounds can be separated from the salt and the salt can be recycled to the incinerator for reuse, as shown in Fig. 17-1.

After startup, the waste stream may have sufficient heat content to maintain the salt bed in a hot, fluid condition without the addition of supplemental fuel.

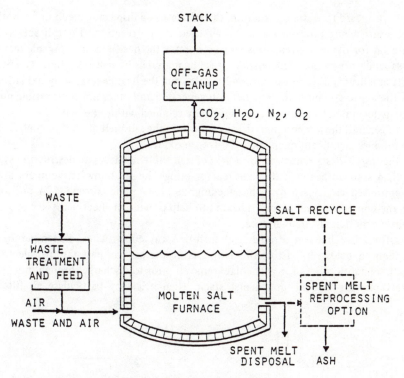

Fig. 17-1. Molten salt incinerator. *Source:* Rockwell International, Energy Systems Group, Canoga Park, CA.

Table 17-1. Molten Salt Combustion Process Combustion Chemistry.

Salt: Sodium Carbonate
Oxidizing Agent: Air

Element in Feed	Combustion Product
Carbon	carbon dixoide
Hydrogen	steam
Chlorine, fluorine	sodium chloride, fluoride
Phosphorus	sodium phosphate
Arsenic	sodium arsenate
Sulfur silicon	sodium sulfate
Silicon (glass)	sodium silicate
Iron (stainless steel)	iron oxide
Silver (photo film)	silver metal

Source: Rockwell International, Energy Systems Group, Canoga Park, CA

With regard to waste destruction, the salt acts as a dispersing medium for both the waste being processed and the air used in the reaction. The salt acts as a catalyst for oxidation reactions and accelerates the destruction of organic materials while preventing the emission of acidic materials by neutralization. The bed acts as a heat sink for absorbing and distributing the heat generated by oxidation of the waste components. Further, ash generated and other noncombustible materials generated by the waste are physically retained within the melt.

In general, the temperatures developed within the salt melt are not high enough to produce significant quantities of nitrogen oxides.

Figure 17-2 is a schematic of a generalized molten salt waste destruction system. A start-up heater within the reactor brings the salt to molten temperature. Compressed air is used for solids feeding, i.e., pneumatic conveying. A portion of the compressed air supply is heated to help provide the heat necessary to sustain the molten salt temperature.

Off-gas is exhausted through a quench or spray cooler where its temperature is reduced below 300°F. At this temperature the gas can pass through electrostatic or fabric filters for particulate removal prior to discharge through induced draft blowers feeding an exhaust stack. High-efficiency particulate air filters

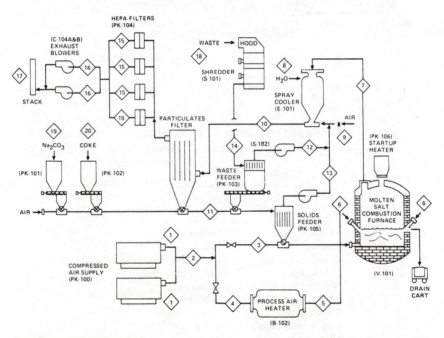

Fig. 17-2. Molten salt system flow schematic. *Source:* Rockwell International, Energy System Group, Canoga Park, CA.

(HEPA filters) may or may not be necessary, depending on the particulate loading and size distribution.

Liquid and sludge waste can be taken directly to the incinerator and charged beneath the melt surface. Solid waste, as shown as item 18 in Fig. 17-2, is shredded to uniform size and is then transferred to a waste feeder. Shredded solid waste, residue from the particulate filter, fresh salt and coke are conveyed, through a single pipeline, to a solids feeder which in turn feeds the furnace.

As waste is added the melt level increases and overflows to a drain cart. When the ash and other inerts build up to 20% of the melt, as sampled in the melt overflow, the melt must be tapped and fresh salt added. The purpose of the coke is to provide a source of carbon, i.e., combustion, to aid in startup and to act as a supplemental fuel when the waste heating value is insufficient.

Spurious emissions or leakages from the waste feeder and solid feeder are collected and passed through the off-gas particulate removal system to provide essentially "zero-discharge" from the system.

FLASH DRYING

Flash drying is a system used for disposal of various sludge wastes. Sludge waste is dried and can then be burned or disposed of by another method or within the flash drying system. Fig. 17-3 is a schematic of a flash drying system. Its principal elements are a hot gas heater, sludge mixer, cage mill, cyclone collector, vapor fan and dry product conveyor.

The operation of this system is as follows:

Wet sludge is blended with a small quantity of dried sludge in the mixer to improve its transportability. This mixed sludge is fed to a windbox on the cage mill where hot gas, from 1000 to 1400°F (depending upon the nature of the sludge), contacts the sludge. Moisture begins to evaporate into steam. The mixture of sludge, hot gas, and steam is ground together in the cage mill. By the time the sludge leaves the mill it is virtually dry, with only 8–10% moisture. The dried sludge is pneumatically conveyed to a cyclone where separation of dry sludge from gas and moisture vapor occurs, at approximately 300°F.

A dry sludge is discharged by the dry product conveyor. This material can be burned to produce the hot gas required for initiation of the process or the dry product can be incinerated in another furnace or it can be taken away for other uses. The furnace shown in Fig. 17-3 utilizes conventional fuel for generating hot gas; the dry sludge is removed from the process as a dry material.

The spent gases from the cyclone must be deodorized and discharged. In this scheme a vapor fan blows these gases from the cyclone to a vertical heat exchanger (recuperator) within the furnace. The spent gas picks up heat from the

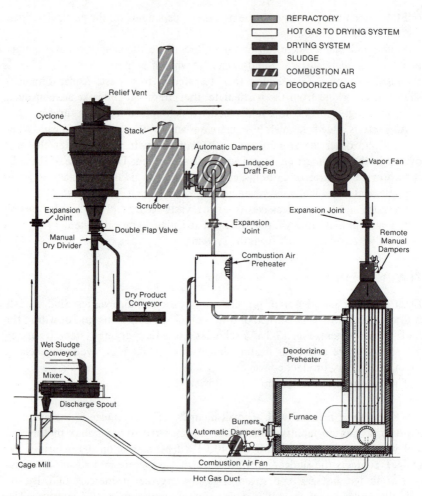

Fig. 17-3. Flash dryer system. *Source:* C-E Raymond, Combustion Engineering, Inc. Chicago, IL.

recuperator and is heated further, within the furnace, to 1400°F. At this temperature, with an appropriate residence time, the odor components will destruct. The hot gas will pass over the tubes in the recuperator, losing some of its heat to the incoming gas. The gas continues to another heat exchanger, a preheater, which heats combustion air prior to its injection into the furnace, adjacent to the supplemental fuel burner. The spent gas will exit the preheater and will discharge the stack at approximately 500°F.

The flash drying process is a rapid process. Drying is practically instantaneous in the cage mill. In the cyclone, separation resulting from the flashing of the hot moisture occurs in from six to ten seconds.

THE CARVER-GREENFIELD PROCESS

The Carver-Greenfield process is a method of drying moisture-laden liquid or sludge through heating and evaporation. Fig. 17-4 illustrates a typical system.

After the feed is passed through a grinder to give it a uniform consistency, it is pumped to a fluidizing tank where an oil is added to it. The feed/oil mixture passes through a series of evaporators where the moisture is evaporated leaving the solids suspended in the oil as a fluid slurry. Four stages of evaporation are shown in Fig. 17-4; however, the number of stages required is a function of the feed properties and as few as two stages may be sufficient for some materials.

The feed is pumped through a heat exchanger prior to each stage of evaporation and ultimately is discharged to a centrifuge. The centrifuged solids contain less than 1% of the original oil added to the feed and less than 2% by weight of moisture. The dry solids can be burned to provide steam for the evaporators, or, as with flash dryers, they can be utilized external to the system.

If additional oil removal is required a hydroextractor can be used, as shown in Fig.17-4. This is a steam operated stripper where higher-temperature superheated steam is used to boil off the residual oil within the dried sludge.

The balance of the equipment used in this process is involved in the capture and recovery of the carrier oil, its reuse, and the generation of steam.

Petroleum based oils, animal oils, or vegetable oils can be used in this process, depending on the nature of the sludge and its means of ultimate disposal.

The off-gases, including spent steam, from this system are normally burned in the steam boiler as combustion air. The only atmospheric discharge is the stack from the steam boiler. The only other discharge is the dried product.

WET AIR OXIDATION

Wet air oxidation is a process whereby a sludge will release bound water upon application of heat. Sludges which are colloidal gels, such as organic sludges, are composed of minute particles bound by sheaths which contain water as well as sludge solids. Upon application of heat the sheath will dissolve, releasing the bound water within the sludge particle. Upon release of bound water the sludge solids can be dewatered or can be further processed.

Thermal conditioning is a low pressure wet air oxidation process where the end product, thermal conditioned sludge, is dewaterable. High pressure wet air oxidation produces a sludge that is not only dewaterable but that has been oxidized.

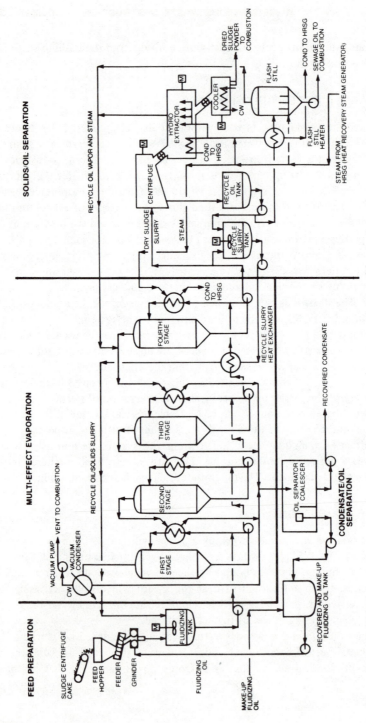

Fig. 17-4. CARVER-GREENFIELD PROGRESS. *Source:* Foster Wheeler/Carver Greenfield, Livingston, NJ.

Referring to Fig. 17-5, raw sludge is first ground to uniform consistency. A low-pressure pump feeds this sludge to a high-pressure positive displacement pump which develops sufficient pressure to pump to the reactor. For thermal conditioning the reactor pressure will be 175–200 psig and for high-pressure oxidation the reactor pressure can be as high as 1200–2400 psig.

Compressed air is injected into the sludge feed. Air is required for both thermal conditioning and wet air oxidation to provide turbulence within the sludge stream, increasing heat transfer efficiency through the heat exchanger(s). It is also required for oxidation of some of the malodorous gaseous compounds released with the release of bound water. High-pressure wet air oxidation, however, requires additional air for oxidation of the sludge solids. Depending on the amount of air injected into the sludge, and system pressure, the conditioned sludge can be partially oxidized or can be fully oxidized to a sterile ash.

Steam is injected within the reactor to provide the required reactor pressure and residence time. Temperatures range from 300 to 650°F and normal residual time within the reactor is from 10 to 30 minutes.

The hot sludge exiting the reactor passes through one or a series of heat exchangers where most of the available heat is transferred to the entering raw sludge to reduce the total system heating requirements. Raw sludge is pumped through the heat exchanger tubes and the hot conditioned sludge passes through the shell side of the heat exchanger.

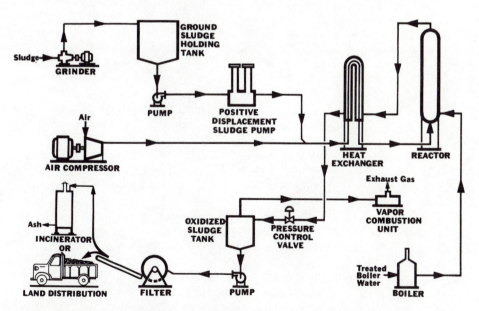

Fig. 17-5. Flowsheet for high pressure/high temperature wet air oxidation. *Source:* ZIM-PRO, Division of Sterling Drug, Inc., Rothschild, WI.

From the heat exchanger the conditioned sludge goes to a separator, or storage tank, where a quiescent period is provided. Gases are released and these gases are highly odorous. They are normally incinerated by direct flame or catalytic incineration before discharge to the atmosphere.

When a low pressure thermal conditioning process is used relatively little air is added in the system and the resultant sludge has approximately the same heating value (volatile content) as the raw sludge. Steam (or in some cases hot water) must be provided to maintain this reaction.

The high pressure oxidation process, where relatively large quantities of air are introduced into the sludge stream, can produce a sterile ash, where the initial volatile content is completely oxidized and the resultant sludge has no heating value. Other than for initial startup, this process will often generate enough heat by oxidation of the volatiles to maintain itself without steam (or hot water) injection.

Chapter 18
Energy Recovery

Incineration produces heat which can often be reclaimed and utilized. In one step, i.e., incineration, a waste can be destroyed and energy reclaimed by converting the waste heat to fuel-saving energy. Techniques for determining the energy available from waste heat will be presented in this chapter.

RECOVERING HEAT

Steam is used for incinerator heat recovery far more frequently that hot water or hot air (gas) generation. Steam is more versatile in its application and one pound of steam contains significantly more energy than does one pound of water or air. In general, while hot water is normally of use only for building heat during winter months, or can be used in limited quantities for feedwater heating, steam can be used for process requirements and for equipment loads, which are year-round loads. Further, steam can be converted to hot water or used for air heating when these needs arise.

The calculations presented in this chapter are for steam generation.

APPROACH TEMPERATURE

With t the temperature of the heated medium, steam or hot water; t_i the temperature of the entering flue gas; and t_o the temperature of the flue gas entering the boiler (see Fig. 18-1) the heat available can be calculated.

For any heat exchanger there is an approach temperature t_x. This temperature is the difference between the temperatures of the heated medium (t) and of the exiting flue gas (t_o). Therefore,

$$t_x = t_o - t$$

The more efficient the heat exchanger, the less will be the approach temperature. The larger the heat exchanger, the less will be t_x until, in the extreme case,

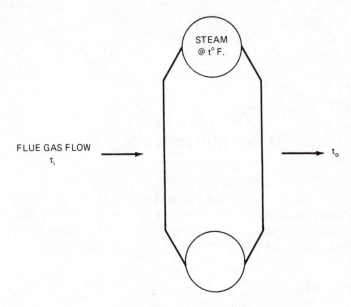

Fig. 18-1. Waste heat boiler.

with an infinitely large heat exchanger, t_x will be zero and the steam (or hot water) will be at the same temperature as the exiting flue gas.

In practice the approach temperature of a waste heat boiler is on the order of 100°F for very efficient and 150°F for standard, economical construction.

AVAILABLE HEAT

The heat available in exhaust or flue gas is equal to that heat at the boiler inlet less the gas heat content at the boiler outlet.

With Q the heat available from the flue gas stream, in BTU/lb, note the following:

$$Q = W(h @ t_i - h @ t_o)$$

The flue gas will have a dry and a wet component. Considering the dry gas component to have the properties of air (W_{dg}, h_a), and W_m the moisture component, this equation becomes:

$$Q = W_{dg}(h_{ai} - h_{ao}) + W_m(h_{mi} - h_{mo})$$

The inlet temperature t_i is the temperature of the incinerator outlet. The outlet temperature of the heat exchanger (t_o) is defined by the approach temperature (t_x) and the temperature of the heated medium (t):

$$t_o = t_x + t$$

The enthalpy at the outlet of the heat exchanger must be evaluated at t_o.

Example

Consider a gas flow, 1400°F, of 15000 lb/hr of dry gas plus 2000 lb/hr of moisture. Let the available heat be calculated for the generation of saturated steam at 100 psia, 200 psia and 400 psia, with a 150°F approach temperature:

Inlet condition:
$$t_i = 1400°F$$
$$h_a = 341.85 \text{ BTU/lb}$$
$$h_{mi} = 1719.82 \text{ BTU/lb}$$
$$W_{dg} = 15000 \text{ lb/hr}$$
$$W_m = 2000 \text{ lb/hr}$$

Outlet condition (note Table 1):

With $p = 100$ psia,
$$t = 328°F$$
$$t_o = 150 + 328 = 478°F$$

by interpolation,
$$h_{ao} = 101. \text{ BTU/lb}$$
$$h_{mo} = 1249. \text{ BTU/lb}$$

$$Q = 15000(341.85 - 101.) + 2000(1719.82 - 1249)$$
$$= 4.554 \text{ MBH}$$

With $p = 200$ psia,
$$t = 382°F$$
$$t_o = 150 + 382 = 532°F$$

by interpolation,
$$h_{ao} = 115. \text{ BTU/lb}$$
$$h_{mo} = 1274. \text{ BTU/lb}$$

$$Q = 15000(341.85 - 115.) + 2000(1719.82 - 1274)$$
$$= 4.294 \text{ MBH}$$

With $p = 400$ psia,
$$t = 445°F$$
$$t_o = 150 + 445 = 595°F$$

by interpolation,
$$h_{ao} = 130. \text{ BTU/lb}$$
$$h_{mo} = 1304. \text{ BTU/lb}$$

$$Q = 15000(341.85 - 130.) + 2000(1719.82 - 1305.)$$
$$= 4.007 \text{ MBH}$$

Table 18-1.

Inlet, MBH	°F	Steam Pressure, psia	Steam Temp., °F	Flue Gas Temp., °F	Δt, °F	Available Heat, MBH	Effic., %
8.567	1400	100	328	478	922	4.554	53
8.567	1400	200	382	532	868	4.294	50
8.567	1400	400	445	595	805	4.007	47

These calculations are summarized in Table 18-1. The inlet is the total heat in the flue gas, related to 60°F:

$$15000 \times 341.85 + 2000 \times 1714.82 = 8.567 \text{ MBH}$$

The column Δt is the difference in flue gas temperature entering and leaving the boiler. The efficiency noted is the available heat divided by the total heat in the flue gas entering the boiler.

Of significance is the relationship between available heat and Δt, the temperature difference of the flue gas across the boiler. The available heat is proportional to Δt.

For example: Q at $\Delta t = 805$ versus $\Delta t = 922$,

$$Q @ 805 = \frac{805}{922} \times 4.554 = 4.0 \text{ MBH}$$

Q at $\Delta t = 868$ versus $\Delta t = 922$,

$$Q @ 868 = \frac{868}{922} \times 4.554 = 4.3 \text{ MBH}$$

Comparing these values to the calculated values for Q in Table 19-1 it is clear that the available heat Q is directly proportional to Δt, the temperature loss in the flue gas.

STEAM GENERATION

Knowing the heat availability, the amount of steam that can be generated can be calculated. Fig. 18-2 shows typical flow through a waste heat boiler producing steam. Makeup water temperature is raised to feedwater temperature by steam flow from the boiler and by the heat contained in return condensate. Condensate is returned to the deaerator.

Beside raising the feedwater temperature prior to injection into the boiler,

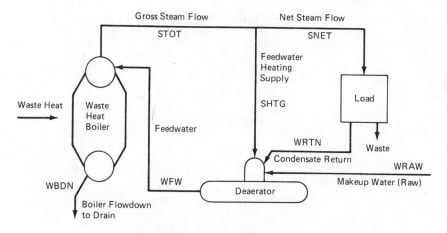

Fig. 18-2. Waste heat boiler, steam flow.

the deaerator acts to help release dissolved oxygen from feedwater. Additional feedwater treatment is usually employed to reduce, or prevent, scaling and corrosion of boiler surfaces. Water softeners are used to remove most of the calcium and magnesium hardness from raw water. Chemical addition is also used, typically as follows:

Sodium sulfite. This is an oxygen-scavenging chemical that chemically removes the dissolved oxygen component not removed in the deaerator. Hydrazine is another oxygen scavenger that is used in high pressure (over 1200 psia) boiler applications.

Amine. There are a number of amines in use for feedwater treatment. They are used for boiler pH or alkalinity control. Excess alkalinity (pH greater than 11) will result in accelerated scale build-up while low pH (below 6) can cause excessive boiler tube corrosion. Normally boiler water pH is maintained in the range of 8.0–9.5, slightly alkaline.

Phosphates. This treatment is used to precipitate residual calcium and magnesium hardness remaining in feedwater after softening. Certain phosphates will act as dispersants preventing adhesion of the precipitate to tube walls.

These chemicals will form a sludge, or mud, which will accumulate in the lower drum of a boiler. The boiler water must have a blowdown on a regular basis to prevent a build-up of mud within the boiler. This blowdown will normally represent from 2 to 5% of the boiler steam generation.

CALCULATING STEAM GENERATION

Using the steam tables (Table 18-2), calculations will be performed for obtaining steam, makeup, blowdown, and feedwater flows.

Table 18-2. Saturated Steam Properties.

Pressure, psia	Temperature, °F	Pressure, psia	Temperature, °F
14.7	212	45	274
15	213	50	281
16	216	55	287
17	219	60	293
18	222	65	298
19	225	70	303
20	228	75	308
21	231	80	312
22	233	90	320
23	235	100	328
24	238	125	344
25	240	150	358
26	242	175	371
27	244	200	382
28	246	250	401
29	248	300	417
30	250	350	432
32	254	400	445
34	258	450	456
36	261	500	467
38	264	600	486
40	267	700	503

Source: Ref. 18-17.

Using an approach temperature of 150°F, generating 100 psia steam, dry and saturated (h_{stm} = 1187 BTU/lb, h_{bdn} = 298 BTU/lb), the heat available (from page 307) is 4.554 MBH. For this illustration, blowdown is 4% of the feedwater flow, feedwater is provided to the boiler at 220°F (h_{fw} = 188 BTU/lb) and 20% of the steam generation is returned as condensate at 170°F (h_{rtn} = 138 BTU/lb). In addition, raw water enters the deaerator at 60°F (h_{mu} = 27 BTU/lb) and radiation loss from the boiler is 1% of the total boiler input.

The heat available for generating steam is the heat lost by the flue gas less the heat lost by boiler radiation:

Waste Heat = 4.554 MBH – 0.01 X 4.554 MBH = 4.508 MBH

Referring to Fig. 18-2:

Waste Heat = Heat in Steam + Heat in Blowdown – Heat in Feedwater

Heat in Blowdown = WBDN X h_{bdn}
WBDN = 0.04 X WFW
Heat in Blowdown = 0.04 X 298 X WFW = 11.92 WFW

$$\text{Heat in Feedwater} = \text{WFW} \times h_{fw}$$
$$= 188 \text{ WFW}$$

$$\text{Heat in Steam} = \text{STOT} \times h_{stm} = \text{STOT} \times 1187 \text{ BTU/lb}$$
$$\text{STOT} = \text{WFW} - \text{WBDN} = \text{WFW} - 0.04 \times \text{WFW} = 0.96 \text{ WFW}$$
$$\text{Heat in Steam} = 0.96 \text{ WFW} \times 1187 = 1139.52 \text{ WFW}$$

$$\text{Waste Heat} = 4.508 \text{ MBH} = 11.92 \text{ WFW} + 1139.52 \text{ WFW} - 188 \text{ WFW}$$
$$= 963.44 \text{ WFW}$$

Therefore: $\quad\quad\quad\quad$ WFW = 4679 lb/hr
and: $\quad\quad\quad\quad\quad\quad$ STOT = 0.96 × 4679 = 4492 lb/hr
also: $\quad\quad\quad\quad\quad\quad$ WBDN = 0.04 × 4679 = 187 lb/hr

To calculate steam required for feedwater heating, make-up, and condensate return flows, a material balance and a heat balance must be performed around the deaerator.

Material (flow) balance:

$$\text{WFW} = \text{SHTG} + \text{WRTN} + \text{WRAW}$$

From above: $\quad\quad\quad$ WRTN = 0.2 STOT = 0.2 × 4492 = 898 lb/hr
$\quad\quad\quad\quad\quad\quad\quad$ WFW = 4679 lb/hr

Therefore:

$$\text{SHTG} = \text{WFW} - \text{WRTN} - \text{WRAW} = 4679 - 898 - \text{WRAW} = 3781 - \text{WRAW}$$

Heat balance:

$$\text{WFW} \times h_{fw} = \text{SHTG} \times h_{stm} + \text{WRTN} \times h_{ret} + \text{WRAW} \times h_{mu}$$

Therefore:

$$4679 \times 188 = (3781 - \text{WRAW}) \times 1187 + 898 \times 138 + \text{WRAW} \times 27$$

$$\text{WRAW} = 3218 \text{ lb/hr}$$
$$\text{SHTG} = 3781 - 3218 = 563 \text{ lb/hr}$$

Where condensate is not returned to the deaerator, i.e., to the boiler system, the steam required for feedwater heating will increase:

Material balance:

$$\text{SHTG} = \text{WFW} - \text{WRAW} = 4679 - \text{WRAW}$$

Heat balance:

$$4679 \times 188 = (4679 - \text{WRAW}) \times 1187 + \text{WRAW} \times 27$$

Therefore:

$$\text{WRAW} = 4030 \text{ lb/hr}$$
$$\text{SHTG} = 649 \text{ lb/hr}$$

In general, with no separate source of heat for feedwater heating (such as returned condensate), 12–15% of generated steam is required.

Table 18-3 summarizes the above calculations.

The net flow of steam (SNET) is that quantity of steam available for useful work. As can be seen, use of condensate return for feedwater heating increases the quantity of steam available for a load.

WATERWALL SYSTEMS

Larger incinerators dedicated to destruction of refuse or other paper-type waste materials are often designed with "waterwall" construction, as described in a Chapter 11. These installations can be provided with a variety of features including the following:

- Convection boiler section. Boiler tubes placed perpendicular to the flow of gas as it exits the incinerator. A major portion of available heat is captured by these tubes, producing saturated steam.
- Economizer. Used to heat feedwater by extracting heat from gases as they leave the convection boiler section.
- Superheater. A tubular section normally placed upstream of the convection section. Hot incinerator gases superheat steam generated from the convection section of the boiler.
- Air preheater. Used in lieu of, or directly downstream of, the economizer. It produces heated combustion air from the relatively low-temperature gas flow at this location.

Table 18-3.

Avail. Heat, MBH	STOT, lb/hr	WBDN, lb/hr	WFW, lb/hr	WRET, lb/hr	WRAW, lb/hr	SHTG, lb/hr	WNET, lb/hr
4.554	4492	187	4679	898	3218	563	3929
4.554	4492	187	4679	0	4030	649	3843

Calculations of steam generation from each of these sections are a complex task and will not be detailed here. Tables 11-8 and 11-9 indicate steam generation for typical waterwall incinerators for a variety of waste quality.

To calculate available heat by the methods of this chapter, the exit gas temperature (the temperature of flue gas exiting the boiler sections and entering the air emissions control system) can be assumed to be in the range of 350–550°F.

Chapter 19
Air Pollution Control

Air discharges from incinerators include particulate matter, gases, odor and noise. In this chapter equipment available for air pollution control will be discussed.

THE CONTROL PROBLEM

There are numerous types and sizes of air pollution control equipment and schemes on the market today. They range from the unsophisticated, a series of baffles to separate large particulate matter from a gas stream, to the highly sophisticated high energy water scrubbing devices utilizing alkalai to clean gas streams of solid, liquid and gaseous pollutants.

As presented in Chapter 7, and referenced in Chapters 2 and 3, incinerators will necessarily produce air pollutants which must be removed from the gas stream prior to atmospheric discharge. These pollutants must be removed to provide discharge levels below those mandated by governmental statutes. In many cases incinerator discharges must be free of additional components in order to safeguard downstream equipment from excessive wear from erosion (physical wear of materials) and/or corrosion (chemical degradation of materials.)

The generation of pollutants is a function of the following factors:

- Waste compositon.
- Charging rate.
- Method of charge.
- Furnace type.
- Furnace design.
- Burning conditions (three T's, i.e., temperature, turbulence, time).
- Excess air introduced.

The more common pollutants discharged from incinerators which must be substantially removed from the gas stream include the following:

- Particulate matter.
- Chlorides.
- Sulfur oxides.

Other pollutants, as discussed in Chapter 7, are not normally present in incinerator discharges to the extent that they warrant particular attention. If they are present in any significant amounts, however, they will often be removed by the systems installed for control of the three pollutants identified above.

PARTICLE SIZE

The efficiency of any control device will vary as a function of particle catch size. Particulate matter is normally measured by mean physical size in units of one millionth of a meter (microns), represented by the greek letter μ.

Particle size cannot be calculated; it can be measured and estimates can only be inferred from prior data based on measurement. As would be expected, therefore, data on particle sizes is rare. It is a function of waste properties, furnace design and furnace operation.

Table 19-1 lists particle size data measured at the outlet of a multiple hearth

Table 19-1. Sewage Sludge Incineration, Airborne Particle Size.

	Percent By Weight Less Than Indicated Size	
	Location 1[a]	Location 2[b]
Particulate loading,		
grains/SCF	1.88	0.01
pounds/hour	217.01	1.48
pounds/ton	52.08	0.36
Size distribution		
18.7 μ	37.9	100.0
11.7 μ	30.6	98.0
8.0 μ	16.4	94.9
5.4 μ	6.6	93.4
3.5 μ	2.6	92.8
1.8 μ	1.6	83.3
1.1 μ	0.9	67.7
0.76 μ	0.1	54.6

[a]Measured at incinerator outlet without controls, burning 100 TPD wet sludge cake.
[b]Measured after Venturi and tray scrubber with a total pressure drop of 30 in. WC.

Source: Envisage Environmental Laboratory, Cleveland, OH, "Tests of the Cleveland Southerly Wastewater Treatment Center Multiple-Hearth Incinerators, May 1981."

**Table 19-2. Refuse Incineration, Airborne
Particle Size.**

	Unit No. 1 (250 TPD)	Unit No. 2 (120 TPD)
Particle specific, gravity, lb/CF	2.70	3.77
Particle bulk density, lb/CF	30.9	9.4
Loss on ignition at 1400° F, %	8.2	30.4
Size distribution (% by weight less than indicated size):		
30 μ	40.4	50.0
20 μ	34.6	45.0
15 μ	31.1	42.1
10 μ	26.8	38.1
8 μ	24.8	36.3
6 μ	22.3	33.7
4 μ	19.2	30.0
2 μ	14.6	23.5
Particulate emission rate,		
lb/ton	24.6	9.1
lb/hour	256.3	45.5

Source: Ref. 19-21.

incinerator burning thermally conditioned sewage sludge. It lists the particulate size and loading both before and after the gas scrubbing system.

Listed in Table 19-2 are particle size and emissions data from each of two incinerators burning municipal refuse.

Note that with sludge or refuse over half the particulate is larger than 30 μ. The scrubber, as shown in Table 19-1, removes practically all of the particulate matter over 5 μ in diameter.

Table 19-3 lists chemical elements found in a typical particulate catch after organic (carbonaceous) components present were removed.

Particle diameters and their general properties and characteristics are shown in Fig. 19-1. Some of the information in this table can be used to calculate the minimum size chamber required for settling of particulate. For example:

A 30 μ particle has a settling velocity of approximately 5 cm/sec, equivalent to $5 \div 2.54 \div 12 = 0.164$ ft/sec. Passing through a 2-ft-high chamber at 20 ft/sec the chamber length would have to be long enough to provide sufficient time for settling:

$$\frac{2 \text{ ft high}}{0.164 \text{ ft/sec}} = 12.2 \text{ sec to settle}$$

20 ft/sec velocity × 12.2 sec = 244 ft long

Table 19-3. Inorganic Component Analysis: Refuse Burning Incinerator.

Component	Weight Percent
SiO_2	53.0
CaO	14.8
MgO	9.3
Al_2O_3	6.2
Na_2O	4.3
TiO_2	4.2
K_2O	3.5
Fe_2O_3	2.6
P_2O_5	1.5
ZnO	0.4
BaO	0.2

Source: Ref. 19-32.

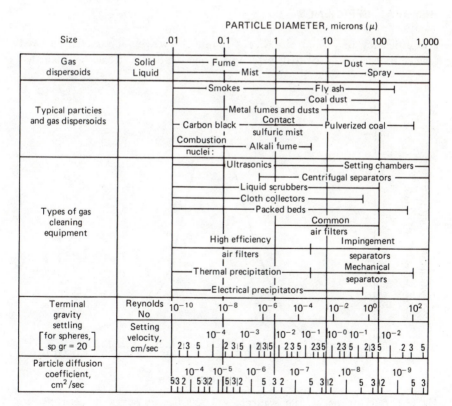

Fig. 19-1. Particle size data. *Source:* Ref. 19-35.

The time necessary for settling of a 30 μ particle would require a chamber length of 244 ft, assuming no turbulence, which would increase settling velocity.

This calculation indicates the impracticality of utilizing chambers for removal of small size particulate matter. As previously noted, half of the particulate generated from incineration processes may be less than 30 μ in diameter. The smaller the diameter, the larger the settling time, therefore, the larger the settling chamber required. Using the above example of a 244-ft-long chamber, the size of such equipment makes their use of questionable value.

Of interest is the fact that particles greater than 10 μ can be seen as individual particles by the naked eye.

Below 10 μ, although the eye cannot distinguish individual particles, there is a net effect of particle size on atmospheric clarity. As shown in Fig. 19-2, particulate in the range of 0.3–1.0 μ will severely affect visual clarity.

REMOVAL EFFICIENCY

The efficiency of a control device is simply the weight of material removed divided by the weight of that material entering the device. It can be calculated

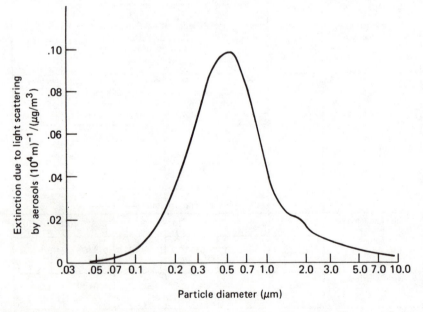

Fig. 19-2. Effectiveness of light scattering as a function of particle diameter. *Source:* Ref. 19-27.

for a specific size of particulate or can be based on total particulate loading. For example, using the particle size information in Table 19-1:

Total flow into control devices, 217.01 lb/hr for particles less than 1.1 μ:

$$0.9\% \times 217.01 \text{ lb/hr} = 1.95 \text{ lb/hr}$$

Total flow out of control device, 1.48 lb/hr for particles less than 1.1 μ:

$$67.7\% \times 1.48 \text{ lb/hr} = 1.00 \text{ lb/hr}$$

Removal efficiency of particles less than 1.1 μ:

$$\frac{1.95 - 1.00}{1.95} = 48.72\%$$

Removal efficiency of total particulate load:

$$\frac{217.01 - 1.48}{217.01} = 98.32\%$$

As would be expected, in this case the removal efficiency of small particulate matter is significantly less than the removal of larger particles—less than 50% of the 1.1 μ particles are removed, while over 98% of the total particulate input is removed.

AIR POLLUTION CONTROL EQUIPMENT

The choice of air pollution equipment depends not just on the emission quality and quantity but on conditions external to the incinerator system. For instance, if water is not readily available on a site wet scrubbing systems cannot be considered; limited sewer facilities may preclude the use of electrostatic precipitators where the precipitator soot loading is sewered; space limitations will impact choice of a baghouse for air pollution control; and so on.

EQUIPMENT SELECTION

There are a number of industries where air emissions control has been found to be effective with only one type of control device. This is particularly true with the incineration of municipal waste:

Municipal solid waste. In the vast majority of large-scale installations of refuse burning equipment electrostatic precipitators have been found to provide effective and reliable gas cleaning. Particulate matter is the more critical of emissions from refuse incinerators and the ability of electrostatic precipitators to remove particles down to fractions of a micron is unique. Other control devices have been used, including wet scrubbers and baghouses, but their efficiency (in the case of scrubbing equipment) and reliability (in the case of fabric filters) have not been found to be satisfactory. In the United States there are a limited number of installations utilizing new technology in control devices that show some promise, such as the Electroscrubber® in Pittsfield, MA, and the Teller Dry Scrubbing System, in Framingham, MA. However, as noted above, the electrostatic precipitator has a proven track record for this application.

Municipal sludge. The incineration of sewage sludge requires not only the control of particulate but the removal of gaseous components from the exhaust stream. In general, a sludge incinerator does not operate at the higher temperatures that are developed within a refuse incinerator. As a result, additional unburned organics may be present in the gas stream exiting a sludge incinerator as compared to the exhaust of a refuse incinerator. Wet scrubbers have been found to be effective in removal of gaseous as well as particulate components from sludge incinerator exhaust. In the majority of installations these scrubbers are high-energy devices, requiring in excess of 20 in. WC differential; however, their water flow requirement is usually not a limitation. Sewage sludge incinerators are located at wastewater treatment plants where vast quantities of water are available.

Other incinerator usage may be required to provide wet scrubbing if, for instance, acid components are generated and the waste is considered hazardous. In general, however, except for the above two instances, the variability of industrial waste streams precludes generalization of the type of control equipment to provide. Industrial applications each require their own analysis.

The design of air emissions control equipment for a particular installation requires not only selection of the appropriate type of device, flues, dampers, and air moving equipment must also be selected and sized. Sizing of ancillary equipment, and the control device itself, is in large measure a function of particle size distribution. As noted previously in this chapter, particle size cannot be calculated. Estimates can be made from previous measurements, such as the distribution of particle size listed in Tables 19-1 and 19-2. There is no certainty, however, that these data are representative of any other installation. The data are based on grab sampling, where a discrete number of samples have been obtained and each of these samples has been analyzed. At another time of day, this distribution could be different. If there were any change in waste quality, such as moisture content or heating value (which normally will vary from one lot to another), the size distribution will most likely change. These tables can be used as guides to relative magnitudes of particulate in off-gas, but they cannot necessarily be used as a design basis. Accurate size distribution can be determined, unfortunately, only by full-scale testing of actual equipment. Control device designers and manufacturers will normally be required to make their own estimates of size data to determine equipment sizing, based on their own experience and related empirical factors. An incinerator designer is usually not able to provide this information.

SETTLING CHAMBERS

As discussed previously, settling chambers (expansion chambers), which are basically long, boxlike structures, are ineffective for all but the heaviest, i.e., largest, particles. They are normally not used where the particulate size is 40 μ or less.

Some attempts have been made to utilize gravity settling by decreasing the height of the chamber, i.e., by placing horizontal plates in the direction of flow. With the previous calculations, if the vertical drop were reduced from 2 ft to 1 in., maintaining the same velocity of the gas stream (20 ft/sec) a length of $\frac{1}{24}$ of 260, or less than 11 ft, would be required for settling of larger particles.

In general, settling chambers are not practical for use as pollution control devices other than as part of a larger system to help remove larger particle sizes from the gas stream.

DRY IMPINGEMENT SEPARATORS

Impingement separators are essentially a series of baffles placed in the gas stream. The relatively high inertia of particulate within the gas stream will maintain their direction of flow while the gaseous component of the stream will change its direction of flow around a baffle. Particulate matter will drop and the gas flow will continue through the process stream.

This method of particulate removal can be effective for larger particles, above 15 μ, but smaller particulate matter will continue to flow with the gas stream.

DRY CYCLONIC SEPARATORS

The cyclone is an inertial separator. As shown in Fig. 19-3, gas entering the cyclone forms a vortex which eventually reverses and forms a second vortex leaving the cyclonic chamber. Particulate, because of their inertia, tend to move toward the outside wall. They will drop from this wall, the sides of the cylcone, to an external receiver for ultimate disposal.

Cyclones will remove larger-size particulate matter from the gas stream (greater than 15 μ), but will have negligible effect on smaller particles. A cyclone collector is often placed before another control device such as an electrostatic precipitator or a baghouse. The cyclone will remove larger particles from the gas stream,

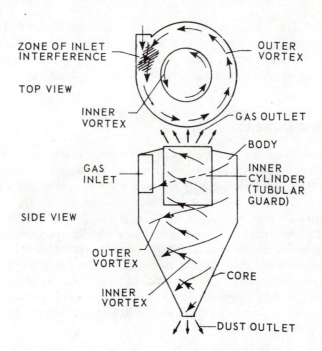

Fig. 19-3. Conventional reverse-flow cyclone. *Source:* Ref. 19-17.

and in many cases the cyclone solids discharge is tied directly to the incinerator, feeding collected particulate back to the furnace for complete burnout. Removal of these larger particles from the gas stream results in increased efficiency for the downstream control equipment, which now have only the smaller-sized particles to deal with. Of note is that the larger particles are normally the greatest weight component of particulate in a gas stream. For instance, 85% of the emissions weight may be over 15 μ, with only 15% below 15 μ. Use of a cyclonic collector, which has a good removal efficiency for larger particles, will allow provision of less efficient downstream control devices while providing a high overall removal efficiency, based on the total particulate size loading.

With D_{pc} the particle cut size, the ratio D_p/D_{pc} is the actual particle size D_p related to the particle cut size. The cut size is the diameter of those particles collected by a particular piece of equipment with 50% efficiency. Fig. 19-4 shows the relationship between particle size and collection efficiency. Collection efficiency drops off rapidly as the particle size decreases.

One danger in the use of cyclonic separators is the reluctance of collected particulate to drop from the cyclone walls. This condition, agglomeration, may occur if the dust is fibrous, sticky, or hygroscopic (water-absorbing), or if the gas stream contains too much particulate matter (100 grains per cubic foot is a practical maximum, with 7000 grains per pound).

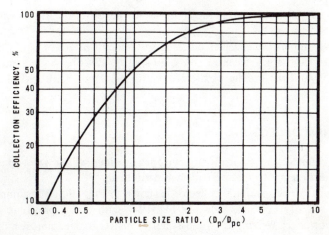

Fig. 19-4. Cyclone efficiency versus particle size ratio. *Source:* Ref. 19-28.

WET COLLECTION

Wet devices for particulate removal utilize two mechanisms. First, particles are wetted by contact with a liquid droplet. Second, the wetted particles impinge on a collecting surface where they are subsequently removed from that surface by a liquid flush.

One of the simplest devices for wet collection of particles is wetting a collecting baffle. Particulate is continually washed from the baffle, or wall, preventing agglomeration and presenting a continual clean liquid surface to aid in catch and removal.

LIQUID SPRAYING

A spray directed along the path of dust or other particulate matter will impinge upon these particles with an efficiency directly proportional to the number of droplets and to the force imparted to the droplets. This results in a droplet size range, i.e., the smaller the size of the individual droplets the greater their number. However, the smaller the droplet particle size the less impingement, or inertial force is associated with them. The optimum water droplet particle size is approximately 100 μ. Above 100 μ there are too few particles, and below 100 μ the droplets have insufficient inertia.

The mechanism of diffusion promotes deposition of dust particles on water droplets. Diffusion or Brownian movement, is that property of materials of different diameters to intermingle although the materials may be at rest, much as natural gas diffusing within a contained room although the air within that room is at rest. Diffusion helps the particulate and droplets come in contact and is inversely proportional to the size of water and/or solid particulate. The smaller the particle or droplet size, the more rapid the diffusion or the quicker the wetting process will be.

Spray chambers are at times utilized as a low-cost means of removing heavier particulate from a gas stream. Water is sprayed at rates of 3–8 gpm per 100 CFM of gas flow and the heavier particulate matter is wetted and drops out of the gas stream.

CYCLONE COLLECTORS

Cyclone collectors are basically dry cyclones provided with a water spray to promote collection and to remove collected particulate.

One type of wet cyclonic collector is shown in Fig. 19-5. The equipment is a conventional dry cyclone, shown in Fig. 19-3, with water sprays installed near the gas entrance. Wetting of the particulate matter occurs almost immediately upon entering the cyclone and the spray water washes the cyclone walls clean.

Figure 19-6 shows a cyclonic collector where gases enter tangentially from

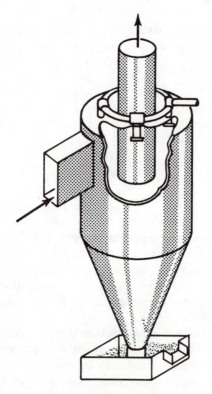

Fig. 19-5. Conventional cyclone converted to a scrubber. *Source:* Ref. 19-7.

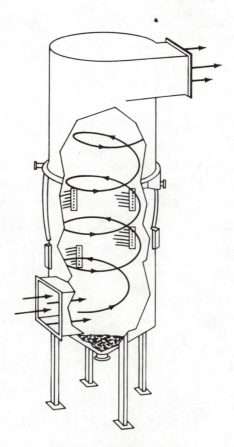

Fig. 19-6. Cyclonic scrubber.

the bottom of the unit. The placement of sprays within the scrubber promotes a cyclonic effect as the gas rises to the scrubber exit. As with the other cyclones, water collects from the walls of the scrubber and exits from the bottom of the unit.

Cyclone collectors require 4–10 gpm of water per 1000 CFM of entering gas flow. Their pressure drop can be as low as 2 or as high as 8 in. WC.

Dimensions of a typical wet cyclone are shown in Fig. 19-7.

WET CORROSION

Gases exiting an incinerator will contain sulfide and halogen components from the waste feed. These components, in addition to organic residuals, can result in significant acid formation in the gas scrubbing liquid. The use of nonmetallic chambers or nonmetallic-coated carbon steel is a growing method of corrosion

SATURATED GAS VOLUME (ACFM)*	MIST ELIM. SEP.		CYCLONE SEP.	
	A	B	A	B
2,500	2'-2"	7'-8"	3'-0"	8'-0"
5,000	3'-0"	8'-11"	4'-0"	10'-0"
10,000	4'-3"	10'-6"	5'-4"	15'-1"
25,000	6'-8"	14'-0"	7'-8"	20'-6"
50,000	9'-5"	17'-3"	10'-4"	27'-4"
75,000	11'-7"	19'-9"	12'-3"	31'-9"
100,000	13'-4"	21'-11"	13'-10"	37'-0"
150,000	16'-4"	25'-8"		
200,000	18'-10"	28'-10"	NOT RECOMMENDED	
300,000	23'-1"	33'-10"	(Parallel Units	
500,000	29'-9"	41'-3"	can be used)	

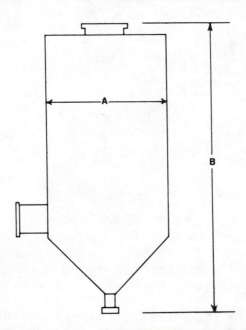

Fig. 19-7. Wet cyclone. *Source:* Koch Engineering Company, Inc., Pollution Control Division, New York, NY.

control. These materials, such as flake polyester, fiberglass, neoprene, and epoxy, have good anticorrosion properties; however, they are temperature sensitive. In most cases these materials cannot withstand temperatures above 160–200°F. Use of such materials must include provisions for emergency quenching to provide cooling if the normal cooling water supply is lost.

Table 19-4 lists the resistance of selected alloys to incinerator scrubber solu-

Table 19-4. Evaluation of Alloys for Incinerator
Scrubber.

Alloy	Corrosion Results	
	Scrubber Solutions	Fan Deposits
Ti6Al-4V	good resistance	good resistance
Hastelloy C	good	good
Inconel 625	good	good
Hastelloy F	good	–
Hastelloy C-276	good	–
Hastelloy G	good	–
Ti75A	good	–
S-816	good	pitted
Carpenter 20	pitted	pitted, SCC
Incoloy 825	pitted	pitted
Incoloy 800	–	pitted
316L	pitted, SCC	pitted, SCC
310	pitted	–
446	pitted	–
Inconel 600	trenches	–
Inconel 601	trenches	–
Armco 22-13-5	pitted	pitted, SCC
USS 18-18-2	pitted	pitted, SCC
Type 304	pitted, SCC	pitted, SCC

SCC: Stress corrosion cracking.
Source: Ref. 19-26.

tions having relatively high acidity. Also listed is the resistance of these alloys to corrosion when fabricated as a fan (induced draft fan) downstream of the scrubber. Note that stainless steel (316L, 310, 446 and USS 18-18-2) has poor corrosion resistance to acid attack.

MIST ELIMINATORS

Wet gas cleaning equipment floods a gas with scrubbing liquid. Water droplets usually are carried off by the gas stream exiting the scrubber. A mist eliminator is a stationary piece of equipment that removes most of the entrained water droplets from the gas. This is desirable to reduce the plume exiting the stack (excess water in off-gas will appear as a white plume), and will reduce the moisture collected at the induced-draft fan and other downstream equipment. Decreasing this liquid accumulation will also decrease attendant corrosion.

The most common mist eliminator in use with scrubbing equipment is the chevron type, Fig. 19-8. The moisture-laden gas flow travels through a series of baffles. Gas exits the baffles and the heavier moisture particles, because of their inertia, impact the chevron-shaped baffles and fall to the bottom of the chamber, leaving the gas stream.

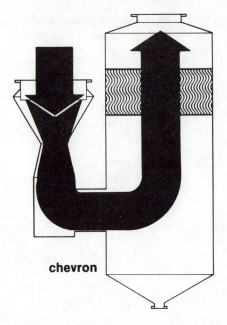

Fig. 19-8. Mist eliminator. *Source:* Peabody-Galion Co., Princeton, NJ.

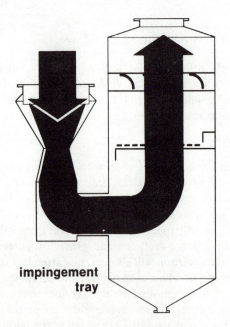

Fig. 19-9. Mist eliminator. *Source:* Peabody-Galion Co., Princeton, NJ.

Figure 19-9 is a diagrammatic representation of an impingement tray tower which functions as a mist eliminator. The wet gas stream passes through a perforated plate and then impacts on a series of small plates, impingement plates. As with the chevron scrubber, inertial forces allow the gas to flow around the impingement plates while the entrained moisture falls to the bottom of the tower. The use of impingement plates for gas scrubbing is discussed later in this chapter.

A packed tower used as a demister is shown in Fig. 19-10. Moisture laden gas passes through a bed of spherical balls (it can be other shapes) which tends to allow passage of gas. The entrained moisture collects on the packing and eventually drops to the bottom of the tower.

There are many other types of demister equipment, unique to particular manufacturers. They typically have a pressure drop in the range of 1–3 in. WC at rated flow. They are often used as part of another item of gas cleaning (or scrubbing) equipment as will be seen in the discussion of high-energy scrubbing systems. Dimensions of a typical cyclone separator with a chevron-type demister are listed in Fig. 19-7 for various gas flow rates.

GAS SCRUBBING

The wet collectors previously discussed were basically washing devices used in conjunction with centrifuges or other inertial collection devices. Scrubbing of a

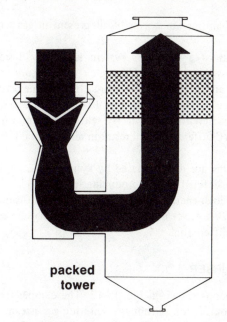

packed tower

Fig. 19-10. Mist eliminator. *Source:* Peabody-Galion, Princeton, NJ.

gas stream, however, involves more than inertial effects. The mechanisms for gas scrubbing (bringing particulate matter in contact with liquid droplets) includes the following:

Interception. Interception of a solid particles with a liquid particle occurs when the two particles have relative motion towards each other and are within one radius of each other. This is the radius of the smaller of the particles.

Gravitational force. Gravitational force causes a particle, when passing an obstacle, to fall from its streamline and settle upon the surface of the obstacle.

Impingement. Impingement occurs when an object, placed in the path of a particle-laden gas stream, causes the gas to flow around the obstacle. Larger particles will tend to continue in a straight path because of their inertia and may impinge on the obstacle and be collected.

Contraction/expansion. Contraction of a gas stream will tend to produce condensation of the moisture within the stream. High turbulence in a contracted area will result in good contact between solid particulate and the liquid particles. The dust-laden liquid particles will have the same velocity as the rest of the gas stream and then passing through an area of expansion, these particles will continue their direction of flow while the balance of the gas stream can be directed to flow in another direction. In effect, this process of contraction/expansion produces good separation of particulate matter from the gas stream.

The above mechanisms are normally all present in gas scrubbing equipment at the same time.

The effectiveness of a scrubbing system is usually directly related to the pressure drop across the scrubber. The higher the pressure drop, the greater the turbulence/mixing and, therefore, the more effective the scrubbing action. This feature is illustrated by the graph in Fig. 19-11. For a 2 μ size particle, for instance, a pressure differential of 8 in. WC will have a removal efficiency of 95%, whereas a 35 in. WC differential will result in almost total (99.9%) removal from the gas stream.

Scrubbing systems are differentiated by system pressure drop. A low-energy system is normally defined as one producing less than 12 in. WC for particulate removal, whereas high-energy systems will have significantly higher pressure differential, from 20 to over 60 in. WC. Likewise, medium-energy systems produce 7–20 in. WC differential.

CYCLONIC SCRUBBER

A cyclonic scrubber is shown in Fig. 19-12. The damper (referred to as a spin damper) creates contraction within the entering gas stream. Normally water is injected immediately upstream of the spin damper. The flow expands within the

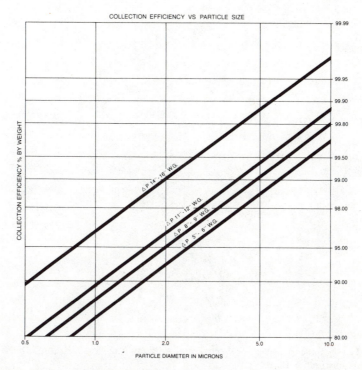

Fig. 19-11. Collection efficiency vs. particle size. *Source:* JOY Manufacturing Co., Western Precipitator Div., Los Angeles, CA.

cyclonic chamber and the expansion/contraction effect plus the inertial effects within the cyclonic tower scrub the gas and separate the dust-laden water from the gas stream.

This scrubber is termed low-energy. The pressure drop across the entire unit ranges from 6 to 12 in. WC. The spin damper position can usually be controlled to provide a lesser or greater inlet velocity to the cyclone. This control also provides the necessary adjustment when passing low flow versus high or rated flow. The water consumption will vary from 5 to 15 gallons per 1000 cubic feet of gas.

VENTURI SCRUBBER

Venturi scrubbers are widely used, where water is readily available, as high-efficiency, high energy gas cleaning devices. The heart of this system is a venturi throat where gases pass through a contracted area, reaching velocities of 200–600 feet per second, and then pass through an expansion section. From the expansion section the gas enters a large chamber for separation of particles or

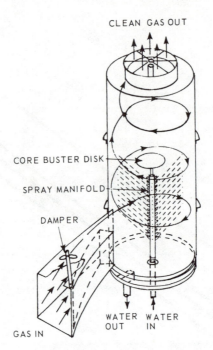

Fig. 19-12. Cyclonic scrubber. *Source:* Pease Anthony Construction Corp, Division of Chemical Construction Corp.

for further scrubbing. Water is injected at the venturi throat or just upstream of the venturi section. Fig. 19-13 illustrates a venturi scrubber where water is injected at its throat. For this design, 5–7 gallons per 1000 cubic feet of gas is required.

An adjustable venturi throat is shown in Fig. 19-14. Two throat flaps are illustrated. However, it is often designed with a single flap. Water is injected into a precooler section immediately before the throat. The throat area is adjustable and is normally controlled (manually or automatically) to maintain a desired pressure drop. Note that in this illustration a vane or chevron demister (water separator) is included in the outlet chamber to remove entrained moisture particles.

A flooded-disk (or plumb-bob) scrubber design is illustrated in Fig. 19-15. The conical plug is positioned to increase or decrease the gas flow area, similar to the throat flap design. Typical dimensions for this unit are given for a range of gas flows entering the stystem, through flange A. Water is injected above the Venturi section, tangentially to the inlet gas flow.

A serious problem associated with venturi sections is the erosive effect of the gas/liquid mixture passing through the throat. As indicated, the throat velocity

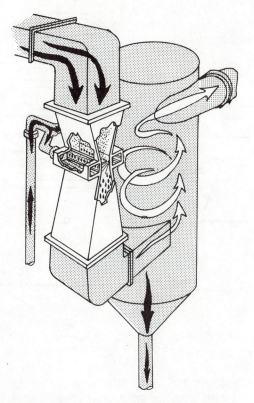

Fig. 19-13. Venturi scrubber. *Source:* UOP Air Correction Division, Des Plaines, IL.

is extremely high, creating the potential for wear of the throat surface through the mechanism of erosion. In addition, the corrosive effect of the wet, acidic gas flow is heightened by its turbulent action.

TRAY SCRUBBERS

Impingement tray scrubbers are essentially perforated plates with target baffles. Tray scrubbers have no large gas-directing baffles but are simply perforated plates within a tower, usually immediately downstream of a Venturi. Fig. 19-16 illustrates a typical impingement plate. A water level is maintained above the trays (there are usually two or more trays). The geometrical relationship of the tray thickness, hole size, and spacing, as well as the impinger details, result in a high-efficiency device for the removal of small-size particulate, less than 5 μ.

As many as 300 openings are provided per square foot of tray area. The openings can be from $\frac{1}{16}$ inch to $\frac{3}{8}$ inch in diameter. Gas flows up through the open-

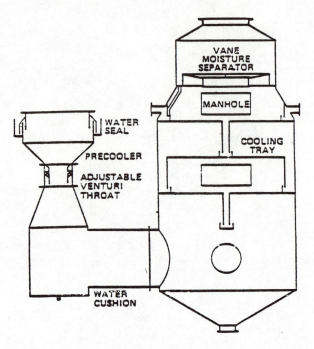

Fig. 19-14. Venturi-scrubber.

ings, while water is trying to flow, by gravity, counter to the gas flow. Highly effective turbulence with attendant mixing of, or scrubbing of, the solid particulate with the water effectively catches the small-micron particulate and removes them from the gas stream.

Tray scrubbers will have a pressure drop of 2–3 in. WC per tray. Impingement plates will add another $\frac{1}{2}$–1 in. WC differential per tray.

SELF-INDUCED SCRUBBER

There are a number of scrubbers that employ a unique geometry, utilizing the gas flow to generate scrubbing action. For instance, in the unit shown in Fig. 19-17 the water level controls the scrubbing action of the gas. The higher the water level, the greater, or longer, the contact between the gas stream and the water particles. As the water level decreases, however, the gas discharge into the water bath will generate a surface effect, atomizing some of this water surface. The net effect of this system is to obtain relatively good efficiency while utilizing relatively low water flows and low gas differential pressure.

ELECTROSTATIC PRECIPITATOR

Electrostatic precipitators (ESPs) are effective devices for the removal of airborne particulate matter. The first commercial ESP was designed and constructed

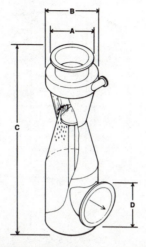

Saturated Gas Volume (ACFM)	A	B	C	D
2,500	1'-0"	3'-6"	6'-9"	1'-1"
5,000	1'-4"	3'-10"	7'-11"	1'-6"
10,000	1'-10"	4'-4"	8'-10"	2'-0"
25,000	3'-0"	5'-6"	10'-4"	3'-3"
50,000	4'-2"	7'-2"	13'-7"	4'-5"
75,000	5'-2"	8'-2"	14'-10"	5'-7"
100,000	5'-11"	8'-11"	16'-3"	6'-0"
150,000	7'-3"	10'-3"	18'-3"	7'-6"
350,000	11'-0"	13'-6"	25'-0"	11'-0"

Fig. 19-15. Flooded disc scrubber.

by Fredrick G. Cottrell earlier in this century, and they are sometimes referred to as *Cottrell process equipment*. In Europe they are called *electro-filters*.

An ESP (a typical unit is shown in Fig. 19-18, with dimensional information listed in Table 19-5), operates as follows:

- The gas stream passes through a series of discharge electrodes. These electrodes are negatively charged, in the range of 1000–6000 volts. This voltage creates a corona around the electrode. A negative charge is induced in the particulate matter passing through the corona.
- A grounded surface, or collector electrode, surrounds the discharge electrode. Charged particulate will collect on the grounded surface, usually in the form of plate surfaces.
- Particulate matter will be removed from the collector surface for ultimate disposal.

Typical discharge electrode and collector plate designs are shown in Fig. 19-19.

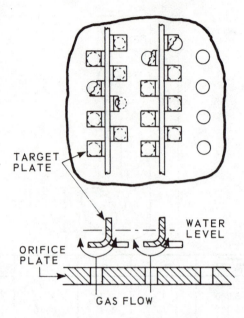

TARGET
PLATE

WATER
LEVEL

ORIFICE
PLATE

GAS FLOW

Fig. 19-16. Arrangement of "target plates" in impingement scrubber. *Source:* Ref. 19-17.

A variation of this design is the two-stage ESP, where the gas passes through a corona discharge prior to entering the collector plate area.

The ESP is extremely efficient in the collection of small-size particulate, down to the sub-micron range. It can be designed for temperatures as high as 750°F; however, its efficiency is sensitive to variation in temperature and flue

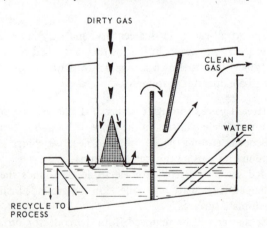

DIRTY GAS

CLEAN
GAS

WATER

RECYCLE TO
PROCESS

Fig. 19-17. Self-induced spray scrubber. *Source:* JOY Manufacturing Co., Western Precipitator Div., Los Angeles, CA.

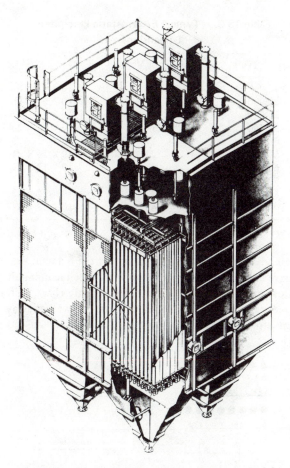

Fig. 19-18. Parallel plate precipitator. *Source:* Ref. 19-24.

gas humidity. As shown in Fig. 19-20, ESP efficiency can have substantial variance when operated beyond its design point.

Removal of particulate from the collecting surface is the key to the success of an ESP installation, If it is not removed it will act as an insulator, preventing the required electrostatic action from occurring, reducing the ability of the ESP to function.

Various methods have been developed for removing particulate. The most common device is use of "rappers," members that are sequenced to "rap" each plate section at regular intervals. Particulate will fall off the plates by this action, collecting in a hopper or series of hoppers beneath the ESP for eventual final disposal. The ESP pictured in Fig. 19-18 is provided with four such collection hoppers.

Table 19-5. Typical Electrostatic Precipitator
Dimensions.

Capacity, CFM	Number of Modules	W × L × H (ft)
12,000	5	10 × 4 × 18
20,000	8	10 × 6 × 19
25,000	11	10 × 8 × 19
35,000	14	10 × 10 × 20
40,000	17	10 × 12 × 20
50,000	21	10 × 14 × 21
60,000	25	10 × 17 × 22
70,000	29	10 × 20 × 23
80,000	33	10 × 22 × 23
100,000	43	10 × 19 × 24

Source: Compiled from Ref. 19-24.

Another method of cleaning the plates is to wet them down. Wet electrostatic precipitators have been developed where sprays wet the incoming flue gas stream to a saturated or supersaturated condition. The electric charge is transferred to the liquid droplets and the liquid charges, collects, and washes away the particulate from the gas stream.

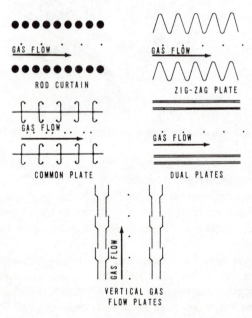

Fig. 19-19. Some special collecting electrodes used in electrical precipitators. *Source:* Ref. 19-24.

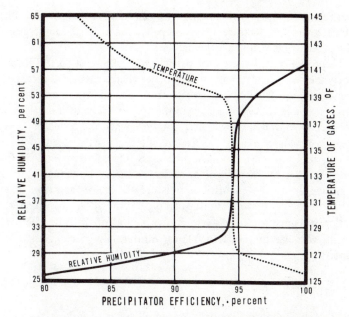

Fig. 19-20. Curves showing effects of variation of the gas stream's temperature and humidity upon efficiency of a specific electrical-precipitator installation. *Source:* Ref. 19-22.

The resistivity of the particulate matter is a significant parameter in ESP design and in the ability of an ESP to collect that specific material. If a particle, or dust, has a high resistivity it is unable to give up its electric charge to the collecting electrode. The dust will therefore build up on the collector, acting as an insulating layer. As this layer increases in depth the surface of the dust layer will develop a significant electric charge relative to the collector. This charge will generate a back discharge, or back corona, which can destroy the collecting electrode. Eventually the insulating dust layer will prevent the ESP from collecting particulate.

With too small a resistivity a dust will readily relinquish its negative charge to the collector and will assume a positive charge. With the collecting electrode at a positive potential, the particle is repelled back to the gas stream. In the gas stream, which is saturated with negatively charged particles, the dust will pick up a negative charge again and will eventually return to the collector plate and be repelled. The low-resistivity particle, therefore, will be successively repelled by the collecting electrode and will not be collected. It will pass through the ESP system.

Electrical resistivity is measured in units of ohm-centimeter. The optimum resistivity for particulate matter to be effectively collected within an ESP is from 10,000 to 10,000,000,000 ohm-centimeters. The resistivity of most materials varies significantly with temperature, and the use of an ESP may very well mandate the temperature range of collection to that temperature range where the resistivity is within acceptable limits.

With the proper sensitivity a dust particle will relinquish part of its charge to the collecting electrode. The rate at which the charge dissipates increases as the dust layer builds up. When the weight of the collected dust exceeds the electrostatic force available to hold the layer to the collector, the dust particles will fall off under their own weight or will be jarred loose when the collectors are rapped.

Another particulate quality associated with resistivity is the tendency to agglomerate, to form a hard or a tarlike mass impossible to remove from the collector electrodes by rapping or by water washing. The tendency to agglomerate is influenced by the quantity of moisture within the gas stream and by the temperature of the gas stream, as discussed previously within this chapter.

Collection is a function of gas velocity, as well as of the other factors noted. The velocity through the collector plates will normally range from 2 to 4 feet per second.

FABRIC FILTERS

Fabric filters, or baghouses, are prevalent in all types of industrial applications. They are essentially a series of permeable bags which allow the passage of gas but not particulate matter. They are effective for particle sizes in the sub-micron range.

Filter fabrics will usually be woven with relatively large spaces, in excess of 50 μ. The filtering process, therefore, is not just simple sieving, since particles less than 1 μ are caught. Filtering occurs as a result of impaction, diffusion, gravitational attraction, and electrostatic forces generated by interparticle friction. The dust layer itself acts as a filter medium.

As dust collects on a bag filter surface the resistance of the filter increases. In addition to this dust mat resistance there is a clean cloth resistance which is a function of the type of cloth fiber and its weave. Clean cloth resistance is measured by an ASTM permeability test procedure: Permeability is measured as that air volume, in standard cubic feet per minute, that will produce a pressure differential of 0.5 in. WC across 1 square foot of cloth area. The usual values of permeability are 7-70 CFM per square foot.

Typical filter fabrics in common use are listed in Table 19-6. Operating temperatures for continuous (long) and intermittent (short) duty are provided as well as information on flammability, permeability, and resistance to corrosive attack.

The filtering ratio of a fabric is the ratio of gas flow, in cubic feet per minute, to filter area in square feet. The more efficient the filter cloth, the greater the filtering ratio. Fig. 19-21 shows the filtering ratio for various materials as a func-

Table 19-6. Filter Fabric Characteristics.

Fiber	Composition	Operating Exposure °F		Supports Combustion	Air Permeability[a] cfm/ft²	Resistance[b]			
		Long	Short			Abrasion	Mineral Acids	Organic Acids	Alkali
Cotton	Cellulose	180	225	yes	10–20	G	P	G	G
Wool	Protein	200	250	no	20–60	G	F	F	P
Nylon	Polyamide	200	250	yes	15–30	E	P	F	G
Orlon	Polyacrylonitrile	240	275	yes	20–45	G	G	G	F
Dacron	Polyester	275	325	yes	10–60	E	G	G	G
Polypropylene	Olefin	200	250	yes	7–30	E	E	E	E
Nomex	Polyamide	425	500	no	25–54	E	F	E	E
Fiberglass	Glass	550	600	yes	10–70	P-F	E	E	G
Teflon	Polyfluoroethylene	450	500	no	15–65	F	E	E	E

[a] cfm/ft² at 0.5 in. W.C.
[b] P = Poor, F = Fair, G = Good, E = Excellent.
Source: Ref. 19-20.

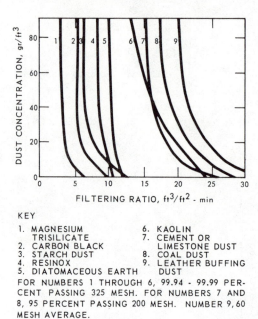

KEY
1. MAGNESIUM
   TRISILICATE
2. CARBON BLACK
3. STARCH DUST
4. RESINOX
5. DIATOMACEOUS EARTH

6. KAOLIN
7. CEMENT OR
   LIMESTONE DUST
8. COAL DUST
9. LEATHER BUFFING
   DUST

FOR NUMBERS 1 THROUGH 6, 99.94 - 99.99 PER-
CENT PASSING 325 MESH. FOR NUMBERS 7 AND
8, 95 PERCENT PASSING 200 MESH. NUMBER 9, 60
MESH AVERAGE.

Fig. 19-21. Typical performance of fabrics on a variety of dusts (dust load versus filtering ratio at 3.5 in. w.c. pressure drop). *Source:* Ref. 19-29.

tion of dust loading, grains of particulate per cubic foot of gas. Particulate from an incinerator would most closely resemble carbon black or diatomaceous earth, curve numbers 2 and 5 respectively.

A major feature of a baghouse is its ability to discharge collected particulate. A number of different methods have been developed for particulate removal including the following:

Shaker mechanism (Fig. 19-22). An eccentric rod physically shakes a bag section and the falling particles drop to the bottom of a silo, by gravity, for eventual disposal. The shaker motor is sequenced to operate in conjunction with operation of the fresh air dampers. Fresh air is admitted to that damper section that has its bags shaken, or agitated. The fresh, clean air aids in discharging the collected dust.

Compressed air (Fig. 19-23). A blast of compressed air is directed into the inside of each bag discharging the dust accumulated on the external surface of the bags. Wire retainers are provided to help reinforce the bags against the abrupt action of the air blast.

Repressurization. The filter sections are independent of each other. Through a series of inlet and exhaust valves, as shown in Fig. 19-24, the dirty gas flow passes through the inside of the bags. On a timing sequence the flow will be

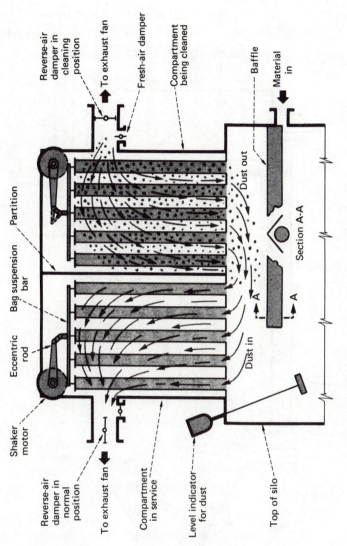

Fig. 19-22. Baghouse with shaker mechanism. *Source:* Ref. 19-34.

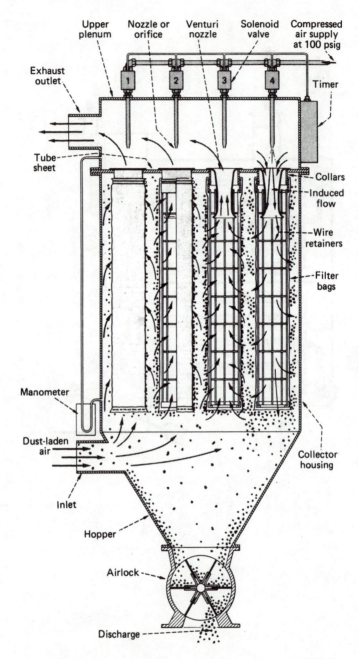

Fig. 19-23. Compressed air utilization in baghouse. *Source:* Ref. 19-34.

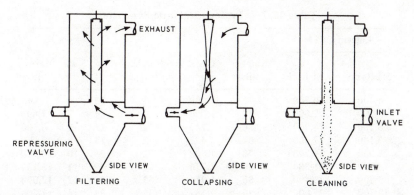

Fig. 19-24. Reverse air flexing to clean dust collector bags by repressuring. *Source:* Ref. 19-17.

reversed to pressurize the outside of the bags of a specific baghouse section. Under external pressure the internal dust loading will fall from the bag surfaces to a collection hopper for disposal.

Sonic cleaning. A source of intense sound tuned to the resonant frequency of the bags will create sympathetic vibrations of the bags. Under this vibration the dust will fall from the bags for collection and eventual removal.

A typical baghouse assembly is shown in Fig. 19-25 with dimensions and operating parameters listed in Table 19-7.

ODOR EMISSIONS

Odor from incinerators can best be controlled by an alteration of the burning process. In general, odors emanating from combustion processes are composed

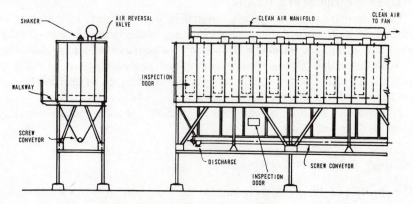

Fig. 19-25. Fully automatic, compartmented baghouse with hopper discharge screw conveyor. *Source:* Buell Engineering Company, Cleveland, OH.

Table 19-7. Typical Baghouse Parameters.

Capacity, CFM		Number of Bags	Cloth, ft^2	Number of Hoppers	A, ft	B, ft	Weight, lb
5:1 Ratio	12:1 Ratio						
4,400	10,700	40	888	1	23	6	6,900
5,300	12,800	48	1,066	2	24	7	7,800
6,200	14,900	56	1,243	2	25	8	8,500
7,100	17,100	64	1,421	2	27	9	9,600
8,000	19,200	72	1,598	2	28	10	10,300
8,900	21,300	80	1,776	2	29	11	11,100
9,800	23,400	88	1,954	3	30	12	12,000
10,700	25,600	96	2,131	3	31	14	12,700
11,500	27,700	104	2,309	3	32	15	13,500
12,400	29,800	112	2,486	3	33	16	14,300
13,300	32,000	120	2,664	3	34	17	15,000
14,200	34,100	128	2,842	4	35	18	16,200
15,100	36,200	136	3,019	4	36	19	17,000
16,000	38,400	144	3,197	4	37	20	17,800
16,900	40,500	152	3,374	4	39	21	18,600
17,800	42,600	160	3,552	4	40	22	19,400
18,600	44,800	168	3,730	5	41	23	20,300
19,500	46,900	176	3,907	5	42	25	21,100
20,400	49,000	184	4,085	5	43	26	21,900
21,300	51,100	192	4,262	5	44	27	22,700

Source: Buell Engineering Company, Cleveland, OH; A, length; B, width.

of unburned hydrocarbons and to a lesser degree inorganic compounds. By proper combustion equipment design and operation the generation of odor can be effectively reduced or eliminated.

Unfortunately, waste streams are normally not consistent in chemistry or quality. As waste quality changes the efficiency of the burning process for that waste will change and burning will have to be adjusted to reduce the possibility of odor generation. Changes in the combustion process may not be practical with a waste that changes in character within a short period of time. Additional mechanisms of odor control may be required.

A number of methods of odor control, including incineration, are currently in use, as described below.

Dilution

The principle of dilution is to reduce the odor concentration to a level where it is no longer detectable, or if detectable, is no longer considered noxious. Methods of dilution include provision of tall stacks to disperse odor by the time the odorous compounds reach a populated surface. Another method of diluting odorous discharges is to increase the temperature of the gas exiting the stacks, thereby increasing its outlet velocity. Also, increasing the velocity of discharge from the stack by decreasing the stack outlet diameter will promote dilution. These latter methods, increasing the gas exit velocity, will increase the time until the odorous compounds return to earth, at which time they will be increasingly dilute.

Yet another method of odor control by dilution is the addition of clean air to the odorous gas exiting the stack. Clean air can be added by providing a flue damper just upstream of the induced draft fan, or by a barometric damper where there is no such fan in the system.

Absorption

In this method of odor control the odorous constituent is absorbed into a solution by either chemical bond or solubility. In the case of solubility an odorous gas may be absorbed by a liquid. However, if the odorous component of the gas does not condense and dissolve, the odorous gas will eventually be released.

Chemical absorption is generally an oxidative process. As noted previously, many odors are unburned, i.e., unoxidized hydrocarbons. By mixing, or scrubbing with chemicals with the ability to release oxygen (such as potassium permanganate ($KMNO_4$), sodium hypochlorite (NaOCl), caustic (NaOH), or chlorine dioxide (ClO_2), the odorous constituent can be effectively oxidized and lose its odor characteristic. Fig. 19-26 shows a typical packed tower used for odor absorption.

Adsorption

A number of materials have the physical property of high adsorptive capacity, the ability to adsorb or attract gases or vapors onto their surface. Adsorptive media commonly used include activated carbon, silica gel, aluminum oxide, and magnesium silicate. The particles of each of these materials have extremely high area-to-weight ratios, providing high retentive capacity.

Activated carbon is the most widely used adsorbant. One reason for this is that it has a low affinity for moisture. Moisture will not compete with contaminants in searching out the carbon surface. Other adsorbants will attract moisture and will have, therefore, a short useful life in wet gas environments.

Another property of an adsorbant material is its ability to regenerate. Normally, with application of heat, the contaminant will be released and the adsorbant will be reactivated for reuse. Reactivation may also require sparging of the spent adsorbant with steam.

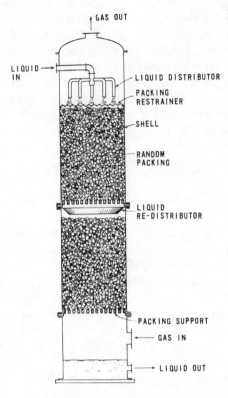

Fig. 19-26. Schematic diagram of a packed tower. *Source:* Ref. 19-30.

Fume Incineration

The ultimate odor destruction technique is burning. Separate fume incinerators are often utilized for odor destruction when other techniques are not effective. Destruction of odor by incineration requires the use of supplemental fuel and with current energy costs and availability this technique is avoided whenever possible. The proper use of fume incinerators or afterburners, that is, provision of the appropriate temperature and residence time, will completely destroy odor, whether organic or inorganic in nature. A discussion of fume or gas burning equipment is included in Chapter 15.

Other Techniques

A number of other methods of odor control have been developed which have application to specific odor types.

Masking is a technique where one odor is added to another to mask its unpleasantness. If an inappropriate masking agent is applied or if the original odor changes in characteristic, the resultant odor can be far worse than the original odor.

Catalytic oxidation utilizes a catalyst such as palladium, rhodium, or platinum, to induce oxidation at lower temperatures than direct flame oxidation. The catalyst can be sensitive to the constituents of the odor and can easily foul or permanently lose its potency with unanticipated odor sources. Catalytic incineration is discussed in a previous chapter.

Techniques for odor control from incineration processes are normally one of the following:

- Fume incineration (afterburner).
- Packed tower (absorber).

Table 19-8.   Gas-Absorption Systems of Commercial Importance.

Solute	Solvent	Reagent	Degree of Commercial Importance		
			High	Moderate	Low
CO_2, H_2S	water		X	–	–
CO_2, H_2S	water	monoethanolamine	X	–	–
CO_2, H_2S	water	diethanolamine	X	–	–
CO_2, H_2S	water	triethanolamine	–	–	X
CO_2, H_2S	water	diaminoisopropanol	–	–	X
CO_2, H_2S	water	methyldiethanolamine	–	–	X
CO_2, H_2S	water	K_2CO_3, Na_2CO_3	X	–	–
CO_2, H_2S	water	NH_3	–	X	–
CO_2, H_2S	water	NaOH, KOH	–	X	–
CO_2, H_2S	water	K_3PO_4	–	X	–
HCl, HF	water		X	–	–
HCl, HF	water	NaOH	X	–	–
Cl_2	water		X	–	–
SO_2	water		–	–	X
SO_2	water	NH_3	–	X	–
SO_2	water	Xylidine	–	X	–
SO_2	water	Dimethylaniline	–	X	–
SO_2	water	$Ca(OH)_2$, oxygen	–	–	X
SO_2	water	Aluminum hydroxide-sulfate	–	X	–
NH_3	water		X	–	–
NO_2	water		X	–	–
HCN	water	NaOH	X	–	–
CO	water	Copper ammonium salts	X	–	–

Source: Ref. 19-33.

Dilution is not a positive means of odor control, and is dependent on atmospheric conditions, which are highly variable. Adsorption requires lower temperatures than that generated by incineration systems in order to maintain the adsorption property of the active media.

ACID ABSORPTION

Wet scrubbing systems and packed towers are often used to absorb gases as well as to capture particulate. As shown in Table 19-8 the solute is that gas or gas component that is to be removed. Water is sued as the solvent for gas removal in each case; however, a chemical reagent is often added to the water for increased removal efficiency.

The mechanism of removal includes solubility plus chemical reactivity. For reactivity, the reaction between hydrogen chloride in the gas stream and sodium hydroxide in the scrubbing liquid is a typical reaction:

$$HCl + NaOH \longrightarrow H_2O + NaCl$$

36.46	40.00	18.02	58.44
1.00	1.09	0.49	1.60

The first line beneath the equation is the atomic weight of each element and the second line is the atomic weight of each component normalized to 1 lb of

Fig. 19-27. HEPA filter. *Source:* American Air Filter Co., Louisville, KY.

Table 19-9. Typical HEPA Filter Sizes.

Rated SCFM at 1″ W.G.	Actual Dimensions (Less Gaskets)		
	A (in.)	B (in.)	C (in.)
900	24	36	$5\frac{7}{8}$
1230	24	48	$5\frac{7}{8}$
1550	24	60	$5\frac{7}{8}$
1900	24	72	$5\frac{7}{8}$
750	30	24	$5\frac{7}{8}$
925	30	30	$5\frac{7}{8}$
1150	30	36	$5\frac{7}{8}$
1550	30	48	$5\frac{7}{8}$
1975	30	60	$5\frac{7}{8}$
2350	30	72	$5\frac{7}{8}$
900	36	24	$5\frac{7}{8}$
1150	36	30	$5\frac{7}{8}$
1400	36	36	$5\frac{7}{8}$
1900	36	48	$5\frac{7}{8}$
2350	36	60	$5\frac{7}{8}$
2850	36	72	$5\frac{7}{8}$

Source: American Air Filter Company, Louisville, KY.

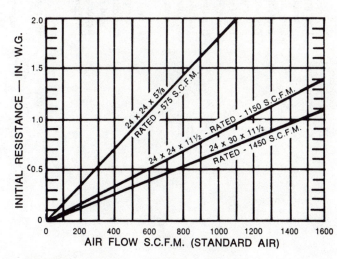

Fig. 19-28. Typical HEPA filter characteristics. *Source:* American Air Filter Co., Louisville, KY.

Table 19-10. Advantages and Disadvantages of Collection Devices.

Collector	Advantages	Disadvantages
Gravitational	Low pressure loss, simplicity of design and maintainance.	Much space required. Low collection efficiency.
Cyclone	Simplicity of design and maintenance	Much head room required.
	Little floor space required	Low collection efficiency of small particles.
	Dry continuous disposal of collected dusts	Sensitive to variable dust loadings and flow rates.
	Low to moderate pressure loss.	—
	Handles large particles.	—
	Handles high dust loadings.	—
	Temperature independent.	—
Wet collectors	Simultaneous gas absorption and particle removal.	Corrosion, erosion problems.
		Added cost of wastewater treatment and reclamation.
	Ability to cool and clean high-temperature, moisture-laden gases.	Low efficiency on submicron particles.
		Ineffective for hydrophobic materials.
	Corrosive gases and mists can be recovered and neutralized.	Contamination of effluent stream by liquid entrainment.
	Reduced dust explosion risk	Freezing problems in cold weather.
	Efficiency can be varied	Reduction in buoyancy and plume rise.
		Water vapor contributes to visible plume under some atmospheric conditions.
Electrostatic precipitator.	99+ percent efficiency obtainable	Relatively high initial cost.
	Very small particles can be collected	Precipitators are sensitive to variable dust loadings or flow rates.
	Particles may be collected wet or dry	Resistivity causes some material to be economically uncollectable.

Method	Advantages	Limitations
	Pressure drop and power requirement are small compared to other high-efficiency collectors.	Precautions are required to safeguard personnel from high voltage.
	Maintenance is nominal unless corrosive or adhesive materials are handled.	Collection efficiencies can deteriorate gradually and imperceptibly.
	Few moving parts	—
	Can be operated at high temperatures (550° to 850°F.).	—
Fabric filtration	Dry collection possible	Sensitivity to filtering velocity.
	Decrease of performance is noticeable	High-temperature gases must be cooled to 200° to 550°F.
	Collection of small particles possible	Affected by relative humidity (condensation).
		Susceptibility of fabric to chemical attack.
Afterburner, direct flame.	High efficiencies possible	High operational cost. Fire hazard.
	High removal efficiency of submicron odor-causing particulate matter.	Removes only combustibles.
	Simultaneous disposal of combustible gaseous and particulate matter.	—
	Direct disposal of non-toxic gases and wastes to the atmosphere after combustion.	—
	Possible heat recovery	—
	Relatively small space requirement	—
	Simple construction	—
	Low maintenance	
Afterburner, catalytic	Same as direct flame afterburner	High initial cost.
	Compared to direct flame: reduced fuel requirements, reduced temperature, insulation requirements, and fire hazard.	Catalysts subject to poisoning.
		Catalysts require reactivation.

Source: Ref. 20-17, Page XXIV.

**Table 19-11. Average Collection Efficiencies of
Gas Cleaning Equipment.**

Equipment Type	Percentage Efficiency at		
	50 μm	5 μm	1 μm
Inertial collector	95	16	3
Medium-efficiency cyclone	94	27	8
Low-resistance cellular cyclones	98	42	13
High-efficiency cyclone	96	73	27
Impingement scrubber	98	83	38
Self-induced spray deduster	100	93	40
Void spray tower	99	94	55
Fluidized bed scrubber	>99	99	60
Irrigated target scrubber	100	97	80
Electrostatic precipitator	>99	99	86
Irrigated electrostatic precipitator	>99	98	92
Flooded disk scrubber, low energy	100	99	96
Flooded disk scrubber, medium energy	100	99	97
Venturi scrubber, medium energy	100	>99	97
High-efficiency electrostatic precipitator	100	>99	98
Venturi scrubber, high energy	100	>99	98
Shaker-type fabric filter	>99	>99	99
Reverse-jet fabric filter	100	>99	99

Source: Ref. 19-23.

HC1. Therefore, 109 lb of sodium hydroxide is theoretically required to neutralize one pound of hydrogen chloride, producing 1.60 lb of sodium chloride.

HEPA FILTERS

HEPA (High-Efficiency Particulate Air) filters are extremely efficient filters that have been originally developed for control of particulate from nuclear energy facilities. They can remove over 99.97% of particles 0.3 μ and greater.

The standard test for evaluating removal efficiency is the DOP (dioctylphalate) smoke test. DOP particles are generated in a standard size of 0.3 μ. The percent removal of DOP smoke with a filter differential pressure drop of 1.0 in. WC is defined as the filter efficiency.

A typical HEPA filter, as shown in Fig. 19-27 is constructed of a glass fiber mat pleated to increase its unit surface area. The filter is mounted in a frame and a series of frames are mounted in a filter band to provide the required flow capacity. Table 19-9 lists typical HEPA filter sizes and Fig. 19-28 shows a series of curves relating flow to clean filter resistance.

Table 19-12. Average Control Efficiency of Air Pollution Control (APC) Systems.

APC Type	APC System Removal Efficiency (wt %)								
	Mineral Particulate	Combustible Particulate	Carbon Monoxide	Nitrogen Oxides	Hydrocarbons	Sulfur Oxides	Hydrogen Chloride	Polynuclear Hydrocarbons	Volatile Metals
None (flue settling only)	20	2	0	0	0	0	0	10	2
Dry expansion chamber	20	2	0	0	0	0	0	10	0
Wet bottom expansion chamber	33	4	0	7	0	0	10	22	4
Spray chamber	40	5	0	25	0	0.1	40	40	5
Wetted wall chamber	35	7	0	25	0	0.1	40	40	7
Wetted, close-spaced baffles	50	10	0	30	0	0.5	50	85	10
Mechanical cyclone (dry)	70	30	0	0	0	0	0	35	0
Medium-energy wet scrubber	90	80	0	65	0	1.5	95	95	80
Electrostatic precipitator	99	90	0	0	0	0	0	60	90
Fabric filter	99.9	99	0	0	0	0	0	67	99

Source: Ref. 19-25, page 260.

Table 19-13. Maximum Demonstrated
Collection Efficiency of Incinerator
Control Equipment.

Collection Device	Collection Efficiency, Percent
Settling chamber	35
Wetted baffles	53
Cyclones	75 to 80
Impaction scrubbers (with pressure drop less than ten inches of water)	94 to 96
Electrostatic precipitators	99+
Bag filters	99+

Source: Ref. 19-19.

SUMMARY TABLES

As a concise guide to the types of air pollution control equipment available and their performance Tables 19-10 through 19-13 have been included in this chapter.

Table 19-10 summarizes the advantages and disadvantages of the equipment discussed in this chapter. Typical collection efficiencies of control equipment as a function of particle size are listed in Table 19-11. Table 19-12 relates collection efficiency to gas stream components, and Table 19-13 is an overview of incinerator air pollution control devices.

Chapter 20
Comprehensive Design Example

The analytical methods developed in this text will be tied together in this chapter. Design parameters will be developed for an incinerator system burning solid and liquid waste. Refractory design will also be included.

EXAMPLE

A multi-purpose incinerator is required to burn plant trash (rubbish) and an organic liquid. The trash is normally contaminated with chlorides in relatively small quantities. The organic liquid, cycloheptane (C_7H_{14}), contains trace amounts of solid contaminants. The facility must be designed for destruction of 2000 lb/hr of rubbish and 475 lb/hr of liquid cycloheptane, either burning concurrently or firing the trash alone.

Future plant expansion may result in the burning of solid organic powders. Additional capacity shall be obtained by operating additional hours during the week.

CRITERIA

It will be assumed that the wastes charged are hazardous and the incinerator will have to comply with RCRA requirements for hazardous wastes containing chlorides. Not knowing the specific nature of the waste stream with regard to parameters of destruction the following criteria will be used:

- A temperature zone of 2200°F should be provided with a retention time of 2 seconds.
- A minimum of 3% oxygen should be found in the off-gas stream.
- A maximum particulate loading of 0.03 grains per dry standard cubic foot should be allowed.

Additional criteria shall include the following:

- A maximum external temperature of 200°F at the surface of the primary chamber.
- A temperature limit of 250°F at the surface of the secondary chamber.

INCINERATOR EQUIPMENT

The incinerator system must handle solid waste and/or liquid waste. A rotary kiln with an afterburner or a modular combustion unit (controlled air incinerator) would have application. The provision that powders must be incinerated in the future requires that turbulence must be provided within the incinerator. Without sufficient turbulence to agitate the powder waste and expose the powder surfaces to air and heat, the powder may not burn out properly. The rotary kiln appears to be the favorable option because it provides turbulence to the solid waste charge.

A rotary kiln will be used for solid waste incineration. The off-gas will pass through an afterburner (secondary chamber) where the required temperature and residence time will be maintained.

Fuel oil (No. 2 grade) will be used to obtain the required temperature if the waste will not maintain combustion. The liquid waste stream will be fired in the afterburner.

SOLID WASTE

The incinerator must burn 2000 lb/hr of rubbish, type 1 waste as noted in Table 8-2. The total heat release is as follows:

$$6500 \text{ BTU/lb} \times 2000 \text{ lb/hr} = 13.0 \text{ MBH*}$$

Assuming a kiln heat release of 25000 BTU/hr-ft³, from Chapter 15, the required volume can be calculated:

$$\frac{13,000,000 \text{ BTU/hr}}{25,000 \text{ BTU/hr-ft}^3} = 520 \text{ ft}^3$$

With a 4:1 ratio of length to diameter, following the methods of Chapter 15,

$$L = 4D$$

$$V = \frac{\pi D^2 L}{4} = \frac{\pi D^2}{4} (4D) = \pi D^3$$

---

*MBH = million BTU per hour.

$$D = \left(\frac{V}{\pi}\right)^{1/3} = \left(\frac{520}{\pi}\right)^{1/3} = 5.491 \text{ ft}$$

$$L = 4D = 4 \times 5.491 = 21.964 \text{ ft}$$

The kiln inside dimensions are, therefore:

$$D = 5'6'' \text{ inside diameter}$$

$$L = 22'0'' \text{ inside length}$$

For an external temperature of 200°F with an internal temperature of 1600°F a refractory will be selected. The kiln is continually rotating and convection heat transfer can be best approximated as a vertical wall. The kiln surface is oxidized steel. Using the resistance graph, Fig. 5-5, a resistance R of 5.3 is required. From Chapter 5:

$$x = RK$$

Using a dense castable, $K = 5$, over 2 ft of refractory is required. 2 ft of refractory is not practical, (too large) and a two-wall refractory system will be used, insulating castable against the kiln surface and 6 in. of firebrick adjacent to the hot stream. Following the methods of Chapter 5.

$$R = R_1 + R_2 = \frac{X_1}{K_1} + \frac{X_2}{K_2}$$

with

$$X_1 = 6 \text{ in. firebrick}$$

$$K_1 = 10.0 \text{ BTU-in./ft}^2\text{-}°\text{F-hr for firebrick}$$

and

$$K_2 = 1.3 \text{ BTU-in./ft}^2\text{-}°\text{F-hr for insulating castable}$$

To find the thickness of castable required, X_2:

$$X_2 = \left(R - \frac{X_1}{K_1}\right)K_2$$

$$= \left(5.3 - \frac{6}{10.0}\right) \times 1.3 = 6.11 \text{ in.}$$

Therefore the kiln shall be lined with 7 inches of insulating castable protected from the gas stream by 6 inches of firebrick.

From the resistance graph the heat loss Q from the kiln, which will have a surface temperature of 200°F, will be 263 BTU/hr-ft².

The internal kiln dimensions are 5'6" diameter by 22'0" long and with a total lining of 13 inches the kiln external dimensions are:

Diameter: 5'6" + 2 × 13" = 7'8"
Length: 22'0" + minimum of wall diameter (13") = 23'1" *minimum*

Heat loss through the kiln wall, per the resistance graph, is 263 BTU/hr-ft², as noted above. Now that the external kiln size is determined the total heat loss can be calculated.

Kiln external cylindrical area:

$$\pi DL = \pi \times (7'8") \times (23'1") = 556 \text{ ft}^2$$

Heat loss:

$$556 \text{ ft}^2 \times 263 \text{ BTU/hr-ft}^2 = 0.15 \text{ MBH}$$

LIQUID WASTE

Cycloheptane will be fired in the secondary chamber, or afterburner, at its supply rate of 475 lb per hour. To determine the heating value of cycloheptane the methods of Chapter 8 will be used.

Using the heat of formation method, with Tables 8-3 and 8-5:

-37.5   -94.1   -68.3(g)

$$C_7H_{14} + 10.5\, O_2 + 3.3197(10.5)N_2 \longrightarrow 7CO_2 + 7H_2O + 34.857\, N_2$$

98.21   44.01   18.02   28.02
1.00   3.1368   1.2843   9.9449

$$
\begin{aligned}
\Delta H_f\ H_2O(g): &\quad -68.3 \times 7 = -478.1 \\
\Delta H_f\ CO_2: &\quad -94.1 \times 7 = -658.7 \\
\Delta H_f\ C_7H_{14}: &\quad -(37.5) \times 1 = \underline{+37.5} \\
&\quad \sum \Delta H_f = -1099.3
\end{aligned}
$$

$$\frac{1}{98.21 \text{ gram/mole}} \times (-1099.3)(\text{kcal/mole}) \times 1802 = -20170 \text{ BTU/lb (exothermic)}$$

Using the hydrocarbon approximation:

$$\frac{10.5 \text{ lb mole } O_2}{\text{mole } C_7H_{14}} \times \frac{\text{mole } C_7H_{14}}{98.21 \text{ lb}} \times \frac{184{,}000 \text{ BTU}}{1 \text{ lb mole } O_2} = 19{,}672 \text{ BTU/lb}$$

These two heating values are relatively close:

$$\frac{20{,}170 - 19{,}672}{20{,}170} \times 100 = 2.5\%$$

The more accurate number, 20170 BTU/lb, will be used for further calculations.

The generation of off-gas is calculated from the above equation, noting the second line of values beneath the equation is the normalized quantity of each component:

$$\frac{3.1368 \text{ lb } CO_2}{\text{lb } C_7H_{14}} + \frac{9.9449 \text{ lb } N_2}{\text{lb } C_7H_{14}} = \frac{13.0817 \text{ lb DG}}{\text{lb } C_7H_{14}}$$

$$\frac{13.0817 \text{ lb DG}}{\text{lb } C_7H_{14}} \times \frac{\text{lb } C_7H_{14}}{20.170 \text{ KBTU}} = \frac{6.486 \text{ lb DG}}{10 \text{ KBTU}}$$

$$\frac{1.2843 \text{ lb } H_2O}{\text{lb } C_7H_{14}} \times \frac{\text{lb } C_7H_{14}}{20.170 \text{ KBTU}} = \frac{0.637 \text{ lb } H_2O}{10 \text{ KBTU}}$$

The size of the afterburner will be calculated using an external low-pressure atomizing burner, as described in Chapter 14. This burner is utilized because of the presence of solids in the liquid waste stream and the low-pressure compressed air required. With a heat release of 60000 BTU/hr-ft^3 and 475 lb/hr of cycloheptane, the *minimum* chamber size is calculated as follows:

$$475 \text{ lb/hr} \times 20170 \text{ BTU/lb} = 9.581 \text{ MBH}$$

$$\frac{9581000 \text{ BTU/hr}}{60000 \text{ BTU/hr-ft}^3} = 159.683 \text{ ft}^3$$

For a 4-ft-diameter chamber the required length will be calculated:

Area: $\qquad\qquad\qquad \dfrac{\pi}{4}(4)^2 = 12.566 \text{ ft}^2$

Length: $\qquad\qquad \dfrac{159.683 \text{ ft}^3}{12.566 \text{ ft}^2} = 12.708 \text{ ft}$

Use a length of 12'9"

As noted above, this is the *minimum* chamber size to take care of the heat release of the cycloheptane. Additional fuel may be required to bring the 1600°F gas stream leaving the kiln to the required 2200°F.

The afterburner refractory will be selected on the basis of 2200°F inside temperature and 250°F external temperature. Using an outside surface of steel, with conduction averaged as a vertical wall, from the resistance graph, (Fig. 5-5) the required wall thickness is calculated as follows:

$$R = \frac{X}{K} = 4.7 \text{ hr-ft}^2\text{-}°\text{F/BTU (from resistance graph)}$$

Using insulating castable

$$K = 1.9 \text{ BTU-in./hr-ft}^2\text{-}°\text{F}$$

$$X = KR = 1.9 \times 4.7 = 8.93 \text{ in.}$$

A thickness of 9.0 in. will be used.

The heat loss, again from the resistance graph, is 418 BTU/hr-ft^2.

INCINERATOR CALCULATIONS

For rubbish, from Table 8-3, the ash is 10% of the as received weight and moisture is 25% of the as received weight. For the mass balance, Table 20-1, the wet feed is 2000 lb/hr. The moisture content is 25% which is 500 lb/hr. The dry feed is the difference between wet feed and moisture, or 1500 lb/hr.

The ash content is 10% of the wet feed, or 200 lb/hr and the volatile content is therefore 1500 − 200 = 1300 lb/hr. This is 200 ÷ 1500 = 13.3% of the dry weight. From the tabular data in Chapter 8, Type 1 waste has an as-received heating value of 6500 BTU/lb. This calculates to 6500 × 2000 ÷ 1300 = 10000 BTU/lb volatile.

Dry gas and moisture generation will be taken as 7.50 lb/10 kB and 0.51 lb/10 kB respectively, as noted in Chapter 11 for refuse.

The mass balance is computed using 150% excess air for burning the solid waste.

In the heat balance, 120 BTU/lb is assumed for ash losses and the kiln radiation loss has been calculated for a 200°F external kiln temperature, 0.15 MBH. The radiation related to heat release is 0.15 ÷ 13 = 1.2%.

The temperature developed within the kiln, to obtain the 12.61 MBH released, is calculated from the methods of Chapter 4, as follows:

Temperature   MBH

$1700°F$	$425.08 \times 23{,}422 + 1{,}890.11 \times 1{,}391 = 12.59$
x	13.05
$1800°F$	$453.24 \times 23{,}422 + 1{,}948.02 \times 1{,}391 = 13.33$

$$x = 1700 + (1800 - 1700) \frac{(13.05 - 12.59)}{(13.33 - 12.59)} = 1762°F$$

The off-gas from the solid waste must be raised to $2200°F$ and 63.78 gph F.O. is required, from the heat balance, to obtain this heat increase.

At this point the gas volume will be evaluated. Using Table 4-5, at $2200°F$ the gas flow is calculated as follows:

Dry gas: $67.0 \text{ ft}^3/\text{lb} \times 30{,}275 \,(\text{lb/hr}) \times (\text{hr}/3600 \text{ sec}) = 563 \text{ ft}^3/\text{sec}$

Moisture: $107.9 \text{ ft}^3/\text{lb} \times 2071 \,(\text{lb/hr}) \times (\text{hr}/3600 \text{ sec}) = \underline{62 \text{ ft}^3/\text{sec}}$

Total: $= 625 \text{ ft}^3/\text{sec}$

For a 2 second retention time a volume of $2 \times 625 = 1250 \text{ ft}^3$ is required.

Table 20-1.   Mass Flow.

	Kiln	Aft.	Kiln w/Aft.	12% CO_2
Wet Feed, lb/hr	2,000	475		
Moisture, %	25			
lb/hr	500			
Dry Feed, lb/hr	1,500	475		
Ash, %	13.3			
lb/hr	200			
Volatile, lb/hr	1,300	475		1,775
Vol. Htg. Value, BTU/lb	10,000	20,170		
MBH	13.00	9.58	22.58	
Dry Gas, lb/10 kB	7.50	6.49		
lb/hr	9,751	6,217		15,968
Comb. H_2O, lb/10 kB	0.51	0.64		
lb/hr	663	613		1,276
DG + Comb. H_2O, lb/hr	10,414	6,830		17,244
100 Air, lb/hr	9,114	6,355	15,469	15,469
Total Air Fract.	2.5	1.1		2.29
Total Air, lb/hr	22,785	6,991	29,776	35,424
Excess Air, lb/hr	13,671	636	14,307	19,955
Humid/Dry Gas (Air), lb/lb	0.01	0.01		
Humidity, lb/hr	228	67	295	
Total H_2O, lb/hr	1,391	680	2,071	
Total Dry Gas, lb/hr	23,422	6,853	30,275	35,923

This volume is 8 times greater than that minimum volume necessary for firing the liquid waste stream, cycloheptane as initially calculated. It is also twice the volume required for the rotary kiln at the head of the system.

A reasonable chamber would have an internal diameter of 10 feet and a length of $16'6''$. As calculated previously the wall thickness is $9''$. The external diameter would be therefore, $10'0'' + (2)(9'') = 11'6''$.

The radiation and convection loss from the kiln is also as noted previously, 418 BTU/hr-ft^2.

The total heat loss from the afterburner is calculated as follows:

$$A = \pi DL = \pi(11.5 \text{ ft})(16.5 \text{ ft}) = 596 \text{ ft}^2$$

Heat Transfer: $\qquad$ 596 ft^2 $\times$ 418 BTU/hr-ft^2 = 0.25 MBH

The percent oxygen, as noted previously, must be at least 2% within the afterburner. The total gas flow, from the heat balance, is 30,275 lb/hr dry gas plus 2,071 lb/hr H_2O, a total of 32,346 lb/hr wet gas flow. The oxygen component of this flow is the sum of the excess oxygen injected in the primary chamber (kiln) plus the excess oxygen injected into the afterburner:

Kiln: $\qquad$ 13671 lb/hr excess air $\times$ 0.2315 lb O_2/lb air

$\qquad\qquad$ = 3165 lb/hr oxygen

Afterburner: $\qquad$ 8220 lb/hr total air $\times \dfrac{0.1 \text{ excess air}}{1.1 \text{ total air}}$

$\qquad\qquad$ = 636 lb/hr excess air $\times$ 0.2315 lb O_2/lb air

$\qquad\qquad$ = 147 lb/hr oxygen

The total oxygen in the gas flow is therefore:

$$3165 + 147 = 3{,}312 \text{ lb/hr } O_2$$

The percent oxygen is:

$$\frac{3{,}312 \text{ lb/hr } O_2}{32{,}346 \text{ lb/hr gas flow}} \times 100 = 10.24\% \text{ oxygen, by weight}$$

This is well in excess of the minimum oxygen requirement, 2% of the gas flow.

The above calculations, as noted in the first column of Tables 20-1 and 20-2, the mass flow and heat balance sheets, represent burning solid waste, using No. 2 fuel oil in the afterburner to reach the desired temperature. The second column is used to calculate combustion characteristics of the waste liquid, cycloheptane. Cycloheptane will be used in lieu of fuel oil if it has sufficient heating value to raise the temperature of the kiln off-gas to at least 2200°F.

Table 20-2.   Heat Balance.

	Kiln	Aft.	Kiln w/Aft.
Cooling Air Wasted, lb/hr			
°F			
BTU/lb			
MBH			
Ash, lb/hr	200		
BTU/lb	120		
MBH	0.02		
Radiation, %	1.2	2	
MBH	0.15	0.25	
Humidity, lb/hr	228	67	
Correction (@970 BTU/lb), MBH	-0.22	-0.06	
Losses, Total, MBH	-.05	.19	.14
Input, MBH	13.00	9.58	22.58
Outlet, MBH	13.05	9.39	22.44
Dry Gas, lb/hr	23,422	6,853	30,275
H_2O, lb/hr	1,391	680	2,071
Temperature, °F	1,762		2,272
Desired Temp., °F	2,200		2,272
MBH	16.34		
Net, MBH	3.54*		
Fuel Oil, Air Fraction	1.1		
Net BTU/gal	55,507		
gph	63.78		
Air, lb/gal	114.64		
lb/hr	7,312		
Dry Gas, lb/gal	115.12		
lb/hr	7,342		
H_2O, lb/gal	8.62		
lb/hr	550		
DG w/FO, lb/hr	30,764		30,275
H_2O w/FO, lb/hr	1,941		2,071
Air w/FO, lb/hr	30,097		29,776
Outlet, MBH	21.96*		22.58
Reference t, °F	60	60	60

*Including 0.25 heat loss in the afterburner.

The mass flow sheet calculates the air feed and products of combustion of cycloheptane. The heat losses when burning cycloheptane are inserted in the heat balance sheet, second column.

The third column of the mass flow and heat balance sheets represents the conditions within the afterburner when burning cycloheptane in the afterburner, and trash in the kiln. The temperature developed within the afterburner is 2272°F, more than the 2200°F minimum required for complete destruction of the combustible components within the off-gas.

Table 20-3.   Flue Gas Discharge.

	Kiln	Kiln w/Aft.
Inlet, °F	2,200	2,272
Dry Gas, lb/hr	30,764	30,275
Heat, MBH	21.96	22.44
BTU/lb dry gas	714	741
Adiabatic t, °F	178	179
H_2O Saturation, lb/lb dry gas	.6008	.6279
lb/hr	18,483	19,260
H_2O Inlet, lb/hr	1,941	2,071
Quench H_2O, lb/hr	16,542	17,189
gpm	33	34
Outlet Temp., °F	120	120
Raw H_2O Temp., °F	70	70
Sump Temp., °F	148	149
Temp. Diff., °F	78	79
Outlet, BTU/lb dry gas	111.65	111.65
MBH	3.43	3.42
Req'd Cooling, MBH	18.95	19.06
H_2O, lb/hr	242,949	241,266
gpm	486	483
Outlet, ft^3/lb dry gas	16.515	16.515
ft^3/min	8,468	8,443
Fan Press., in. WC	15	15
Outlet, ACFM	8,780	8,754
Outlet, H_2O/lb dry gas	0.08128	0.08128
H_2O, lb/hr	2,500	2,493
Recirc. (Ideal), gpm	33	34
Recirc. (Actual), gpm	265	275
Cooling H_2O, gpm	486	483

The total off-gas flow is slightly less when burning cycloheptane (30,275 + 2,071 = 32,346 lb/hr) than when burning fuel oil (30,764 + 1,941 = 32,705 lb/hr) in the afterburner. The flue gas flow parameters are calculated for each of these two cases in Table 20-3, the flue gas discharge table.

PARTICULATE EMISSIONS

Allowable particulate emissions are related to a 12% CO_2 discharge standard, i.e., 0.03 grain per dry standard cubic foot corrected to 12% CO_2. This correction is calculated as follows:

Trash: assume $C_6H_{10}O_5$ (cellulose)
Cycloheptane: C_7H_{14}

Equilibrium equation:

$$44.01 \qquad\qquad 28.2 \qquad 32.00$$
$$C_6H_{10}O_5 + C_7H_{14} + 16.5xO_2 + 16.5(3.3197)xN_2 \longrightarrow 13CO_2 + 12H_2O + 54.775xN_2 + 16.5(x-1)$$
$$572.13 \qquad\qquad 1534.80 \qquad 528x - 528$$

With the unit molecular weight above the dry gas components and the total molecular weight beneath the components the total dry gas flow is:

$$572.13 + 1534.80x + 528.00x - 528.00 = 44.13 + 2062.80x$$

The CO_2 fraction present is 0.12 (12% CO_2):

$$0.12 = \frac{572.13}{44.13 + 2062.80x}$$

$$x = 2.290$$

This calculation indicates that 229% total air, or 129% excess air, will produce a CO_2 concentration of 12% of the exiting dry gas stream. The actual dry gas flow must be calculated for this condition in order to determine allowable emissions.

The mass flow sheet, Table 20-1, column 4, is calculated for 129% excess air, i.e., 12% CO_2. The total dry gas flow is 35923 lb/hr.

Emissions, as stated above, are .03 grains per dry standard cubic foot at 12% CO_2, and one dry cubic foot of air (dry gas) at standard conditions weighs 0.075 lbs:

$$(35{,}923 \text{ lb/hr}) \times (\text{DSCF}/0.075 \text{ lb}) \times (0.03 \text{ gr/DSCF}) \times (\text{lb}/7000 \text{ gr})$$

$$= 2.05 \text{ lb/hr emissions}$$

CONTROL EQUIPMENT

Air pollution control equipment must be provided to reduce particulate emissions to an acceptable level. From Table 7-6 the particulate discharge from a multiple chamber incinerator is approximately 7 lb/ton of charge. For the solid waste incinerated in this example:

$$2000 \frac{\text{lb}}{\text{hr}} \times 7 \frac{\text{lb}}{\text{ton}} \times \frac{\text{ton}}{2000 \text{ lb}} = 7 \frac{\text{lb}}{\text{hr}}$$

Allowable emissions: $2.05 \dfrac{\text{lb}}{\text{hr}}$

Required collection efficiency:

$$\frac{7 - 2.05}{7} \times 100 = 71\%$$

Examining Table 12 in Chapter 20 a low efficiency wet scrubber has a particulate removal efficiency of up to 90%. This is more than adequate to provide the 71% removal efficiency required. A pressure drop of 12 inches WC will be assumed for the scrubber. The ID fan will be sized for an additional 3 inches WC drop through the exhaust gas system.

PSD EVALUATION

As noted in Chapter 2, when emissions exceed de minimus values PSD procedures are triggered. For particulate matter the trigger is 25 tons per year. For this example emissions of 2.05 lb/hr of particulate is anticipated. On a yearly basis, with 5000 hour per year operations:

$$2.05 \frac{lb}{hr} \times 5000 \frac{hr}{yr} \times \frac{ton}{2000\ lb} = 5.13 \frac{tons}{year}$$

This is below the de minimus value for particulates.

The nitrogen oxide de minimis value is 40 tons per year. Using Fig. 7-2:

For the Kiln: 13.00 MBH heat release
150% excess air
1700°F

Generation of NO_x is 0.3 lb/MBTU.

In one year:

$$13.00 \frac{MBTU}{hr} \times \frac{0.3\ lb}{MBTU} \times 5000 \frac{hr}{yr} \times \frac{ton}{2000\ lb} = 9.75 \frac{ton}{year}$$

For the Afterburner: 9.58 MBH heat release
10% excess air
2216°F

Generation of NO_x is 0.2 lb/MBTU.

In one year:

$$9.58 (MBTU/hr) \times (0.2\ lb/MBTU) \times 5000 (hr/yr) \times (ton/2000\ lb) = 4.79\ ton/yr$$

Total NO_x generation:

$$9.75 + 4.79 = 14.54\ ton/year\ NO_x$$

This is less than the 40 ton/year de minimis value.

The carbon monoxide de minimis value is 100 tons per year. Examining Table 7-3, which is not directly applicable to trash burning (Table 7-3 applies to pure hydrocarbon compounds, whereas trash contains oxygen as an elemental component) indicates an order of magnitude of CO generation. At $2200°F$, 50% excess air, 3×10^{-6} lbs of CO is generated per lb of stoichiometric air (15469 lb/hr):

$$3 \times 10^{-6} (lb\ CO/lb\ air) \times 15,469 (lb\ air/hr) \times 5,000 (hr/yr) \times (ton/2000\ lb)$$

$$= 0.12\ tons/yr\ CO$$

This value of CO is insignificant compared to the de minimis value, 100 tons per year.

From the above calculations a PSD analysis will not be triggered by the quantities of particulate, NO_x, and/or CO generated by this system.

CONCLUSIONS

The previous analysis resulted in a determination of the following equipment items and parameters:

- Use of a rotary kiln, 7'8" outside diameter by 23'1" long lined with 7" insulating firebrick and 6" of firebrick.
- An afterburner 11'6" outside diameter by 16'6" long lined with 9" of insulating castable refractory.
- Use of a wet scrubber with 71% (minimum) collection efficiency.
- At 5000 hours of operations per year this system will not require a PSD new source review.
- The solid waste incinerated will not generate sufficient heat to maintain $2200°F$ in the afterburner. Fuel oil must be provided, 63.78 gph, for $2200°F$ operation.
- The liquid waste has sufficient heating value to replace all of the fuel oil otherwise required to maintain $2200°F$.

- Make-up water at a temperature of 70°F is required at a flow of 486 gpm to cool the incinerator off-gas.
- A recirculation flow of 275 gpm is required for adiabatic cooling (quencher flow). For an assumed pump head of 30 psig, the pump horsepower required is calculated as follows, assuming 85% pump efficiency:

$$\frac{\dfrac{275 \text{ gal}}{\text{min}} \times \dfrac{8.34 \text{ lb}}{\text{gal}} \times \dfrac{30 \text{ lb}}{\text{in.}^2} \times \dfrac{\text{ft}^3}{62.4 \text{ lb}} \times \dfrac{144 \text{ in.}^2}{\text{ft}^2}}{33000 \dfrac{\text{ft-lb}}{\text{min-HP}} \times 0.85 \text{ efficiency}} = 5.66 \text{ HP}$$

A 7.5 HP recirculation pump motor is required.
- Combustion air for the kiln is 22,785 lb/hr (22,785 ÷ 0.075 ÷ 60 = 5,063 SCFM). Assuming a fan pressure of 8 in. WC, with 75% efficiency, the fan motor horsepower calculation is as follows:

$$\frac{\dfrac{5063 \text{ ft}^3}{\text{min}} \times 8 \text{ in.} \times \dfrac{62.4 \text{ lb}}{\text{ft}^3} \times \dfrac{\text{ft}}{12 \text{ in.}}}{33000 \dfrac{\text{ft-lb}}{\text{min-HP}} \times 0.75 \text{ efficiency}} = 8.51 \text{ HP}$$

A 10 HP fan motor is required.
- Combustion air required in the afterburner is 6991 lb/hr (1554 SCFM) when burning liquid waste and 7312 lb/hr (1625 SCFM) when burning fuel oil. Using the larger flow value the fan horsepower required, at 12 in. WC fan pressure with 80% efficiency, is as follows:

$$\frac{\dfrac{1625 \text{ ft}^3}{\text{min}} \times 12 \text{ in.} \times \dfrac{62.4 \text{ lb}}{\text{ft}^3} \times \dfrac{\text{ft}}{12 \text{ in.}}}{33000 \dfrac{\text{ft-lb}}{\text{min-HP}} \times 0.80 \text{ efficiency}} = 3.84 \text{ HP}$$

Use a 5 HP fan motor.
- The induced draft fan must provide a maximum of 8780 CFM gas flow at 15 in. WC. The required horsepower, at 80% efficiency, is:

$$\frac{\dfrac{8780 \text{ ft}^3}{\text{min}} \times 15 \text{ in.} \times \dfrac{62.4 \text{ lb}}{\text{ft}^3} \times \dfrac{\text{ft}}{12 \text{ in.}}}{33000 \dfrac{\text{ft-lb}}{\text{min-HP}} \times 0.80 \text{ efficiency}} = 25.94 \text{ HP}$$

A 30 HP fan motor (minimum) is required.

- The scrubber system will produce a net spent water discharge which must be drained. To calculate this discharge:

	w/Fuel Oil	w/Liquid Waste
Entering in flue gas, lb/hr	1,941	2,071
in scrubber, lb/hr	242,949	241,266
total entering lb/hr	244,890	243,337
Exiting in flue gas	2,500	2,493
to drain, lb/hr	242,390	240,844
gpm	485	482
Exiting temperature, °F	178	179

Glossary

ABRASION. The wearing away of surface material by the scouring action of moving solids, liquids or gases.

ABSORPTION. The penetration of one substance into or through another.

ACTIVATED CARBON. A highly absorbent form of carbon used to remove odors and toxic substances from gaseous emissions or to remove dissolved organic material from wastewater.

ACUTE LC(50). A concentration of a substance, expressed as parts per million parts of medium, that is lethal to 50% of the test population of animals under specified test conditions.

ACUTE TOXICITY. Any poisonous effect produced within a short period of time, usually up to 24–96 hours, resulting in severe biological harm and often death.

ADHESION. Molecular attraction which holds the surfaces of two substances in contact, such as water and rock particles.

ADSORPTION. The attachment of the molecules of a liquid of gaseous substance to the surface of a solid.

AEROSOL. A particle of solid or liquid matter that can remain suspended in the air because of its small size.

AFTERBURNER. A device that includes an auxiliary fuel burner and combustion chamber to incinerate combustible gas contaminants.

AGRICULTURAL SOLID WASTE. The solid waste that is generated be the rearing of animals and the producing and harvesting of crops or trees.

AIR CLASSIFIER. A system that uses a forced air stream to separate mixed material according to size, density and aerodynamic drag of the pieces.

AIR COOLED WALL. A refractory wall with a lane directly behind it through which cool air flows.

AIR EMISSIONS. For stationary sources, the release or discharge of a pollutant by an owner or operator into the ambient air either by means of a stack or as a fugitive dust, mist, or vapor as a result inherent to the manufacturing or forming process.

AIR HEATER. A heat exchanger through which air passes and is heated by a medium of a higher temperature, such as hot combustion gases in metal tubes.

AIR POLLUTANT. Dust, fumes, smoke and other particulate matter, vapor, gas, odorous substances or any combination thereof. Also any air pollution agent or combination of such agents, including any physical, chemical, biological, radioactive substances or matter which is emitted into or otherwise enters the ambient air.

AIR POLLUTION. The presence of any air pollutant in sufficient quantities and of such characteristics and duration as to be, or likely to be, injurious to health or welfare, animal or plant life, or property, or as to interfere with the enjoyment of life or property.

AMBIENT AIR. That portion of the atmosphere external to buildings to which the general public has access.

ANIMAL WASTE. The high organic waste that is generated by the breeding, maintenance, use and slaughter of animals.

ASH. Inorganic residue remaining after ignition of combustible substances determined by definite prescribed methods.

AUTOGENOUS (AUTOTHERMIC) COMBUSTION. The burning of wet organic material where the moisture content is at such a level that the heat of combustion of the organic material is sufficient to vaporize the water and maintain combustion; no auxiliary fuel is required except for startup of the process.

BAFFLES. Deflector vanes, guides, grids, grating, or similar devices constructed or placed in air or gas flow systems, flowing water or slurry systems to effect a more uniform distribution of velocities, absorb energy, divert, guide, or agitate fluids, and check eddies.

BAGASSE. An agricultural waste material consisting of the dry pulp residue that remains after juice is extracted from sugar cane or sugar beets.

BAGHOUSE. An air pollution abatement device used to trap particulates by filtering gas streams through large fabric bags usually made of cloth or glass fibers.

BAROMETRIC DAMPER. A hinged or pivoted plate that automatically regulates the amount of air entering a duct, breeching, flue connection or stack, thereby maintaining a constant draft within an incinerator.

BATCH-FED INCINERATOR. An incinerator that is periodically charged with waste; one charge is allowed to burn out before another is added.

BATTERY WALL. A double or common wall between two incinerator combustion chambers; both faces are exposed to heat.

BILGE OIL. Waste oil which accumulates, usually in small quantities, in the lower spaces in a ship, just inside the shell plating, and usually mixed with larger quantities of water.

BIODEGRADABLE. Any substance that decomposes chemically and/or physically through the action of microorganisms.

BIOLOGICAL WASTE. Waste derived from living organisms.

BIOMASS. The amount of living matter in a given unit of the environment.

BLOWDOWN. The minimum discharge of recirculating water for the purpose of discharging materials contained in the process, the further buildup of which would cause concentrations or amounts exceeding limits established by best engineering practice.

BLOWER. A fan used to force air or gas under pressure.

BOTTOM ASH. The solid material that remains on a hearth or falls off the grate after thermal processing is complete.

BREECHING. A passage that conducts the products of combustion to a stack or chminey.

BRIDGE WALL. A partition between chambers over which the products of combustion pass.

BRITISH THERMAL UNIT. The amount of heat required to raise the temperature of one pound of water one degree Farenheit.

BULKY WASTE. Large items of solid waste such as household appliances, furniture, large auto parts, trees, branches, stumps, and other oversize wastes whose large size precludes or complicates their handling by normal solid waste collection, processing or disposal methods.

BUNKER "C" OIL. A general term used to indicate a heavy viscous fuel oil.

BURIAL GROUND. A disposal site for unwanted radioactive materials that uses earth or water for a shield.

BURNING AREA. The horizontal projection of a grate, a hearth, or both.

BURNING HEARTH. A solid surface to support the solid fuel or solid waste in a furnace and upon which materials are placed for combustion.

BURNING RATE. The volume of solid waste incinerated or the amount of heat released during incineration. The burning rate is usually expressed in pounds of solid waste per square foot of burning area per area or in British Thermal Units per cubic foot of furnace volume per hour.

BUTTERFLY DAMPER. A plate or blade installed in a duct, breeching, flue connection, or stack that rotates on an axis to regulate the flow of gases.

CAKE. The solids discharged from a dewatering apparatus.

CALCINATION. The process of heating a waste material to a high temperature without fusing in order to effect useful change (e.g., oxidation, pulverization.)

CARBONACEOUS MATTER. Pure carbon or carbon compounds present in the fuel or residue of a combustion process.

CARBON SORPTION. The process in which a substance (the sorbate) is brought into contact with a solid (the sorbent), usually activated carbon, and held there either by chemical or physical means.

CARCINOGENIC. Capable of causing the cells of an organism to react in such a way as to produce cancer.

CATALYTIC COMBUSTION SYSTEM. A process in which a substance is introduced into an exhaust gas stream to burn or oxidize vaporized hydrocarbons or odorous contaminants; the substance itself remains intact.

CAUSTIC SODA. Sodium hydroxide (NaOH), a strong alkaline substance used as an acid neutralizer.

CENTRIFUGAL COLLECTOR. A mechanical system using centrifugal force to remove areosols from a gas stream or to dewater sludge.

CHAMBER. An enclosed space inside an incinerator.

CHAIN GRATE STOKER. A stoker with a moving chain as a grate surface. The grate consists of links mounted on rods to form a continuous surface that is generally driven by a shaft with sprockets.

CHARGE. The amount of solid waste introduced into a furnace at one time or, in the steel industry, the addition of iron and steel scrap or other materials into the top of an electric arc furnace.

CHARGING CHUTE. An overhead passage through which waste materials drop into an incinerator.

CHARGING GATE. A horizontal, movable cover that closes the opening on a top-charging furnace.

CHARGING HOPPER. An enlarged opening at the top of a charging chute.

CHECKER WORK. A pattern of multiple openings through which the products of combustion pass to accelerate the turbulent mixing of gases.

CHUTE-FED INCINERATOR. An incinerator that is charged through a chute that extends two or more floors above it.

CLAMSHELL BUCKET. A vessel used to hoist and convey materials; it has two jaws that clamp together when it is lifted by specially attached cables.

CLASSIFICATION. The separation and rearrangement of waste materials according to composition (e.g., organic or inorganic), size, weight, color, shape, and the like, using specialized equipment.

CLINKERS. Hard, sintered, or fused pieces of residue formed in a furnace by the agglomeration of ash, metals, glass and ceramics.

COAL REFUSE. Waste products of coal mining, cleaning, and coal preparation operations and containing coal, matrix material, clay, and other organic and inorganic material.

CODISPOSAL. The technique in which sludge is combined with other combustible materials (e.g., refuse, refuse derived fuel, coal), to form a furnace feed with a higher heating value than the original sludge.

COFFIN. A thick walled container (usually lead) used for transporting radioactive materials.

COLD DRYING HEARTH. A surface upon which unheated waste materials are placed to dry or to burn. Hot combustion gases are then passed over the materials.

COMBUSTIBLES. Materials that can be ignited at a specific temperature in the presence of air to release heat energy.

COMBUSTION. The production of heat and light energy through a chemical process, usually oxidation.

COMBUSTION AIR. The air used for burning a fuel.

COMBUSTION GASES. The mixture of gases and vapors produced by burning.

COMMERCIAL WASTE. All types of solid wastes generated by stores, offices, restaurants, warehouses, and other nonmanufacturing activities, excluding residential and industrial wastes.

COMPOST. A relatively stable mixture of decomposed organic waste materials, generally used to fertilize and condition the soil.

CONDUCTION. The transfer of heat by physical contact between substances.

CONICAL BURNER. A hollow, cone-shaped combustion chamber with an exhaust vent at its point and a door at its base through which waste materials are normally charged; air is delivered to the burning waste inside the cone. Also called a teepee burner.

CONSERVATION. The protection, improvement and use of natural resources according to principles that will insure their highest economic or social benefits.

CONSTRUCTION AND DEMOLITION WASTE. The waste building materials, packaging and rubble resulting from construction, remodeling and demolition operations on pavements, houses, commercial buildings, and other structures.

CONSUMER WASTE. Materials used and discarded by the buyer, or consumer, as opposed to wasted created and discarded in-plant during the manufacturing process.

CONTAMINANT. Any physical, chemical, biological or radiological substance in water.

CONTINUOUS FEED INCINERATOR. An incinerator into which solid waste is charged almost continuously to maintain a steady rate of burning.

CONTROLLED-AIR INCINERATOR. An incinerator with two or more combustion chambers in which the amounts and distribution of air are controlled. Partial combustion takes place in the first zone and gases are burned in a subsequent zone or zones.

CORROSION. The gradual wearing away of a substance by chemical action.

CURIE. A measure of radioactivity.

CURTAIN WALL. A refractory construction or baffle that deflects combustion gases downward.

CYCLONE SEPARATOR. A separator that uses a swirling air flow to sort mixed materials according to the size, weight and density of the pieces.

DAIRY WASTE. The waste generated by dairy plants in their processing of milk to produce cream, butter, cheese, ice cream, and other dairy products.

DECIBEL. A unit of sound measurement.

DECONTAMINATION/DETOXIFICATION. Processes which will convert pesticides into nontoxic compounds or the selective removal of radioactive material from a surface or from within another material.

DEEP-WELL INJECTION. Disposal of raw or treated hazardous wastes by pumping them into deep wells for filtration through porous or permeable subsurface rock and then containment within surrounded layers of impermeable rock or clay.

DEWATERING. A physical process which removes sufficient water from sludge so that its physical form is changed from that of a fluid to that of slurry or damp solid.

DISPERSION TECHNIQUE. The use of dilution to attain ambient air quality levels including any intermittent or supplimental control of air pollutants varying with atmospheric conditions.

DOMESTIC WASTE OR HOUSEHOLD WASTE. Solid waste, comprised of garbage and rubbish, which normally originates in the residential private household or apartment house.

DOWNPASS. A chamber or gas passage placed between two combustion chambers to carry the products of combustion downward.

DRAFT. The difference between the pressure within an incinerator and that in the atmosphere.

DRAG CONVEYOR. A conveyor that uses vertical steel plates fastened between two continuous chains to drag material across a smooth surface.

DROP ARCH. A form of construction that supports a vertical refractory furnace wall and serves to deflect gases downward.

DRYING HEARTH. A solid surface in an incinerator upon which wet waste materials, liquids, or waste matter that may turn to liquid before burning are placed to dry or to burn with the help of hot combustion gases.

DUCT. A conduit, usually metal or fiberglass, round or rectangular in cross section, used for conveyance of air.

DUMP. A site used to dispose of solid wastes, or other containerized wastes, without environmental controls.

DUST. Fine grain particles light enough to be suspended in the air.

EFFLUENT. Waste materials, usually waterborne, discharge into the environment, treated or untreated.

ELECTROSTATIC PRECIPITATOR. An air pollution control device that imparts an electric charge to particles in a gas stream causing them to collect on an electrode.

ELUTRIATION. The reduction of the concentration of an impurity on a solid by repeated washings.

EMISSION RATE. The amount of pollutant emitted per unit of time.

ENCAPSULATION. The complete enclosure of a waste in another material in such a way as to isolate it from external effects such as those of water or of air.

ENVIRONMENT. Water, air, land, and all plants and man and other animals living therein, and the interrelationships which exist among them.

ENVIRONMENTALLY PERSISTENT WASTE. Any waste which, if exposed to a natural environment, remains hazardous for an extended length of time.

ERASE STACK. An expanding connection on the outlet of a fan or in an airflow passage to convert kinetic energy into static pressure.

EVAPORATION. The physical transformation of a liquid to a gas at any temperature below its boiling point.

EXHAUST SYSTEM. The system comprised of a combination of components which provides for enclosed flow of exhaust gas from the furnace exhaust port to the atmosphere.

EXPLOSIVE WASTES. Wastes which are unstable and may readily undergo violent chemical change or explode.

EXTRACTION TEST PROCEDURE. A series of laboratory operations ard analyses designed to determine whether, under severe conditions, a solid waste, stabilized waste, or landfill material can yield a hazardous leachate.

FEDERAL REGISTER. Daily publication that is the official method of notice of executive branch actions, including proposed and final regulations.

FEEDLOT WASTE. High concentrations of animal excrement that result from raising large numbers of animals on a relatively small, confined area of land.

FILTER. A porous device through which a gas or liquid is passed to remove suspended particles or dust.

FIRE POINT. The lowest temperature at which an oil vaporizes readily enough to burn at least five seconds after ignition.

FIREBRICK. Refractory brick made from fireclay.

FIRECLAY. A sedimentary clay containing only small amounts of fluxing impurities, high in hydrous aluminum silicate and capable of withstanding high temperatures.

FIXED CARBON. The ash-free carbonaceous material that remains after volatile matter is driven off a dry solid waste sample.

FIXED GRATE. A grate without moving parts, also called a stationary grate.

FLAMMABLE WASTE. A waste capable of igniting easily and burning rapidly.

FLASH DRYING. The process of drying a wet organic material by passing through a high temperature zone at such a rate that the water is rapidly evaporated but the organic material, protected by the boiling point of water, is not overheated.

FLASH POINT. The minimum temperature at which a liquid or solid gives off sufficient vapor to form an ignitable vapor-air mixture near the surface of the liquid or solid.

FLUE. Any passage designed to carry combustion gases and entrained particulates.

FLUE-FED INCINERATOR. An incinerator that is charged through a shaft that functions as a chute for charging waste and as a flue for carrying products of combustion.

FLUE GAS. The products of combustion, including pollutants, emitted to the air after a production process or combustion takes place.

FLUIDIZED BED COMBUSTION. Oxidation of combustible material within a bed of solid, inert (noncombustible) particles which under the action of vertical hot air flow will act as a fluid.

FLY ASH. The airborne combustion residue from burning fuel.

FOOD WASTE. The organic residues generated by the handling, storage, sale, preparation, cooking and serving of foods commonly called garbage.

FORCED DRAFT. The positive pressure created by the action of a fan or blower which supplies the primary or secondary combustion air in an incinerator.

FOSSIL FUEL. Natural gas, petroleum, coal, and any form of solid, liquid or gaseous fuel derived from such materials for the purpose of creating useful heat.

FOULING. The impedance to the flow of fluids or heat that results when material accumulates in flow passages or on heat absorbing surfaces in an incinerator or other combustion chamber.

FUEL. Any material which is capable of releasing energy or power by combustion or other chemical or physical means.

FUGITIVE EMISSIONS. Emissions other than those from stacks or vents.

FUME. Solid particles under one micron in diameter, formed as vapors condense or as chemical reactions take place.

FURNACE. A combustion chamber; an enclosed structure in which heat is produced.

FURNACE ARCH. A nearly horizontal structure that extends into a furnace and serves to deflect gases.

FUSION POINT. The temperature at which a particular complex mixture of minerals can flow under the weight of its own mass.

GARBAGE. Solid waste resulting from animal, grain, fruit or vegetable matter used or intended for use as food.

GRAIN LOADING. The rate at which particles are emitted from a pollution source, in grains per cubic foot of gas emitted, 7000 grains = 1 lb.

GRATE. A piece of furnace equipment used to support solid waste or solid fuel during the drying, igniting and burning process.

GREASE. A group of substances including fats, waxes, free fatty acids, calcium and magnesium soaps, mineral oils, and certain other non-fatty materials.

GUILLOTINE DAMPER. An adjustable plate, used to regulate the flow of gases, installed vertically in a breeching.

HARDNESS. A characteristic of water, imparted by salts of calcium, magnesium, and iron, that causes curdling of soap, deposition of scale in boilers, damage in some industrial processes, and sometimes objectionable taste.

HAZARDOUS WASTE. A waste, or combination of wastes, that may cause or significantly contribute to an increase in mortality or an increase in serious irreversable, or incapacitating reversable illness or that pose a substantial present or potential hazard to human health or the environment when improperly treated, stored, transported, disposed of, or otherwise managed.

HEARTH. The bottom of a furnace on which waste materials are exposed to the flame.

HEAT BALANCE. An accounting of the distribution of the heat input and output of an incinerator or boiler, usually on an hourly basis.

HEAVY METALS. Metallic elements such as mercury, chromium, lead, cadmium and arsenic, with high atomic weights, and which tend to accumulate in the food chain.

HOT DRYING HEARTH. A surface upon which waste materials are placed to dry or to burn. Hot combustion gases first pass over the materials and then under the hearth.

HYDROCARBON. Any of a vast family of compounds containing carbon and hydrogen in various combinations, found especially in fossil fuels.

IGNITABILITY. The characteristic property of liquids having a flash point of less than 140°F; non-liquids liable to cause fires through friction, absorption of moisture, spontaneous chemical change, or retained heat from manufacturing, or are liable, when ignited, to burn so vigorously and persistently as to create a hazard; ignitable compressed gas; oxidizers.

IGNITION ARCH. A refractory furnace arch or surface located over a fuel bed to radiate heat and to accelerate ignition.

IGNITION TEMPERATURE. The lowest temperature of a fuel at which combustion becomes self-sustaining.

INCINERATION. An engineered process using controlled flame combustion to thermally degrade waste materials.

INCINERATOR STOKER. A mechanically operable moving grate arrangement for supporting, burning, or transporting solid waste in a furnace and discharging the residue.

INDUCED DRAFT. The negative pressure created by the action of a fan, blower or other gas moving device located between an incinerator and a stack.

INDUSTRIAL WASTE. Unwanted materials produced in or eliminated from an industrial operation.

INFILTRATION AIR. Air that leaks into the chambers of ducts of an incinerator.

INORGANIC MATTER. Chemical substances of mineral origin, not containing carbon-to-carbon bonding.

INORGANIC REFUSE. Solid waste composed of matter other than plant, animal, and certain carbon compounds, e.g., metals and glass.

INSTITUTIONAL SOLID WASTE. Solid wastes generated by educational, health care, correctional and other institutional facilities.

INSULATED WALL. A furnace wall on which refractory material is installed over insulation.

ISOKINETIC SAMPLING. Sampling in which the linear velocity of the gas entering the sampling nozzle is equal to that of the undisturbed gas stream at the sample point.

JUNK. Unprocessed, discarded materials that are usually not suitable for reuse or recycling, e.g., rags, paper, toys, metal, furniture.

KRAFT PAPER. A comparatively coarse paper noted for its strength and used primarily as a wrapper or packaging material.

LANDFILL. A land disposal site enploying an engineered method of disposing of wastes on land that minimizes envinronmental hazards by spreading wastes in thin layers, compacting the wastes to the smallest practical volume, and applying cover materials at the end of each operating day.

LANTZ PROCESS. A destructive distillation technique in which the combustible components of solid waste are converted into combustible gases, charcoal, and a variety of distillates.

LETHAL DOSE. The quantity of a substance which is fatal to 50% of the population on which it is tested.

LIME. Any of a family of chemicals consisting essentially of calcium oxide or hydroxide made from linestone (calcite).

LINER. The material used on the inside of a furnace wall to insure that a chamber is impervious to escaping gases; the material used on the inside of a sanitary landfill to insure that the basin is impervious to fluids.

MACROENCAPSULATION. The isolation of a waste by imbedding it in, or surrounding it with, a material which acts as a barrier to water or air.

MAGNETIC SEPARATION. The process by which a permanent magnet or electromagnet is used to attract magnetic materials away from mixed waste.

MASKING. Blocking out one sight, sound, or smell with another.

METAL CLEANING WASTES. Any cleaning compounds, rinse waters, or any other water-borne residues derived from cleaning any metal process equipment including, but not limited to, boiler tube cleaning, boiler fireside cleaning and air preheater cleaning.

MICROENCAPSULATION. The isolation of a waste from external effects by mixing it with a material which then cures or converts it to a solid, nonleaching barrier.

MILLED REFUSE. Solid waste that has been mechanically reduced in size.

MINING WASTE. Residues which result from the extraction of raw materials from the earth.

MIST. Liquid particles, measuring 40–500 microns in diameter, that are formed by condensation of vapor.

MIXING CHAMBER. A chamber usually placed between the primary and secondary combustion chambers and in which the products of combustion are thoroughly mixed by turbulence that is created by increased velocities of gases, checkerwork, or turns in the direction of the gas flow.

MIXTURE. Any combination of two or more chemical substances if the combination does not occur in nature and is not, in whole or in part, the result of a chemical reaction.

MODULAR COMBUSTION UNIT. One of a series of incinerator units designed to operate idependently and can handle small quantities of solid waste.

MOVABLE GRATE. A grate with moving parts.

MULTIPLE CHAMBER INCINERATOR. An incinerator that consists of two or more chambers, arranged as in-line or retort types, interconnected by gas passage parts or flues.

MUNICIPAL INCINERATOR. A privately or publicly owned incinerator primarily designed and used to burn residential and commercial solid waste within a community.

MUNICIPAL SOLID WASTES. Garbage, refuse, sludges and other discarded materials resulting from residential and non-industrial operations and activities.

MUTAGENIC. The property of a substance or mixture of substances which, when it interacts with a living organism, causes the genetic characteristics of the organism to change and its offspring to have a decreased life expectancy.

NATURAL DRAFT. The negative pressure created by the height of a stack or chimney and the difference in temperature between flue gases and the atmosphere.

NATURAL GAS. A natural fuel containing methane and hydrocarbons that occurs in certain geologic formations.

NEUTRALIZATION. The chemical process in which the acidic or basic characteristics of a fluid are changed to those of water.

NITROGENOUS WASTES. Animal or plant residues that contain large amounts of nitrogen.

NONFERROUS. Metals that contain no iron, e.g., aluminum, copper, brass and bronze materials.

ODOR THRESHOLD. The lowest concentration of an airborne odor that a human being can detect.

OFFAL. The viscera and trimmings of a slaughtered animal removed from the carcass.

OFFICE WASTES. Discarded materials that consist primarily of paper waste including envelopes, ledgers and brochures.

OIL. Oil of any kind or in any form including, but not limited to, petroleum, fuel oil, sludge, oil refuse, and oil mixed with wastes other than dredged spoil.

ON-SITE INCINERATOR. An incinerator that burns solid waste on the property used by the generator thereof.

OPACITY. Degree of obscuration of light; e.g., a window has zero opacity, while a wall has 100% opacity.

OPEN BURNING. The combustion of any material without control of combustion air or combustion temperature, without containment of the combustion reaction in an enclosure, and/or without provision of a stack or a vent for discharge of combustion products.

OPEN HEARTH FURNACE. A long, wide, shallow reverberatory furnace used to produce steel from cast or pig iron.

OPEN PIT INCINERATOR. A burning apparatus that has an open top and a system of closely spaced nozzles that place a stream of high velocity air over the burning zone.

ORGANIC MATTER. Chemical substances comprised mainly of carbon, covalently bonded. May have its origin in animal or plant life, coal, petroleum, or laboratory synthesis.

ORGANIC NITROGEN. Nitrogen combined in organic molecules such as protein, amines, and amino acids.

ORSAT. An apparatus used to analyze flue gases volumetrically by dissolving the constituent gases selectively in various solvents.

OSCILLATING-GRATE STOKER. A stoker whose entire grate surface oscillates to move the solid waste and residue over the grate surface.

OVERFIRE AIR. Air under control as to quantity and direction, introduced above and beyond a fuel bed by induced or forced draft.

OXIDANT. A substance containing oxygen that reacts chemically to release some of or all of its oxygen component to another substance.

OXIDATION. The addition of oxygen to a compound.

OXIDE. A compound of two elements, one of which is oxygen.

PACKED TOWER. A pollution control device that forces dirty gas through a tower packed with crushed rock, wood chips, or other packing while liquid is sprayed over the packing material. Polutants in the gas stream either dissolve in or chemically react with the liquid.

PAPER. The term for all kinds of matted or felted sheets of fiber laid down on a fine screen from a water suspension. Specifically, as one of the two subdivisions of the general term, paper refers to materials that are lighter in basis weight, thinner and more flexible than paperboard, the other subdivision.

PARTICULATE MATTER. Any material, except water in uncombined form, that is or has been airborne and exists as a liquid or a solid at standard conditions.

PARTICULATES. Fine liquid or solid particles such as dust, smoke, mist, fumes, or smog, found in the air or in emissions.

PATHOGEN. Organism capable of causing disease.

PATHOGENIC WASTE. Discarded materials that contain organisms capable of causing disease.

PERMEABILITY. The property of a solid material which allows fluid to flow through it.

pH. A measure of the acidity (0, most acid) or alkalinity (14, most alkaline) of a liquid or solid on a scale of 0 to 14.

PIG. A container, usually lead, used to ship or store radioactive materials.

PLASTICS. Nonmetallic compounds that result from a chemical reaction and are molded or formed into rigid or pliable structural material.

PLUME. Visible emissions from a flue or chimney.

POLLUTANT. Dredged spoil, solid waste, incinerator residue, sewage, garbage, sewage sludge, munitions, chemical wastes, biological materials, radioactive materials, heat, wrecked or discarded equipment, rock, sand, cellar dirt, and industrial, municipal, and agricultural waste discharged into water.

POLLUTION. The presence of matter or energy whose nature, location or quantity produces undesired environmental effects. Also, the man-made or man-introduced alteration of the chemical, physical, biological, and radiological integrity of water.

POROSITY. The ratio of the volume of pores of a material to the volume of its mass.

POUR POINT. The lowest temperature at which an oil will flow or can be poured under specified conditions of test.

PRECIPITATORS. Air pollution control devices that collect particles from an emission by mechanical or electrical means.

PRIMARY COMBUSTION AIR. The air admitted to a combustion system when the fuel is first oxidized.

PRIMARY POLLUTANT. A pollutant emitted directly from a polluting stack.

PRIMARY STANDARD. A national air emissions standard intended to establish a level of air quality that, with an adequate margin of error, will protect public health.

PROCESS WASTE. Any designated toxic pollutant which is inherent to or unavoidable resulting from any manufacturing process, including that which comes into direct contact with or results from the production or use of any raw material, intermediate product, finished product, byproduct, or waste product.

PROXIMATE ANALYSIS. The analysis of a fuel to determine (on a percentage basis) how much moisture, volatile matter, fixed carbon, and ash the sample contains; usually the fuel's heat value is also obtained.

PUTRESCIBLE. A substance that can rot quickly enough to cause odors and attract flies.

PYROLYSIS. The chemical decomposition of organic matter through the application of heat in an oxygen deficient atmosphere.

QUENCH TANK. A water filled tank used to cool incinerator residues or hot materials during industrial processes.

RADIOACTIVE. Substances that emit rays either naturally or as a result of scientific manipulation.

RATED INCINERATOR CAPACITY. The number of tons of solid waste that can be processed in an incinerator per 24-hour period when specified criteria prevail.

REACTIVITY. The tendency to create vigorous reactions with air or water, tendency to explode, to exhibit thermal instability with regard to shock, ready reaction to generate toxic gases.

RECIPROCATING GRATE STOKER. A stoker with a bed of bars or plates arranged so that alternate pieces, or rows of pieces, reciprocate slowly in a horizontal sliding mode and act to push the solid waste along the stoker surface.

RECYCLED MATERIAL. A material that is used in place of a primary, raw, or virgin material in manufacturing a product and consists of materials derived from post consumer waste, industrial scrap, material derived from agricultural wastes and other items, all of which can be used in the manufacture of new products.

REFRACTORY MATERIAL. Nonmetallic substances used to line furnaces because they can endure high temperatures and resist abrasion, spalling and slagging.

REFRACTORY WALL. A wall made of heat resistant material.

REFUSE. All solid materials which are discarded as useless.

REFUSE DERIVED FUEL. The combustible, or organic, portion of municipal waste that has been separated out and processed for use as fuel.

RESIDUAL OIL. A general term used to indicate a heavy viscous fuel oil.

RESIDUAL WASTES. Those solid, liquid or sludge substances from man's activities in the urban, agricultural, mining, and industrial environments remaining after collection and necessary treatment.

RESIDUE. Solid or semisolid materials such as, but not limited to, ash, ceramics, glass, metal, and organic structures remaining after incineration or processing.

RESIDUE CONVEYOR. Generally a drag or flight conveyor used to remove incinerator residue from a quench through to a discharge point.

RESOURCE CONSERVATION. Reduction of the amounts of wastes generated, reduction of overall consumption, and utilization of recovered resources.

RESOURCE RECOVERY. The extraction of useful materials or energy from waste.

RETORT-TYPE INCINERATOR. A multiple chamber incinerator in which the gases travel from the end of the ignition chamber, then pass through the mixing and combustion chambers.

REVERBERATORY FURNACE. A furnace in which the fuel is not in direct contact with the charge (waste) but the heating effect is basically generated by reflection down from a refractory roof.

ROCKING-GRATE STOKER. A stoker with a bed of bars or plates on axles; when the axles are rocked in a coordinated manner, the solid waste is lifted and advanced along the surface of the grate.

ROTARY KILN STOKER. A cylindrical, inclined device that rotates, thus causing the solid waste to move in a slow cascading and forward motion.

SANITARY LANDFILL. A method of disposing of refuse on the land without creating nuisances or hazards to the public health or safety, by utilizing the principles of engineering to confine the refuse to the smallest practical area, to reduce it to the smallest practical volume, and to cover it with a layer of earth at the conclusion of easch day's operation, or at such more frequencies as may be necessary.

SCRUBBING. The removal of impurities from a gas stream by spraying of a fluid.

SECONDARY BURNER. A burner installed in the secondary combustion chamber of an incinerator to maintain a minimum temperature and to complete the combustion of incompletely burned gas.

SECONDARY COMBUSTION AIR. The air introduced above or below the fuel (waste) bed by a natural, induced or forced draft.

SECONDARY POLLUTANT. A pollutant formed in the atmosphere by chemical changes taking place between primary pollutants and other substances present in the air.

SECONDARY STANDARD. A national air quality standard that establishes that ambient concentration of a pollutant that, with an adequate margin of safety, will protect the public welfare (all parts of the environment other than human health) from adverse impacts.

SETTLING CHAMBER. Any chamber designed to reduce the velocity of the products of combustion and thus to promote the settling of fly ash from the gas stream before it is discharged to the next process or to the environment.

SETTLING VELOCITY. The velocity at which a given dust will fall out of dust-laden gas under the influence of gravity only; also called terminal velocity.

SEWAGE SLUDGE. A semiliquid substance consisting of settled sewage solids combined with varying amounts of water and dissolved materials.

SHREDDER. A machine used to break up waste materials into smaller pieces by cutting or tearing.

SIFTINGS. The fine materials that fall from a fuel bed through its grate openings during incineration.

SILO. A storage vessel, generally tall relative to its cross section, for dry solids; materials are fed into the top and withdrawn from the bottom through a control mechanism.

SLAG. The more or less completely fused and vitrified matter separated during the reduction of metal from its ore.

SLAGGING. Destructive chemical action that forms slag on refractory materials subjected to high temperatures; also a molten or viscous coating produced on refractory materials by ash particles.

SLIDING DAMPER. A plate normally installed perpendicular to the flow of gas in a breeching and arranged to slide across it to regulate the flow.

SLUDGE. Any solid, semisolid or liquid waste generated from a municipal, commercial or industrial wastewater treatment plant, water supply treatment plant, or air pollution control facility, or any other such waste having similar characteristics and effects.

SLURRY. A pumpable mixture of solids and fluid.

SMOKE. Particles suspended in air after incomplete combustion of materials containing carbon.

SOLID WASTE. Any garbage, refuse, sludge from a waste treatment plant, water supply treatment plant, or air pollution control facility and other discarded material, including solid, liquid, semisolid, or contained gaseous material resulting from industrial, commercial, mining and agricultural operations, and from community activities.

SOOT. Carbon dust formed by incomplete combustion.

SOURCE. Any building, structure, facility, or installation from which there is or may be the discharge of pollutants.

SOURCE REDUCTION. Reducing the amount of materials entering the waste stream by voluntary or mandatory programs to eliminate the generation of waste.

SPRAY CHAMBER. A chamber equipped with water sprays that cool and clean the combustion products passing through it.

STACK. Any chimney, flue, vent, roof monitor, conduit or duct arranged to discharge emissions to the ambient air.

STATIONARY SOURCE. Any building, structure, facility, or installation which emits or may emit any air pollutant.

STOICHIOMETRIC COMBUSTION. Combustion with the theoretical air quantity.

STOKER. A mechanical device to feed solid fuel or solid waste to a furnace.

STREET REFUSE. Solid waste picked up when streets and sidewalks are swept manually and mechanically, wastes from public waste receptacles, and material removed from catch basins.

SYNERESIS. The process whereby a colloidal gel releases bound water by the application of heat.

TERATOGENIC. Affecting the genetic characteristics of an organism so as to cause its off-spring to be misshapen or malformed.

THEORETICAL AIR. The quantity of air, calculated from the chemical composition of a waste, that is required to burn the waste completely.

THERMAL EFFICIENCY. The ratio of heat used to total heat generated.

THRESHOLD DOSE. The minimum application of a given substance to produced a measurable effect.

TOXIC SUBSTANCE. A chemical or mixture that may present an unreasonable risk of injury to health or to the environment.

TOXICITY. The degree of danger posed by a substance to animal or plant life.

TRANSURANIUM. Nuclides having an atomic number greater than that of uranium (i.e., greater than 92).

TRAVELING GRATE STOKER. A stoker that is essentially a moving chain belt carried on sprockets and covered with separated, small metal pieces called keys.

TROMMEL. A perforated, rotating, horizontal cylinder that may be used in resource recovery facilities to break open trash bags, to remove glass and such small items as stone and dirt, and to remove cans from incinerator residue.

TUYERES. Openings or ports in a grate through which air can be directed to improve combustion.

ULTIMATE ANALYSIS. The chemical analysis of a solid, liquid or gaseous fuel.

UNDERFIRE AIR. Forced or induced combustion air (quantity and direction are controlled) introduced under a grate to promote burning within a fuel bed.

UNIT-SUSPENDED WALL. A furnace wall or panel that is hung from a steel structure.

UNRECLAIMABLE RESIDUES. Residual materials of little or no value remaining after incineration.

VAPOR. The gaseous phase of substances that are liquid or solid at atmospheric temperature and pressure; e.g., steam.

VAPOR PLUME. The stack effluent consisting of flue gases made visible by condensed water droplets or mist.

VITRIFICATION. A process whereby high temperatures effect permanent chemical and physical change in a ceramic body.

VOLATILE. Any substance that evaporates at a low temperature.

VOLATILITY. The property of a substance or substances to convert into vapor or gas without chemical change.

WASTE. Unwanted materials left over from manufacturing processes, or refuse from places of human or natural habitation.

WASTE REDUCTION. The prevention or restriction of waste generation at its source by redesigning products or the patterns of production and consumption.

WASTEWATER. Water carrying dissolved or suspended solids from homes, farms, businesses, institutions and industries.

WATERWALL INCINERATOR. An incinerator whose furnace walls consist of vertically arranged metal tubes through which water passes and absorbs the radiant energy from burning solid waste.

WINDBOX. A chamber below a furnace grate or surrounding a burner, through which air is supplied under pressure to burn the fuel.

WOOD RESIDUE. Bark, sawdust, slabe, chips, shavings, mill trim and other wood products derived from wood processing and forest management operations.

ZERO GAS. A gas containing less than one part per million parts of sulfur dioxide.

REFERENCES

1. "Common Environmental Terms: A Glossary," USEPA, November 1979.
2. Resource Conservation and Recovery Act, USEPA, May 1980.
3. The Clean Air Act, USEPA, 42USC.
4. Code of Federal Regulations—40, Protection of Environment, Revised July 1, 1978.

Acronyms and Abbreviations

AQCR	Air Quality Control Region
BAAQMD	Bay Area Air Quality Management District
BACT	Best Available Control Technology
BOD	Biochemical Oxygen Demand
BTU	British Thermal Unit
C/A	Cooling Air
CAA	Clean Air Act
CAAA	Clean Air Act Amendments
CE	Combustion Efficiency
CEQ	Council on Environmental Quality
COD	Chemical Oxygen Demand
CPI	Chemical Process Industries
CTG	Control Technology Guideline
DEC	Department of Environmental Conservation
DEP	Department of Environmental Protection
DG	Dry Gas
DOE	United States Department of Energy
DRE	Destruction and Removal Efficiency
DSCF	Dry Standard Cubic Foot
EA	Excess Air
EIS	Environmental Impact Statement
EOP	Emissions Offset Policy
EPA	United States Environmental Protection Agency
ESP	Electrostatic Precipitator
FBF	Fluid Bed Furnace
FGD	Flue Gas Desulfurization
FO	Fuel Oil
GAL	Gallon
GEP	Good Engineering Policy
GLC	Ground Level Concentration
HEPA	High Efficiency Particulate Air Filter
LAER	Lowest Achievable Emission Rate
LEL	Lower Explosive Limit
LLW	Low-Level Radioactive Waste
M/H	Multiple Hearth Incinerator
MBH	Million BTU Per Hour
MCU	Modular Combustion Unit
MSW	Municipal Solid Waste
NA	Non-Attainment

NAAQS	National Ambient Air Quality Standards
NEP	National Environmental Policy Act
NESHAPs	National Emission Standard for Hazardous Air Pollutants
NIOSH	National Institute for Occupational Safety and Health
NPDES	National Pollutant Discharge Elimination System
NRC	Nuclear Regulatory Commission
NSPS	New Source Performance Standards
NSR	New Source Review
NTIS	National Technical Information Service
ORD	Office of Research and Development (EPA)
PIC	Products of Incomplete Combustion
POHC	Principal Organic Hazardous Constituent
POTW	Publicly Owned Treatment Works
PSD	Prevention of Significant Deterioration
RACT	Reasonable Available Control Technology
RCRA	Resource Conservation and Recovery Act
RDF	Refuse Derived Fuel
SAC	Starved-Air Combustion
SIP	State Implementation Plan
SWA	Solid Waste Administration
TRU	Transuranic Material
TSCA	Toxic Substances Control Act
UEL	Upper Explosive Limit
VOC	Volatile Organic Compound
WC	Water Column

References

CHAPTER 1

1-1. *Source Category Survey: Industrial Incinerators*, USEPA 450/3-80-013, May 1980.

1-2. *Incinerator Standards*, Incinerator Institute of America, 1968.

1-3. Black, R., Klee, A., "The National Solid Wastes Survey: An Interim Report," Presented at the 1968 Annual Meeting of the Institute of Solid Wastes of the American Public Works Association, Miami Beach, FL, October 1968.

1-4. Bolton, R., *Public Utilities Regulatory Policies Act (PURPA)*, Malcolm Pirnie, Inc., May 1981.

1-5. *Air Quality Control*, National Association of Manufacturers, 1975.

1-6. *Municipal Incinerator Enforcement Manual*, USEPA 340/1-76-013, January 1977.

1-7. Arubuckle, J., *Environmental Law Handbook*, Government Institutes, Washington, D.C., September 1979.

1-8. Cross, F., *Handbook on Environmental Monitoring*, Technomic, 1974.

1-9. *The Federal Research and Development Plan for Air Pollution Control by Combustion Process Modification*, USEPA Publication CPA/22-69-147, January 1977.

1-10. Binder, R., *Fluid Mechanics*, Prentice Hall, Englewood Cliffs, NJ, January 1956.

1-11. Tchobanoglous, G., *Solid Wastes: Engineering-Principles and Management Issues*, McGraw-Hill, New York, 1977.

1-12. Brunner, C., Schwarz, S., *Energy and Resource Recovery from Waste*, Noyes, 1983.

CHAPTER 2

2-1. "A Review of Standards of Performance for New Stationary Sources: Incinerators," USEPA 60/4-79-009, March 1979.

2-2. "A Review of Standards of Performance for New Stationary Sources: Sewage Sludge Incinerators," USEPA 60/4-79-010, March 1978.

2-3. Devitt, T., *Inspection Manual for Enforcement of New Source Performance Standards: Sewage Sludge Incinerators*, USEPA 340/1-75-004, February 1975.

2-4. Devitt, T., *Inspection Manual for Enforcement of New Source Performance Standards: Municipal Incinerators*, USEPA 340/1-75-003, February 1975.

2-5. Gunther, C., "Resource Recovery and the Clean Air Act," *Waste Age*, May 1981.

2-6. Ambient Air Quality Standards, USEPA, 40CFR50, 1981.

2-7. Dravnieks, A., "Odor Perception and Odorous Air Pollution," *Transactions of the Pulp and Paper Institute*, May 1972.

2-8. Matey, J., "The Clean Air Act," *Pollution Engineering*, November 1981.

2-9. Regulatory Options for the Control of Odors, USEPA 450/5-80-003, February 1980.

2-10. "Swooping Down on Air Pollution," *Chemical Engineering*, June 1, 1981.

2-11. *Source Category Survey: Industrial Incinerators*, USEPA 450/3-80-013, May 1980.

2-12. Berkun, J., Star, A., *Implications of the New Source Review Process*, Malcolm Pirnie, Inc., 1980.

2-13. Niessen, W., Sarofin, A., "Incinerator Air Pollution, Facts and Speculation," National Incinerator Conference, American Society of Mechanical Engineers, 1970.

CHAPTER 3

3-1. Frick, G., "What Wastes Are Hazardous Under RCRA?," *Consulting Engineer*, September 1980.

3-2. Ross, R., "The Burning Issue: Incineration of Hazardous Wastes," *Pollution Engineering*, August 1979.

3-3. *Hazardous Waste Facilities: Standards for Incinerators*, USEPA SW-908, January 1981.

3-4. "Incinerator Standards for Owners and Operators of Hazardous Waste Management Facilities," *Federal Register*, January 23, 1981.

3-5. *Hazardous Waste Management: A Guide to the Regulations*, USEPA, 1980.

3-6. Beck, E., Plehn, S., "EPA's Cradle to Grave Approach to Hazardous Waste Management," *Consulting Engineer*, September 1980.

3-7. Hart, F., Tusa, W., "Organizing to Deal With RCRA," *Waste Age*, April 1981.

3-8. Sittig, M., *Incineration of Industrial Hazardous Wastes and Sludges*, Noyes, Park Ridge, NJ, 1979.

3-9. "Hazardous Waste and Consolidated Permit Regulations (RCRA)," *The Federal Register*, Volume 45, Number 98, Monday, May 19, 1980.

CHAPTER 4

4-1. *Combustion Fundamentals for Waste Incineration*, American Society of Mechanical Engineers, 1974.

4-2. Obert, M., *Thermodynamics*, McGraw-Hill, New York, 1948.

4-3. Norris, E., Therkelson, E., Pratt, L., *Applied Thermodynamics*, McGraw-Hill, New York, 1955.

4-4. Shapiro, A., *The Dynamics and Thermodynamics of Compressible Fluid Flow*, Ronald Press, 1958.

4-5. Lee, J., Sears, F., *Thermodynamics*, Second Edition, Addison Wesley, Reading, MA, 1963.

4-6. Costello, F., "The Second Law of Thermodynamics," American Society of Mechanical Engineers, Winter Annual Meeting, New York, 1979.

4-7. Keenan, J., Keyes, F., *Thermodynamic Properties of Steam*, John Wiley & Sons, New York, 1957.

4-8. *Guidebook for Industrial and Commercial Gas Fired Incineration*, American Gas Association, New York, 1963.

4-9. Keenan, J., Kaye, J., *Gas Tables*, John Wiley & Sons, New York, 1957.

4-10. Zimmerman, O., Lavine, I., *Psychrometric Tables and Charts*, Industrial Research Service, Dover, NH, 1964.

4-11. Baumeister & Marks, *Standard Handbook for Mechanical Engineers*, Seventh Edition, McGraw-Hill, New York.

4-12. Kennan, J., Keyes, F., *Theoretical Steam Rate Tables*, American Society of Mechanical Engineers, New York, 1938.

4-13. Perry & Chilton, *Chemical Engineering Handbook*, Fifth Edition, McGraw-Hill, New York.

4-14. Danielson, J., Air Pollution Engineering Manual, County of Los Angeles, Air Pollution Control District, AP-40, May 1973.

CHAPTER 5

5-1. "Hazardous Waste Laws Put New Strains on Refractories Used in Incineration," *Refractory Magazine*, December 1981.
5-2. Weber, J., "Predict Thermal Conductivity of Pure Gases," *Chemical Engineering*, January 12, 1981.
5-3. Harrison, M., Pelanne, C., "Cost Effective Thermal Insulation," *Chemical Engineering*, December 17, 1977.
5-4. Darroudi, T., "Thermal Shock Damage Resistance Parameters of Refractory Castables," *Industrial Heating*, May 1980.
5-5. Marino, J., "Thermal Design of Refractory and Insulation Systems," *Industrial Heating*, May 1980.
5-6. O'Keefe, W., "Thermal Insulation," *Power Magazine*, August 1974.
5-7. Trinklein, R., "Development of a Lightweight, Thermal Insulating, Corrosion Resistant Refractory Monolith," *Industrial Heating*, April 1979.
5-8. Brunner, C., "Program Calculates Heat Transfer Through Composite Walls," *Chemical Engineering*, June 16, 1980.
5-9. "Saving Heat Energy in Refractory Lined Equipment," *Chemical Engineering*, May 4, 1981.
5-10. Miner, R., "Evaluation of Improved Silicon Refractories," *Industrial Heating*, May 1980.
5-11. Jakob, M., *Elements of Heat Transfer*, Third Edition, John Wiley & Sons, New York, 1957.
5-12. Hess, L., *Insulation Guide for Buildings and Industrial Processes*, Noyes, Park Ridge, NJ, 1979.

CHAPTER 6

6-1. *Combustion Fundamentals for Waste Incineration*, The American Society of Mechanical Engineers, New York, 1974.
6-2. *Guidebook for Industrial and Commercial Gas Fired Incineration*, American Gas Association, New York, 1963.
6-3. Stewart, E., "Determine Excess Air From Wet Stack Gas Analysis," *Pollution Engineering*, April 1979.
6-4. *Fundamentals of Gas Combustion*, American Gas Association, New York, January 1973.
6-5. Eshbach, O., *Handbook of Engineering Fundamentals*, Second Edition, John Wiley & Sons, New York, 1961.
6-6. Cone, C., *Energy Management for Industrial Furnaces*, John Wiley & Sons, New York, 1980.

CHAPTER 7

7-1. Shen, T., "Air Pollutants From Sewage Sludge Incineration," *Journal of the American Society of Civil Engineers*, February 1979.

7-2. *Combustion Fundamentals for Waste Incineration*, The American Society of Mechanical Engineers, 1974.

7-3. Copeland, B., "A Study of Heavy Metal Emissions From Fluidized Bed Incinerators," Industrial Waste Conference, May 1975.

7-4. *Air Pollution Aspects of Sludge Incineration*, USEPA 625/4-75-009, June 1975.

7-5. Rinaldi, G., *An Evaluation of Emission Factors for Waste to Energy Systems*, USEPA, 1979.

7-6. "Dioxin Issue Resolved," *Waste Age*, November 1981.

7-7. "Air Quality Criteria for Particulate Matter," North Atlantic Treaty Organization Committee on the Environment, PB-240-570, Government Printing Office, Washington, D.C., November 1971.

7-8. *Introduction to Manual Methods for Measuring Air Quality*, California State Air Resoruces Board, July 1974.

7-9. *Test Methods to Determine the Mercury Emissions From Sludge Incineration Plants*, USEPA 60/4-79-058, September 1979.

7-10. Greenberg, R., Zoller, W., Gordon, J., "Atmospheric Emissions of Elements From the Parkway Sewage Sludge Incinerator," *Environmental Science and Technology*, January 1981.

7-11. *Dioxin From Combustion Sources*, American Society of Mechanical Engineers, 1981.

7-12. Jahnke, J., "A Research Study of Gaseous Emissions From a Municipal Incinerator," *Journal of the Air Pollution Control Association*, August 1971.

7-13. *Compilation of Air Pollutant Emission Factors*, USEPA AP-42, April 1973.

7-14. *Air Pollution Manual*, American Industrial Hygiene Association, 1972.

7-15. *Industrial Noise Manual*, American Industrial Hygiene Association, undated.

7-16. *Community Air Quality Guides*, American Industrial Hygiene Association, undated.

7-17. Pryde, L., *Environmental Chemistry*, Cummings, Menlo Park, CA, 1973.

7-18. *Source Category Survey: Industrial Incinerators*, USEPA 450/3-80-013, May 1980.

7-19. Dewling, R., Manganelli, R., Baer, G., "Fate and Behavior of Selected Heavy Metals in Incinerated Sludge," *Journal of the Water Pollution Control Federation*, October 1980.

7-20. Peterson, M., Stutzenberger, F., "Microbiological Evaluation of Incinerator Operations," *Applied Microbiological Journal*, July 1967.

7-21. Trijonis, J., Shapland, D., *Existing Visibility Levels in the US*, USEPA 450/5-79-010, 1979.

CHAPTER 8

8-1. *Thermodynamic Data for Industrial Incinerators*, National Bureau of Standards, Government Printing Office, Washington, D.C., 1972.

8-2. Stull, D., Prophet, H., *Thermochemical Tables*, Second Edition, Government Printing Office, Washington, D.C., June 1971.

8-3. Perry & Chilton, *Chemical Engineering Handbook*, Fifth Edition, McGraw-Hill, New York.

8-4. *Incinerator Standards*, Incinerator Institute of America, 1968.

CHAPTER 9

9-1. Achinger, W., Giar, J., *Testing Manual for Solid Waste Incinerators*, USEPA, 1973.

9-2. Corey, R., *Principles and Practice of Incineration*, John Wiley & Sons, New York, 1969.

9-3. *Disposal of Industrial Wastes by Combustion*, American Society of Mechanical Engineers, January 1971.

94. Rimer, A., *Solid Waste Engineering Laboratory Manual*, Duke University, 1978.

9-5. *Steam*, Thirty-Eighth Edition, Babcock & Wilcox, 1972.

CHAPTER 10

10-1. Hitchock, D., "Solid Waste Disposal: Incineration," *Chemical Engineers*, May 21, 1979.

10-2. Rice, R., "Solving the Small Company Solid Waste Problem," *Public Works Magazine*, September 1976.

10-3. Rubel, F., *Incineration of Solid Wastes*, Noyes, Park Ridge, NJ, 1974.

10-4. Archinger, W., Giar, J., *Testing Manual for Solid Waste Incinerators*, USEPA, 1973.

10-5. *Source Category Survey: Industrial Incinerators*, USEPA 450/3-80-013, May 1980.

10-6. *Recommended Methods of Reduction, Neutralization, Recovery or Disposal of Hazardous Waste.* Volume III, Disposal Process Descriptions, Ultimate Disposal, Incineration and Pyrolysis Processes, USEPA 670/2-73-053C, August 1973.

10-7. Danielson, J., *Air Pollution Engineering Manual*, County of Los Angeles, Air Pollution Control District, AP-40, May 1973.

10-8. MacKnight, R., "Controlling the Flue-Fed Incinerator," *Journal of the Air Pollution Control Association*, 10:103–09, April 1960.

CHAPTER 11

11-1. *European Waste-to-Energy Systems: An Overview*, Energy Research & Development Administration, CONS-2103-6, US Department of Commerce, Springfield, VA, June 1977.

11-2. Jahnke, J., "A Research Study of Gaseous Emissions From a Municipal Incinerator," *Journal of the Air Pollution Control Association*, August, 1977.

11-3. Rubel, *Incineration of Solid Wastes*, Noyes, Park Ridge, NJ, 1974.

11-4. Battelle Memorial Institute, *Corrosion Studies in Municipal Incinerators*, USEPA, 1972.

11-5. *The Conversion of Existing Municipal Incinerators to Codisposal*, USEPA SW-743, 1979.

11-6. *A Review of Techniques for Incineration of Sewage Sludge with Solid Wastes*, USEPA 600/2-76-288, December 1976.

11-7. Kennedy, J., "The Disposal of Solid Wastes," *Journal of Environmental Health*, September 1968.

11-8. *Hazardous Material Design Criteria*, USEPA 600/2-79-198, October 1979.

11-9. DeMarco, J., *Municipal Scale Incinerator Design and Operation*, USEPA, 1973.

11-10. *Municipal Incinerator Enforcement Manual*, USEPA 340/1-76-013, January 1977.

11-11. Miller, P., *Corrosion Studies in Municipal Incinerators*, USEPA SW-72-3-3, 1972.

11-12. "World Survey Finds Less Organic Matter," *Refuse Removal Journal*, 10:26, 1967.

11-13. *Municipal Refuse Disposal*, Institute for Solid Wastes, American Public Works Association, 1970.

11-14. Wisely, F., Hinchman, H., "Refuse as a Supplementary Fuel," Proceedings of the Third Annual Environmental Engineering Science Conference, Louisville, KY, March 1973.

11-15. Niessen, W., *Combustion and Incineration Processes*, First Edition, Marcel Dekker, Inc., New York.

11-16. Golembiewski, M., Baladi, E., *Environmental Assessment of a Waste-to-Energy Process: Braintree Municipal Incinerator*, USEPA, 1979.

11-17. Weinstein, N., Toro, R., *Thermal Processing of Municipal Solid Waste for Resource and Energy Recovery*, Ann Arbor Science, Ann Arbor, MI, 1976.

11-18. Popperman, C., "The Harrisburg Incinerator," Proceedings of the National Incinerator Conference, American Society of Mechanical Engineers, Miami, FL, 1974, Pages 247–254.

11-19. Astrom, L., Harris, D., "Comparative Study of European and North American Steam Producing Incinerators," Proceedings of the National Incinerator Conference, American Society of Mechanical Engineers, Miami, FL, 1974, Page 264.

CHAPTER 12

12-1. Sieger, R., Maroney, P., *Incineration, Pyrolysis of WWTP Sludges*, USEPA, 1977.

12-2. Sanner, W., Ortuglio, C., *Conversion of Municipal and Industrial Refuse Into Useful Materials by Pyrolysis*, US Department of the Interior, Bureau of Mines, Report of Investigations 7428, US Government Printing Office, Washington, D.C., August, 1970.

12-3. King, G., *Producing Clean Water and Energy From Pharmaceutical Wastewater*, Midland Ross Report, 1980.

12-4. Galandak, F., Racstain, M., "Design Considerations for Pyrolysis of Sewage Sludge," *Journal of the Water Pollution Control Federation*, February 1979.

12-5. *Recommended Methods of Reduction, Neutralization, Recovery or Disposal of Hazardous Waste*. Volume III, Disposal Process Descriptions, Ultimate Disposal, Incineration and Pyrolysis Processes, USEPA 670/2-73-053C, August 1973.

12-6. Mallan, G., Finney, C., "New Techniques in the Pyrolysis of Solid Waste," *Journal of the Water Pollution Control Federation*, February 1979.

12-7. Drobuy, N., Hull, H., Testin, R., *Recovery and Utilization of Municipal Solid Wastes*, US Public Health Service Publication 1908, 1971.

12-8. "Modular Incinerators Made for the Eighties," *Waste Age*, May 1980.

12-9. Rice, R., "Solving the Small Company Solid Waste Problem," *Public Works Magazine*, July 1976.

12-10. *Source Category Survey: Industrial Incinerators*, USEPA 450/3-80-013, May 1980.

12-11. Ross, R., *Industrial Waste Disposal*, Van Nostrand Reinhold, New York, 1968.

CHAPTER 13

13-1. Kenson, R., "Rotary Kiln Incinerators for Sludge Disposal," *Pollution Engineering*, December 1980.

13-2. *Hazardous Material Incinerator Design Criteria*, USEPA 600/2-79-198, October 1979.

13-3. Brunner, C., *Design of Sewage Sludge Incineration Systems*, Noyes, Park Ridge, NJ, 1980.

13-4. Krindler, E., Youngs, P., Burkhardt, G., "Modifying Existing Multiple Hearth Incinerators to Burn Thermal Conditioned Wastewater Sludge," *Journal of the Water Pollution Control Federation*, September 1980.

13-5. Baturay, A., "Latest Developments in Fluidized Bed Development Technology," Proceedings of the Puerto Rico Water Pollution Control Federation, February 1979.

13-6. *The Conversion of Existing Municipal Sludge Incinerators to Codisposal*, USEPA SW-743, 1979.

13-7. *A Review of Techniques for Incineration of Sewage Sludge with Solid Wastes*, USEPA 600/2-76-288, December 1976.

13-8. Jacknow, J., "Environmental Aspects of Acceptable Sludge Disposal Techniques," Fifth Conference on Acceptable Sludge Disposal Techniques," January 1978.

13-9. Brunner, C., "Sewage Sludge Incineration at the Cleveland Southerly Wastewater Treatment Center," American Society of Mechanical Engineers, Intersociety Energy Conference, Atlanta, GA, August, 1981.

13-10. Jacknow, J., "Thermal Sludge Processing Technology," *Sludge Magazine*, July 1979.

13-11. *Air Pollution Aspects of Sludge Incineration*, USEPA 625/4-75-009, June 1975.

13-12. Lyon, S., "Incineration of Raw Sludges and Greases," *Journal of the Water Pollution Control Federation*, April 1973.

13-13. *Recommended Methods of Reduction, Neutralization, Recovery or Disposal of Hazardous Waste*. Volume III, Disposal Process Descriptions, Ultimate Disposal, Incineration and Pyrolysis Processes, USEPA 670/2-73-053C, August, 1973.

CHAPTER 14

14-1. Shen, T., Chen, M., Lauber, J., "Incineration of Toxic Chemical Wastes," *Pollution Engineering*, October 1978.

14-2. Dunn, K., "Incineration's Role In Ultimate Disposal of Process Wastes," *Chemical Engineering*, October 10, 1975.

14-3. *Engineering Handbook for Hazardous Waste Incineration*, USEPA Draft, November 1980.

14-4. Brunner, C., "Industrial Wastes Incineration at Treatment Plant," *Water Engineering & Management*, June 1981.

14-5. Brunner, C., "Industrial Waste Disposal at Cincinnati," in *The Environmental Professional*, Pergamon Press, Elmsford, New York, 1981.

14-6. Perkins, B., *Incineration Facilities for Hazardous Wastes*, Los Alamos Laboratories, July 1976.

14-7. Brunner, C., Trapp, J., "Progress Report of Industrial Liquid/Fluid Thermal Processing System," *American Society of Mechanical Engineers, Journal of the Solid Waste Processing Division*, Washington, D.C., 1980.

14-8. *At Sea Incineration*, USEPA 600/2-79-137, July 1979.

14-9. Dillon, A., *Hazardous Waste Incineration Engineering*, Noyes, Park Ridge, NJ, 1981.

14-10. *Hazardous Material Incinerator Design Criteria*, USEPA 600/2-79-198, October 1979.

14-11. *Recommended Methods of Reduction, Neutralization, Recovery or Disposal of Hazardous Waste*. Volume III, Disposal Process Descriptions, Ultimate Disposal, Incineration and Pyrolysis Processes, USEPA 670/2-73-053C, August 1973.

14-12. Witt, P., "Disposal of Solid Wastes," *Chemical Engineering*, October 4, 1971.

CHAPTER 15

15-1. Pauletta, C., Hazzard, N., Benfordo, D., "Economics of Heat Recovery Indirect Fume Incineration," *Air Engineering*, March 1967.

15-2. *Energy Savings of Heat Recovery Equipment for Fume Incinerators*, Energy Management Bureau, September 25, 1980.

15-3. Meyers, F., "Fume Incineration with Combustion Air at Elevated Temperature," *Journal of the Air Pollution Control Association*, July 1966.

15-4. Cheremisinoff, P., Young, R., *Pollution Engineering Practice*, First Edition, Ann Arbor, Scientific Publishers, Ann Arbor, MI.

15-5. *Recommended Methods of Reduction, Neutralization, Recovery or Disposal of Hazardous Waste.* Volume III, Disposal Process Descriptions, Ultimate Disposal, Incineration and Pyrolysis Processes, USEPA 670/2-73-053C, August 1973.

15-6. Lund, H., *Industrial Pollution Control Handbook*, McGraw-Hill, New York, 1971.

15-7. Ross, R., *Industrial Waste Disposal*, Van Nostrand Reinhold, New York, 1968.

15-8. Straitz, J., "Flaring with Maximum Energy Conservation," *Pollution Engineering*, February 1980.

15-9. Ross, R., "Incineration of Solvent-Air Mixtures," *Chemical Engineering Progress*, August 1972.

15-10. Cross, F., *Air Pollution Odor Control Primer*, Technomic, 1973.

CHAPTER 16

16-1. Hedahl, T., McCormack, M., *Research and Development Plan for the Slagging Pyrolysis Incinerator*, United States Department of Energy, January 1979.

16-2. Stretz, L., Allem, C., Crippen, M., "Combustible Radioactive Waste Treatment by Incineration and Chemical Digestion," *Journal of the American Institute of Chemical Engineers*, May 28, 1980.

16-3. Oyen, L., Kibbey, A., "Volume Reduction, Solidification and Packaging of Wastes From Nuclear Power Plants, *Nuclear Power Waste Technology*, Chapter 7, American Society of Mechanical Engineers Monograph G00132, New York, 1978.

16-4. Parker, G., "Incineration of Hazardous and Low Level Radioactive Waste," *Pollution Engineering*, August 1981.

16-5. Stretz, L., Koenig, R., "Offgas Treatment for Radioactive Waste Incinerators," Department of Energy Nuclear Air Cleaning Conference, October 1980.

16-6. Aronson, R., "The Search for Safe Nuclear Waste Disposal," *Machine Design*, March 1980.

16-7. Hinga, K., "Disposal of High Level Radioactive Wastes," *Environmental Science and Technology*, January 1982.

16-8. Kibbey, A., *A State of Art Report on Low Level Radioactive Waste Treatment*, Oak Ridge National Laboratory, September 1980.

16-9. Filter, H., Roberson, K., "Solidification of Low Level Radioactive Wastes From Nuclear Power Plants," *Management of Low Level Radioactive Waste*, Volume I, Pergamon Press, Elmsford, New York, 1979.

16-10. Borduin, L., Taboas, A., *US Department of Energy Radioactive Waste Incineration Technology*, Los Alamos Laboratories, March 1980.

CHAPTER 17

17-1. Keller, F., US Patent 3,805,715, April 23. 1974.

17-2. Municipal, Industrial Waste Dehydration System, *Resource Recovery and Energy Review*, January 1976.

17-3. Swets, D., Pratt, L., Metcalf, C., "Thermal Sludge Conditioning in Kalamazoo," *Proceedings of the American Society of Civil Engineers*, March 1974.

17-4. Teletzke, G., "Components of Sludge and Its Wet Air Oxidation Products," *Journal of the Water Pollution Control Federation*, June 1967.

17-5. Sommers, L., Curtis, E., "Wet Air Oxidation," *Journal of the Water Pollution Control Federation*, November 1977.

CHAPTER 18

18-1. Stricker, G., "Twin Incineration Systems with Heat Exchangers Start Up Smoothly," *Chemical Processing*, September 1977.

18-2. Stevens, J., *Energy Recovery and Emissions From Municipal Waste Incineration*, Ontario Air Pollution Control Association, September 1977.

18-3. "Specifying and Operating Reliable Waste Heat Boilers," *Chemical Engineering*, August 13, 1979.

18-4. *Materials and Energy From Municipal Waste*, Office of Technological Assessment, 1979.

18-5. Freeman, H., "Pollutants From Waste-To-Energy Conversion Systems," *Environmental Science and Technology*, November 1978.

18-6. Brown, D., "Energy Recovery From Industrial Waste," *Waste Age Magazine*, April 1981.

18-7. *Energy From Municipal Solid Wastes*, Report of the US House of Representatives, June 1980.

18-8. Recovering the Energy in Wastes," *Mechanical Engineering*, May 1981.

18-9. Merrill, R., *Disposing of Hazardous Wastes in Industrial Boilers*, USEPA, March 1981.

18-10. Hasselriis, F., "Thermal Oxidizers Convert Industrial Waste Into Valuable Energy," *Power*, February 1982.

18-11. *Energy Utilization: Municipal Waste Utilization*, US Department of Energy, March 27, 1981.

18-12. "Heat Recovery," *Compressed Air*, July 1980.

18-13. Boyen, J., *Practical Heat Recovery*, First Edition, John Wiley & Sons, New York.

18-14. Lasday, S., "Production of Energy Pellets From Biomass," *Industrial Heating*, November 1979.

18-15. Rinaldi, G., *An Evaluation of Emission Factors for Waste to Energy Systems*, USEPA, 1979.

18-16. Ross, A., "The Chemical Control of Air Heater Corrosion and Plugging," American Society of Mechanical Engineers, Winter Annual Meeting, November 1973.

18-17. Keenan, J., Keyes, F., *Thermodynamic Properties of Steam*, John Wiley & Sons, New York, 1957.

CHAPTER 19

19-1. Heaney, F., "Air Pollution Controls at Braintree Incinerator," *Journal of the Air Pollution Control Association*, August 1972.

19-2. Grant, A., Tailor, T., Powers, J., "Odor Control Incineration," *Chemical Processing*, February 1972.

19-3. Battelle Memorial Institute, *Process Modifications for Control of Particulate Emissions From Stationary Combustion*, USEPA 650, October 1974.

19-4. *Air Pollution Manual*, American Industrial Hygiene Association, 1972.

19-5. Cross, F., *Air Pollution Odor Control Primer*, Technomic, 1973.

19-6. Brink, J., "Air Pollution Control with Fiber Mist Eliminators," *Canadian Journal of Air Pollution Control*, June 1963.

19-7. Danielson, J., *Air Pollution Engineering Manual*, County of Los Angeles, Air Pollution Control District, AP-40, May 1973.

19-8. *Controlling Airborne Particulates*, Report of the National Academy of Sciences, 1980.

19-9. "Air Pollution Control," *Power*, June 1961.

19-10. Balakrishnan, N., Cheng, C., Patel, M., "Emerging Technologies for Air Pollution Control," *Pollution Engineering*, November 1979.

19-11. *Industrial Guide for Air Pollution Control*, USEPA 625/6-78-004, July 1978.

19-12. Kaplan, N., Maxwell, M., "Removal of Sulfur Dioxide From Industrial Waste Gases," *Chemical Engineering*, October 17, 1977.

19-13. Lasater, R., Hopkins, J., "Removing Particulates From Stack Gases," *Chemical Engineering*, October 17, 1977.

19-14. "Fabric Filters, Dry Scrubbers," *Power*, January 1982.

19-15. Strauss, W., *Industrial Gas Cleaning*, Second Edition, Pergamon Press, Elmsford, New York.

19-16. Pierce, R., "Estimating Acid Dewpoints in Stack Gases," *Chemical Engineering*, November 4, 1977.

19-17. *Control Techniques for Particulate Pollutants*, National Air Pollution Control Administration, AP-51, January 1969.

19-18. Greiner, G., "Latest Developments in Air Pollution Control Technology," *Specifying Engineer*, September 1981.

19-19. Kaiser, E., "Incinerators to Meet New Air Pollution Standards," Presented at the Mid-Atlantic Section Meeting of the Air Pollution Control Association, New York, April 1967.

19-20. Friedrick, H., "Primer on Fabric Dust Collection," *Air Engineering*, Volume 9, May 1967.

19-21. Hagerty, D., Pavoni, J., *Solid Waste Management*, Van Nostrand Reinhold Company, New York, 1973.

19-22. Coulter, R., "Smoke, Dust, Fumes Closely Controlled in Electric Furnaces," *Iron Age*, 173:107–10, January 1954.

19-23. Lapple, C., *Interim Report: Stack Contamination–200 Areas*, HDC-611, August 6, 1948.

19-24. Marchello, J., Kelly, J., *Gas Cleaning for Air Quality Control*, Marcel Dekker, Inc., New York, 1975.

19-25. Niessen, W., *Combustion and Incineration Processes*, Marcel Dekker, Inc., First Edition, New York.

19-26. Miller, P., *Corrosion Studies in Municipal Incinerators*, USEPA SW-72-3-3, 1972.

19-27. Pueschel, R., Noll, K., "Visibility and Aerosol Size Frequency," *Journal of Applied Meteorology*, Volume 6, 1967.

19-28. Lapple, C., "Processes Use Many Collection Types," *Chemical Engineering*, 58:145–51, May 1951.

19-29. Hersey, H., "Reverse-Jet Filters," *Industrial Chemist*, 31:138, March 1955.

19-30. Treybal, R., *Mass Transfer Operations*, McGraw Hill, New York, 1955.

19-31. Teller, A., "Absorption With Chemical Reaction," *Chemical Engineering*, 67:111–24, July 1, 1960.

19-32. Niessen, W., Sarofin, A., "Incinerator Air Pollution: Facts and Speculation," National Incinerator Conference, American Society of Mechanical Engineers, 1970.

19-33. Kohl, R., "Gas Absorption," *Chemical Engineering*, 66:12-127, 1959.

19-34. Kraus, M., "Baghouses," *Chemical Engineering*, April 9, 1979, Pages 94-106.

19-35. First, M., Dreiner, P., "Concentrations of Particulates Found in Air," *Industrial Hygiene & Occupational Medicine*, 5:387, 1952.

Index

(The letter i following a page number indicates an illustration)